Matrix Theory

A SECOND COURSE

THE UNIVERSITY SERIES IN MATHEMATICS

Series Editor: Joseph J. Kohn
Princeton University

THE CLASSIFICATION OF FINITE SIMPLE GROUPS
Daniel Gorenstein
VOLUME 1: GROUPS OF NONCHARACTERISTIC 2 TYPE

ELLIPTIC DIFFERENTIAL EQUATIONS AND OBSTACLE PROBLEMS
Giovanni Maria Troianiello

FINITE SIMPLE GROUPS: An Introduction to Their Classification
Daniel Gorenstein

INTRODUCTION TO PSEUDODIFFERENTIAL AND FOURIER INTEGRAL OPERATORS
François Treves
VOLUME 1: PSEUDODIFFERENTIAL OPERATORS
VOLUME 2: FOURIER INTEGRAL OPERATORS

MATRIX THEORY: A Second Course
James M. Ortega

A SCRAPBOOK OF COMPLEX CURVE THEORY
C. Herbert Clemens

Matrix Theory

A SECOND COURSE

James M. Ortega

University of Virginia
Charlottesville, Virginia

Plenum Press • New York and London

Library of Congress Cataloging in Publication Data

Ortega, James M., 1932-
Matrix theory.

(The University series in mathematics)
Bibliography: p.
Includes index.
1. Matrices. 2. Algebras, Linear. I. Title. II. Series: University series in mathematics (Plenum Press)
QA188.078 1987 512.9'434 86-30312
ISBN 0-306-42433-9

10 9 8 7

A Division of Plenum Publishing Corporation
233 Spring Street, New York, N.Y. 10013

Printed in the United States of America

To Sara and Scott

Preface

Linear algebra and matrix theory are essentially synonymous terms for an area of mathematics that has become one of the most useful and pervasive tools in a wide range of disciplines. It is also a subject of great mathematical beauty. In consequence of both of these facts, linear algebra has increasingly been brought into lower levels of the curriculum, either in conjunction with the calculus or separate from it but at the same level. A large and still growing number of textbooks has been written to satisfy this need, aimed at students at the junior, sophomore, or even freshman levels. Thus, most students now obtaining a bachelor's degree in the sciences or engineering have had some exposure to linear algebra. But rarely, even when solid courses are taken at the junior or senior levels, do these students have an adequate working knowledge of the subject to be useful in graduate work or in research and development activities in government and industry. In particular, most elementary courses stop at the point of canonical forms, so that while the student may have "seen" the Jordan and other canonical forms, there is usually little appreciation of their usefulness. And there is almost never time in the elementary courses to deal with more specialized topics like nonnegative matrices, inertia theorems, and so on.

In consequence, many graduate courses in mathematics, applied mathematics, or applications develop certain parts of matrix theory as needed. At Virginia, we have long had a first semester course in linear algebra for incoming applied mathematics graduate students, a course that is also taken by a number of students in various engineering disciplines. There has been a continuing problem with suitable textbooks at this level. The ones designed for undergraduates are too elementary and incomplete, and the more advanced books are not sufficiently applied, too difficult to read, too specialized, or out-of-print. Hence, the course has been taught primarily from lecture notes, which are the basis for this book.

All students taking the course have some background in linear algebra, but, for most, this background is rather vague and, in many cases, several years in the past. Hence, the book begins with a review of basic concrete matrix theory (Chapter 1) and linear spaces (Chapter 2). These two chapters are largely review and are covered fairly quickly, although much of the material in Chapter 2 is new to many students. Chapter 3 on canonical forms completes the general broad theory and provides the basis for the subsequent more specialized chapters.

Chapter 4 utilizes the basic result of diagonalization of symmetric matrices to study the classification of quadratic forms via inertia and then treats the positive definite and semidefinite cases in relation to minimization with application to least squares problems.

Chapter 5 begins with the development of the matrix exponential to give fundamental solutions of linear differential equations with constant coefficients and to treat stability of solutions. The corresponding results on difference equations are then given with application to iterative methods. Finally, the Lyapunov criterion for stability is discussed with various related results.

Chapter 6 collects several miscellaneous topics. The first section deals primarily with nonnegative matrices or matrices with nonnegative inverses. The next section treats the generalized eigenvalue problem and high-order eigenvalue problems; Section 6.3 gives basic properties of Kronecker products and circulants, and Section 6.4 covers matrix equations and commuting matrices.

Each section has exercises. Each chapter has a References and Extensions section that gives some historical background and references and indicates various extensions of the material.

The numbering system used throughout the book is as follows. Equation numbers are in parentheses: the number (5.3.2) denotes the second equation of Section 3 of Chapter 5. Theorems and definitions are numbered consecutively within each section by, for example, 5.3.2, which denotes the second numbered theorem or definition in Section 5.3. Finally, 5.3-2 denotes Exercise 2 of Section 5.3.

The book assumes a background in calculus and occasionally uses a few facts from elementary analysis, such as the existence of a minimum of a continuous function on a closed bounded set, uniform convergence of series, the Cauchy criterion for convergence of series, the existence of a sequence converging to the supremum of a set, and so on. Other than these results from analysis and the omission of a few proofs, the book is self-contained.

I am deeply indebted to Mrs. B. Ann Turley and Mrs. Carolyn Duprey, who have struggled so valiantly with the twin mysteries of my handwriting and our word processor. I am also indebted to the many students who have suffered through courses based on various drafts of this book and who have made many suggestions. Finally, I acknowledge with sincere thanks the many comments of two anonymous reviewers who read an almost final draft.

Charlottesville

James M. Ortega

Contents

1

Review of Basic Background

In this chapter we review some of the basic ideas of matrix theory. It is assumed that the reader is (or was at one time) familiar with the contents of this chapter, or at least most of it, and a rather quick reading will serve to recall these basic facts as well as to establish certain notation that will be used in the remainder of the book. Some of the topics covered, especially linear equations and eigenvalues, will be expanded upon later in different ways.

The organization of this chapter is as follows. In Section 1.1 we review some properties and classifications of matrices, including partitioned matrices and special types of matrices, such as diagonal, triangular, symmetric, orthogonal, and so on. We also review various properties of vectors, such as linear dependence and orthogonality, and operations on vectors, such as inner and outer products. In Section 1.2 we recall, mostly without proof, the elementary properties of determinants. In Section 1.3 we use the process of Gaussian elimination to establish the existence and uniqueness of a solution of a system of linear equations with a nonsingular coefficient matrix. In Section 1.4 we review the definitions of eigenvalues and eigenvectors and give a few of the simplest properties.

1.1. Matrices and Vectors

A *column vector* is the n-tuple of real or complex numbers

$$\mathbf{x} = \begin{bmatrix} x_1 \\ \vdots \\ x_n \end{bmatrix} \tag{1.1.1}$$

and a *row vector* is $(x_1, \ldots, x_n)$. R^n will denote the collection of all column vectors with n real components, and C^n will denote the collection of all column vectors with n complex components. We will sometimes use the term *n-vector* to denote a vector with n components.

An $m \times n$ *matrix* is the array

$$A = \begin{bmatrix} a_{11} & \cdots & a_{1n} \\ \vdots & & \vdots \\ a_{m1} & \cdots & a_{mn} \end{bmatrix} \tag{1.1.2}$$

with m rows and n columns, where, again, the components a_{ij} may be real or complex numbers. The (i, j) element of the matrix is a_{ij}. The numbers m and n are the *dimensions* of A. When the dimensions are clear, we will also use the notation $A = (a_{ij})$ to denote an $m \times n$ matrix. If $n = m$, the matrix is *square*; otherwise it is *rectangular*. As indicated above, we will generally denote vectors by lowercase boldface letters, and matrices by capital letters.

Note that if $n = 1$ in (1.1.2), then A may be considered as a vector of m components or as an $m \times 1$ matrix. More generally, it is sometimes useful to view the columns of the matrix A as column vectors and the rows as row vectors. In this context we would write A in the form

$$A = (\mathbf{a}_1, \ldots, \mathbf{a}_n) \tag{1.1.3}$$

where each $\mathbf{a}_i$ is a column vector with m components. Likewise, we may also write A as the collection of row vectors

$$A = \begin{bmatrix} \mathbf{a}_1 \\ \vdots \\ \mathbf{a}_m \end{bmatrix} \tag{1.1.4}$$

where the vectors in (1.1.4) are, of course, different from those of (1.1.3). One should keep in mind that matrices and vectors are conceptually quite different, as we shall see more clearly in Chapter 2, but the identifications and partitionings indicated above are useful for manipulations.

A *submatrix* of A is any matrix obtained from A by deleting rows and columns. The partitionings (1.1.3) and (1.1.4) into a collection of vectors are special cases of a partitioning of a matrix A into submatrices. More generally, we consider partitionings of the form

$$A = \begin{bmatrix} A_{11} & \cdots & A_{1q} \\ \vdots & & \vdots \\ A_{p1} & \cdots & A_{pq} \end{bmatrix} \tag{1.1.5}$$

where each A_{ij} is itself a matrix (possibly 1×1), and dimensions of these submatrices must be consistent—that is, all matrices in a given row must have the same number of rows, and all matrices in a given column must have the same number of columns. The partitioned matrix (1.1.5) is sometimes called a *block* matrix. An example is given below for a 5×6 matrix in which the dashed lines indicate the partitioning:

$$A = \left[\begin{array}{cc:c:ccc} 1 & 2 & 3 & 4 & 5 & 6 \\ 1 & 4 & 2 & 1 & 6 & 8 \\ \hdashline 2 & 4 & 5 & 9 & 1 & 3 \\ 8 & 2 & 4 & 3 & 2 & 1 \\ 3 & 7 & 6 & 4 & 1 & 2 \end{array}\right] = \begin{bmatrix} A_{11} & A_{12} & A_{13} \\ A_{21} & A_{22} & A_{23} \end{bmatrix}$$

We next discuss some important special classes of matrices and the corresponding block matrices. An $n \times n$ matrix A is *diagonal* if $a_{ij} = 0$ whenever $i \neq j$, so that all elements off the main diagonal are zero. A is *upper triangular* if $a_{ij} = 0$ for $i > j$, so that all elements below the main diagonal are zero; similarly, A is *lower triangular* if all elements above the main diagonal are zero. The matrix A is *block diagonal, block upper triangular,* or *block lower triangular* if it has the partitioned forms

$$A = \begin{bmatrix} A_{11} & & \\ & \ddots & \\ & & A_{pp} \end{bmatrix}, \qquad A = \begin{bmatrix} A_{11} & A_{12} & \cdots & A_{1p} \\ & & \ddots & \vdots \\ & & & A_{pp} \end{bmatrix},$$

$$A = \begin{bmatrix} A_{11} & & & \\ A_{21} & \ddots & & \\ \vdots & & & \\ A_{p1} & \cdots & & A_{pp} \end{bmatrix}$$

Here, the matrices A_{ii} are all square but not necessarily the same size, and only the (possibly) nonzero matrices are shown. We will sometimes use the notation $\operatorname{diag}(A_{11}, \ldots, A_{pp})$ to denote a block diagonal matrix, and $\operatorname{diag}(a_{11}, \ldots, a_{nn})$ to denote a diagonal matrix.

The *transpose* of the $m \times n$ matrix (1.1.2) is the $n \times m$ matrix

$$A^T = (a_{ji}) = \begin{bmatrix} a_{11} & \cdots & a_{m1} \\ \vdots & & \vdots \\ a_{1n} & \cdots & a_{mn} \end{bmatrix}$$

If A has complex elements, then the transpose is still well defined, but a more useful matrix is the *conjugate transpose* given by

$$A^* = (\bar{a}_{ji}) = \begin{bmatrix} \bar{a}_{11} & \cdots & \bar{a}_{m1} \\ \vdots & & \vdots \\ \bar{a}_{1n} & \cdots & \bar{a}_{mn} \end{bmatrix}$$

where the bar denotes complex conjugate. An example is

$$A = \begin{bmatrix} 2-i & 3 & 2+2i \\ i & 4i & 1 \end{bmatrix}, \qquad A^* = \begin{bmatrix} 2+i & -i \\ 3 & -4i \\ 2-2i & 1 \end{bmatrix}$$

where $i = \sqrt{-1}$. If $\mathbf{x}$ is a column vector, its transpose is the row vector $\mathbf{x}^T = (x_1, \ldots, x_n)$, and similarly, the transpose of a row vector is a column vector. The conjugate transpose is $\mathbf{x}^* = (\bar{x}_1, \ldots, \bar{x}_n)$. We note that another, increasingly standard, notation for the conjugate transpose is A^H or $\mathbf{x}^H$.

A matrix A is *symmetric* if $A = A^T$, and *hermitian* if $A = A^*$. Examples are

$$A = \begin{bmatrix} 2 & 1 \\ 1 & 2 \end{bmatrix}, \qquad A = \begin{bmatrix} 2 & 1+i \\ 1-i & 2 \end{bmatrix}$$

where the first matrix is real and symmetric, and the second is complex and Hermitian. We will see later that symmetric and Hermitian matrices play a central role in matrix theory.

In most of matrix theory it makes no difference whether we work with vectors and matrices with real components or with complex ones; most results are equally true in both cases. However, there are a few places, especially when dealing with eigenvalues, where complex numbers are essential, and the theories are different in the real and complex cases. We shall point out these situations when appropriate. Otherwise, we will work primarily with complex vectors and matrices, keeping in mind that this includes real vectors and matrices as a special case. If the reader is more comfortable thinking in terms of real numbers throughout, no harm is done in doing so, except in those cases that we will delineate.

Basic Operations

Addition of matrices or vectors is allowed whenever the dimensions are the same; it is defined by

$$\begin{bmatrix} x_1 \\ \vdots \\ x_n \end{bmatrix} + \begin{bmatrix} y_1 \\ \vdots \\ y_n \end{bmatrix} = \begin{bmatrix} x_1 + y_1 \\ \vdots \\ x_n + y_n \end{bmatrix}$$

$$\begin{bmatrix} a_{11} & \cdots & a_{1n} \\ \vdots & & \vdots \\ a_{m1} & \cdots & a_{mn} \end{bmatrix} + \begin{bmatrix} b_{11} & \cdots & b_{1n} \\ \vdots & & \vdots \\ b_{m1} & \cdots & b_{mn} \end{bmatrix} = \begin{bmatrix} a_{11}+b_{11} & \cdots & a_{1n}+b_{1n} \\ \vdots & & \vdots \\ a_{m1}+b_{m1} & \cdots & a_{mn}+b_{mn} \end{bmatrix}$$

Multiplication of a matrix $A = (a_{ij})$ by a scalar α is defined by $\alpha A = (\alpha a_{ij})$; similarly for vectors. A sum of the form $\sum_{i=1}^{m} \alpha_i \mathbf{x}_i$, where the α_i are scalars, is a *linear combination* of the vectors $\mathbf{x}_1, \ldots, \mathbf{x}_m$, and $\sum_{i=1}^{m} \alpha_i A_i$ is a linear combination of the matrices A_i.

The product AB of two matrices A and B is defined only if the number of columns of A is equal to the number of rows of B. Then if A is $n \times p$ and B is $p \times m$, the product $C = AB$ is an $n \times m$ matrix C whose ij element is given by

$$c_{ij} = \sum_{k=1}^{p} a_{ik} b_{kj} \tag{1.1.6}$$

The associative and distributive laws

$$(\mathbf{x}+\mathbf{y})+\mathbf{z} = \mathbf{x}+(\mathbf{y}+\mathbf{z}), \qquad (AB)C = A(BC),$$
$$A(B+C) = AB + AC$$

hold for addition and multiplication, as is easily verified. But, whereas addition of matrices is commutative (i.e., $A + B = B + A$), multiplication generally is not, even if A and B are both square. In Chapter 6 we will examine conditions under which $AB = BA$.

The transpose and conjugate transpose operations satisfy

$$(AB)^* = B^*A^*, \qquad (A+B)^* = A^* + B^*$$

where A and B may be either matrices or vectors of the correct dimensions. These relations are easily proved by direct computation (Exercise 1.1-12).

We will see in the next chapter that the definition of matrix multiplication given above arises naturally as a consequence of considering matrices as linear transformations. However, it is also possible to define an elementwise multiplication, analogous to addition, whenever the dimensions of the matrices A and B are the same. This is the *Schur product* (also called the *Hadamard* or *direct* product), which is defined for two $m \times n$ matrices by

$$A \circ B = \begin{bmatrix} a_{11}b_{11} & \cdots & a_{1n}b_{1n} \\ \vdots & & \vdots \\ a_{m1}b_{m1} & \cdots & a_{mn}b_{mn} \end{bmatrix}$$

Unless explicitly stated to the contrary, we will always use the matrix multiplication defined by (1.1.6).

All of the above operations work equally well on block matrices if we assume that all submatrices are of the right dimensions for the indicated operations. For example, for the matrix $A = (A_{ij})$ of (1.1.5),

$$A^* = (A_{ji}^*) = \begin{bmatrix} A_{11}^* & \cdots & A_{p1}^* \\ \vdots & & \vdots \\ A_{1q}^* & \cdots & A_{pq}^* \end{bmatrix}$$

As another example, if $B = (B_{ij})$ is a block matrix of proper dimensions, then $C = AB$ is a block matrix $C = (C_{ij})$ with

$$C_{ij} = \sum_{k=1}^{q} A_{ik}B_{kj}$$

A special case of the matrix multiplication AB is when B is $n \times 1$, a column vector. In this case the product AB is again a column vector. Similarly, if A is $1 \times n$ and B is $n \times m$, then AB is $1 \times m$, a row vector. It is sometimes useful to view matrix multiplication as a collection of matrix-vector multiplications. If $B = (\mathbf{b}_1, \ldots, \mathbf{b}_n)$ is the partitioning of B into its columns, then

$$AB = A(\mathbf{b}_1, \ldots, \mathbf{b}_n) = (A\mathbf{b}_1, \ldots, A\mathbf{b}_n) \tag{1.1.7}$$

where A multiplies each column of B. Similarly, if A is partitioned into its rows, we can view the product as

$$AB = \begin{bmatrix} \mathbf{a}_1 \\ \vdots \\ \mathbf{a}_n \end{bmatrix} B = \begin{bmatrix} \mathbf{a}_1 B \\ \vdots \\ \mathbf{a}_n B \end{bmatrix} \tag{1.1.8}$$

where the rows of A multiply B. The relations (1.1.7) and (1.1.8) are easily verified by applying the definition of matrix multiplication (exercise 1.1-17).

An important special case of matrix multiplication is when A is $1 \times n$ and B is $n \times 1$. Then AB is 1×1, a scalar. On the other hand, if A is $m \times 1$ and B is $1 \times n$, then AB is $m \times n$. These products are given explicitly in terms of vectors by

$$(a_1, \ldots, a_n)\begin{bmatrix} b_1 \\ \vdots \\ b_n \end{bmatrix} = \sum_{i=1}^{n} a_i b_i, \tag{1.1.9a}$$

$$\begin{bmatrix} a_1 \\ \vdots \\ a_m \end{bmatrix} (b_1, \ldots, b_n) = \begin{bmatrix} a_1 b_1 & \cdots & a_1 b_n \\ a_2 b_1 & & a_2 b_n \\ \vdots & & \vdots \\ a_m b_1 & \cdots & a_m b_n \end{bmatrix} \tag{1.1.9b}$$

If $\mathbf{a}$ and $\mathbf{b}$ are column vectors, the above matrix multiplication rules define the products $\mathbf{a}^T\mathbf{b}$ and $\mathbf{a}\mathbf{b}^T$. The former is called the *inner product* or *dot product* of $\mathbf{a}$ and $\mathbf{b}$, if they are real; if they are complex, the proper quantity is $\mathbf{a}^*\mathbf{b}$. The matrix $\mathbf{a}\mathbf{b}^*$ is called the *outer product* of $\mathbf{a}$ and $\mathbf{b}$.

A matrix-vector multiplication $A\mathbf{x}$ may be carried out in two natural ways. In the first let $\mathbf{a}_1, \ldots, \mathbf{a}_m$ denote the rows of A. Then

$$A\mathbf{x} = \begin{bmatrix} \mathbf{a}_1\mathbf{x} \\ \mathbf{a}_2\mathbf{x} \\ \vdots \\ \mathbf{a}_m\mathbf{x} \end{bmatrix} \tag{1.1.10}$$

If A is real, $\mathbf{a}_i\mathbf{x} = (\mathbf{a}_i^T)^T\mathbf{x}$ is the inner product of $\mathbf{x}$ and the column vector obtained by transposing the ith row $\mathbf{a}_i$. In the sequel we will call this simply the inner product of $\mathbf{x}$ and the ith row of A, and refer to (1.1.10) as the inner product form of matrix-vector multiplication. Note, however, that if A is complex, the quantities $\mathbf{a}_i\mathbf{x}$ are not inner products.

In the second way of viewing matrix multiplication, let $\mathbf{a}_1, \ldots, \mathbf{a}_n$ now denote the columns of A. Then it is easy to verify that

$$A\mathbf{x} = \sum_{i=1}^{n} x_i \mathbf{a}_i \tag{1.1.11}$$

so that $A\mathbf{x}$ is a linear combination of the columns of A. Each of the ways, (1.1.10) and (1.1.11), of viewing matrix-vector multiplication has certain computational and conceptual advantages, as we will see later.

Orthogonality and Linear Independence

Two nonzero column vectors $\mathbf{x}$ and $\mathbf{y}$ are *orthogonal* if their inner product, $\mathbf{x}^*\mathbf{y}$, is zero. If, in addition, $\mathbf{x}^*\mathbf{x} = \mathbf{y}^*\mathbf{y} = 1$, the vectors are *orthonormal.* The nonzero vectors $\mathbf{x}_1, \ldots, \mathbf{x}_m$ are orthogonal if $\mathbf{x}_i^*\mathbf{x}_j = 0$ whenever $i \neq j$, and orthonormal if, in addition, $\mathbf{x}_i^*\mathbf{x}_i = 1$, $i = 1, \ldots, m$. A set of orthogonal vectors can always be made orthonormal by scaling in the following way. For any nonzero vector $\mathbf{x}$, $\mathbf{x}^*\mathbf{x} = \sum_{i=1}^{n} |x_i|^2 > 0$. Then $\mathbf{y} = \mathbf{x}/(\mathbf{x}^*\mathbf{x})^{1/2}$ satisfies $\mathbf{y}^*\mathbf{y} = 1$. Thus, if $\mathbf{x}_1, \ldots, \mathbf{x}_m$ are orthogonal vectors, then the vectors $\mathbf{y}_i = \mathbf{x}_i/(\mathbf{x}_i^*\mathbf{x}_i)^{1/2}$, $i = 1, \ldots, m$, are orthonormal.

Row vectors $\mathbf{z}_1, \ldots, \mathbf{z}_m$ may be defined to be orthogonal or orthonormal by applying the above definitions to the column vectors $\mathbf{z}_1^*, \ldots, \mathbf{z}_m^*$.

A real $n \times n$ matrix A is *orthogonal* if the matrix equation

$$A^T A = I = \operatorname{diag}(1, 1, \ldots, 1) = \begin{bmatrix} 1 & & & \\ & 1 & & \\ & & \ddots & \\ & & & 1 \end{bmatrix} = AA^T \quad (1.1.12)$$

holds. This equation may be interpreted as saying that A is orthogonal if its columns (and its rows), viewed as vectors, are orthonormal. If the matrix A is complex with orthonormal columns, then it is called *unitary*, and (1.1.12) is replaced by

$$A^* A = I = AA^* \quad (1.1.13)$$

Examples are

$$\frac{1}{\sqrt{2}}\begin{bmatrix} 1 & 1 \\ -1 & 1 \end{bmatrix}, \quad \frac{1}{\sqrt{3}}\begin{bmatrix} 1 & 1+i \\ -1+i & 1 \end{bmatrix} \quad (1.1.14)$$

in which the first matrix is real and orthogonal, and the second is complex and unitary. On the other hand, the matrices

$$\begin{bmatrix} 1 & 1 \\ -1 & 1 \end{bmatrix}, \quad \begin{bmatrix} 1 & 1+i \\ -1+i & 1 \end{bmatrix} \quad (1.1.15)$$

have orthogonal columns, but the columns are not orthonormal; hence, these are not orthogonal or unitary matrices. We will see later the very special and important role that orthogonal and unitary matrices play in matrix theory.

The matrix I in (1.1.12) is called the *identity matrix* and obviously has the property that $IB = B$ for any matrix or vector B of commensurate dimensions; likewise $BI = B$ whenever the multiplication is defined. We denote the columns of the identity matrix by $\mathbf{e}_1, \ldots, \mathbf{e}_n$, so that $\mathbf{e}_i$ is the vector with a 1 in the ith position and 0's elsewhere. These vectors are a particularly important set of orthogonal (and orthonormal) vectors. Orthogonal vectors, in turn, are important special cases of linearly independent vectors, which are defined as follows. A set of n-vectors $\mathbf{x}_1, \ldots, \mathbf{x}_m$ is *linearly dependent* if

$$\sum_{i=1}^{m} c_i \mathbf{x}_i = 0 \quad (1.1.16)$$

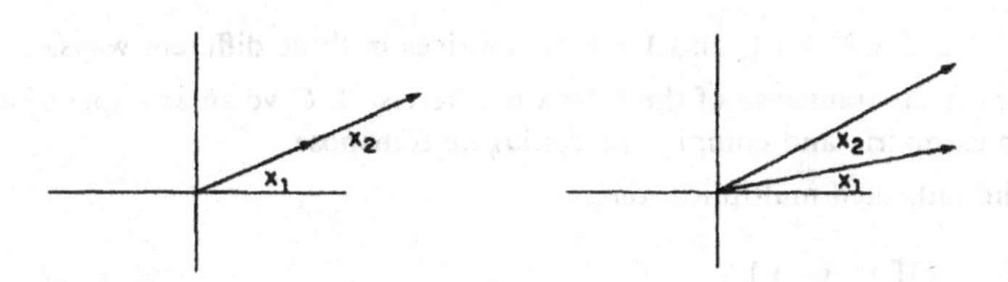

Figure 1.1. (a) Linearly dependent vectors. (b) Linearly independent vectors.

where the scalars $c_1, \ldots, c_m$ are not all zero. If no linear dependence exists—that is, if no equation of the form (1.1.16) is possible unless all c_i are zero—then the vectors are *linearly independent*. For example, if $c_1\mathbf{x}_1 + c_2\mathbf{x}_2 = 0$ with $c_1 \neq 0$, then $\mathbf{x}_1$ and $\mathbf{x}_2$ are linearly dependent, and we can write $\mathbf{x}_1 = -c_1^{-1}c_2\mathbf{x}_2$, so that $\mathbf{x}_1$ is a multiple of $\mathbf{x}_2$. On the other hand, if one vector is not a multiple of the other, then $c_1\mathbf{x}_1 + c_2\mathbf{x}_2 = 0$ implies that $c_1 = c_2 = 0$, and the vectors are linearly independent. This is illustrated in Figure 1.1. Similarly, if $m = n = 3$ and (1.1.16) holds with $c_1 \neq 0$, then

$$\mathbf{x}_1 = \frac{-1}{c_1}(c_2\mathbf{x}_2 + c_3\mathbf{x}_3)$$

which says that $\mathbf{x}_1$ is in the plane containing $\mathbf{x}_2$ and $\mathbf{x}_3$; that is, three 3-vectors are linearly dependent if they lie in a plane; otherwise they are linearly independent.

If $\mathbf{x}_1, \ldots, \mathbf{x}_m$ are linearly dependent, they may be so in different ways: They may all be multiples of one of the vectors, say $\mathbf{x}_m$; in this case the set is said to have only one linearly independent vector. Or they may all lie in a plane, so that they are all linear combinations of two of the vectors, say $\mathbf{x}_m$ and $\mathbf{x}_{m-1}$; in this case we say there are two linearly independent vectors. More generally, if all of the vectors may be written as a linear combination of r vectors but not $r - 1$ vectors, then we say that the set has r linearly independent vectors.

An important property of a matrix is the number of linearly independent columns. We can also consider the rows of A as vectors and determine the number of linearly independent rows. We shall return to this problem in Section 1.3 and also in Chapter 2.

Exercises 1.1

1. Partition the matrix

$$\begin{bmatrix} 1 & 2 & 3 \\ 4 & 5 & 6 \\ 7 & 8 & 9 \end{bmatrix}$$

into 2×2, 1×2, 2×1, and 1×1 submatrices in three different ways.

2. Compute the transpose of the matrix in Exercise 1. Give an example of a 3×3 complex matrix and compute its conjugate transpose.
3. Do the indicated multiplications:

(a) $\begin{bmatrix} 1 & 3 \\ 1 & 4 \end{bmatrix}\begin{bmatrix} 2 & 1 & 1 \\ 3 & 2 & 4 \end{bmatrix}$

(b) $(1, 3)\begin{bmatrix} 2 & 1 \\ 1 & 2 \end{bmatrix}$.

(c) $\begin{bmatrix} 2 \\ 1 \end{bmatrix}(4, 2)$

(d) $(4, 2)\begin{bmatrix} 2 \\ 1 \end{bmatrix}$

4. Let $\mathbf{x}^T = (1, 2, 3)$, and let A be the matrix of Exercise 1. Find the product $A\mathbf{x}$
   (a) by forming the inner products of the rows of A with $\mathbf{x}$;
   (b) as a linear combination of the columns of A.
5. Ascertain if the following pairs of vectors are orthogonal and orthonormal:

(a) $\begin{bmatrix} 1 \\ 0 \end{bmatrix}\begin{bmatrix} 0 \\ 1 \end{bmatrix}$

(b) $\begin{bmatrix} 1 \\ 1 \end{bmatrix}\begin{bmatrix} 1 \\ -2 \end{bmatrix}$

(c) $\begin{bmatrix} 1+i \\ 1-i \end{bmatrix}\begin{bmatrix} 1-i \\ 1+i \end{bmatrix}$

6. Ascertain if the pairs of vectors in Exercise 5 are linearly independent.
7. Let $\mathbf{x}_1, \ldots, \mathbf{x}_m$ be nonzero orthogonal vectors. Show that $\mathbf{x}_1, \ldots, \mathbf{x}_m$ are linearly independent. *Hint*: Assume that a linear dependence (1.1.16) exists and show that orthogonality implies that all c_i must be zero.
8. Show that the following matrix is orthogonal for any real value of θ:

$$\begin{bmatrix} \cos\theta & \sin\theta \\ -\sin\theta & \cos\theta \end{bmatrix}$$

9. Ascertain if the following matrices are orthogonal:

(a) $\begin{bmatrix} 1 & 0 \\ 0 & 1 \end{bmatrix}$

(b) $\begin{bmatrix} 0 & 1 \\ 1 & 0 \end{bmatrix}$

(c) $\begin{bmatrix} 1 & 1 \\ -1 & 1 \end{bmatrix}$

(d) $\begin{bmatrix} \cos\theta & \sin\theta & 0 \\ -\sin\theta & \cos\theta & 0 \\ 0 & 0 & 1 \end{bmatrix}$

10. Show that the product of diagonal matrices is diagonal and that the product of lower (upper) triangular matrices is lower (upper) triangular.
11. Let A and B be $m \times n$ matrices. Show that $(A+B)^T = A^T + B^T$ and $(A+B)^* = A^* + B^*$.
12. Let A and B be $n \times p$ and $p \times m$ matrices, respectively. Show by direct calculation that $(AB)^T = B^T A^T$ and $(AB)^* = B^* A^*$ if A or B is complex.
13. Show that the transpose of an orthogonal matrix is orthogonal and that the conjugate transpose of a unitary matrix is unitary.
14. Show that the product of $n \times n$ orthogonal matrices is orthogonal and that the product of $n \times n$ unitary matrices is unitary.
15. Let A be a real $n \times n$ symmetric matrix, and let P be a real $n \times m$ matrix. Use the result of Exercise 12 to show that the $m \times m$ matrix $P^T AP$ is symmetric. Similarly, show that P^*AP is Hermitian if A and P are complex and A is Hermitian.
16. The *trace* of an $n \times n$ matrix A is defined as the sum of the diagonal elements of A:

$$\operatorname{Tr}(A) = \sum_{j=1}^{n} a_{jj}.$$

For $n \times n$ matrices A and B and scalars α and β, show that
(a) $\operatorname{Tr}(AB) = \operatorname{Tr}(BA)$.
(b) $\operatorname{Tr}(\alpha A + \beta B) = \alpha \operatorname{Tr}(A) + \beta \operatorname{Tr}(B)$.

17. Verify the relations (1.1.7) and (1.1.8).

1.2. Determinants

The *determinant* of an $n \times n$ real or complex matrix A is a scalar denoted by $\det A$ and defined by

$$\det A = \sum_{P} (-1)^{S(P)} a_{\sigma(1)1} a_{\sigma(2)2} \cdots a_{\sigma(n)n} \tag{1.2.1}$$

This rather formidable expression means the following. The summation is taken over all $n!$ possible permutations P of the integers $1, \ldots, n$. Each such permutation is denoted by $\sigma(1), \ldots, \sigma(n)$ so that each of the products in (1.2.1) is made up of exactly one element from each row and each column of A. $S(P)$ is the number of inversions in the permutation, and the sign of the corresponding product is positive if $S(P)$ is even (an even permutation) or negative if $S(P)$ is odd (an odd permutation). The formula (1.2.1) is the natural extension to $n \times n$ matrices of the (hopefully) familiar rules for determinants of 2×2 and 3×3 matrices:

$$n = 2: \qquad \det A = a_{11}a_{22} - a_{12}a_{21}$$

$$n = 3: \qquad \det A = a_{11}a_{22}a_{33} + a_{12}a_{23}a_{31} + a_{13}a_{21}a_{32} - a_{12}a_{21}a_{33} - a_{11}a_{23}a_{32} - a_{13}a_{22}a_{31}$$

Note that the determinant is defined only for square matrices. Note also that a straightforward application of (1.2.1) to the computation of the determinant requires the formation of $n!$ products, each requiring $n - 1$ multiplications; this gives a total of $n!$ additions and $(n - 1)n!$ multiplications. We shall see later that the determinant can actually be computed in about $n^3/3$ additions and multiplications.

We next collect, without proof, a number of basic facts about determinants.

1.2.1. If any two rows or columns of A are equal, or if A has a zero row or column, then $\det A = 0$.

1.2.2. If any two rows or columns of A are interchanged, the sign of the determinant changes, but the magnitude remains unchanged.

1.2.3. If any row or column of A is multiplied by a scalar α, then the determinant is multiplied by α.

1.2.4. $\det A^T = \det A$, $\det A^* = \overline{\det A}$.

1.2.5. If a scalar multiple of any row (or column) of A is added to another row (or column), the determinant remains unchanged.

In the next section we will see the most efficient way to compute a determinant. However, another way that is useful for theoretical purposes is the following. Let A_{ij} denote the $(n - 1) \times (n - 1)$ submatrix of A obtained by deleting the ith row and jth column. Then $(-1)^{i+j} \det A_{ij}$ is called the *cofactor* of the element a_{ij}.

1.2.6. Expansion in cofactors

$$\det A = \sum_{j=1}^{n} a_{ij}(-1)^{i+j} \det A_{ij} \qquad \text{for any } i \tag{1.2.2}$$

$$\det A = \sum_{i=1}^{n} a_{ij}(-1)^{i+j} \det A_{ij} \qquad \text{for any } j \tag{1.2.3}$$

Theorem 1.2.6 shows that det A can be computed as a linear combination of n determinants of $(n-1) \times (n-1)$ submatrices. The two formulas (1.2.2) and (1.2.3) show that det A can be expanded in the elements of a row or the elements of a column. For example, expanding det A in terms of the elements of the first column gives

$$\det A = \sum_{i=1}^{n} a_{i1}(-1)^{i+1} \det A_{i1}$$

$$= a_{11} \det \begin{bmatrix} a_{22} & \cdots & a_{2n} \\ \vdots & & \vdots \\ a_{n2} & \cdots & a_{nn} \end{bmatrix} - a_{21} \det \begin{bmatrix} a_{12} & \cdots & a_{1n} \\ a_{32} & & a_{3n} \\ \vdots & & \vdots \\ a_{n2} & \cdots & a_{nn} \end{bmatrix}$$

$$+ \cdots + (-1)^n a_{n-1,1} \det \begin{bmatrix} a_{12} & \cdots & a_{1n} \\ \vdots & & \vdots \\ a_{n-2,2} & & a_{n-2,n} \\ a_{n,2} & \cdots & a_{nn} \end{bmatrix}$$

$$+ (-1)^{n+1} a_{n1} \det \begin{bmatrix} a_{12} & \cdots & a_{1n} \\ \vdots & & \vdots \\ a_{n-1,2} & \cdots & a_{n-1,n} \end{bmatrix} \tag{1.2.4}$$

Suppose that A is a lower triangular matrix. Then the first row of the last $n-1$ matrices in (1.2.4) is zero. Hence, by 1.2.1, det $A = a_{11}$ det A_{11}. Expanding det A_{11} in terms of elements in its first column and noting that, again, all matrices except the first have a zero row gives

$$\det A = a_{11}a_{22} \det \begin{bmatrix} a_{33} & \cdots & a_{3n} \\ \vdots & & \vdots \\ a_{n3} & \cdots & a_{nn} \end{bmatrix}$$

Continuing in this way, we conclude that $\det A = a_{11}a_{22}\cdots a_{nn}$. An analogous argument holds for an upper triangular matrix. Hence, we have the following result.

1.2.7. The determinant of a triangular matrix is the product of the elements on its main diagonal. In particular, the determinant of a diagonal matrix is the product of its diagonal elements.

Another basic property that will be of continuing use to us concerns the determinant of a product.

1.2.8. If A and B are $n \times n$ matrices, then $\det AB = \det A \det B$.

Exercises 1.2

1. Use (1.2.1) to compute the determinant of a 4×4 matrix.
2. Compute the determinant of the following matrices:

(a) $\begin{bmatrix} 1 & 1 & 1 \\ 1 & 2 & 1 \\ 1 & 1 & 1 \end{bmatrix}$ (b) $\begin{bmatrix} 0 & 1 & 1 \\ 0 & 1 & 2 \\ 0 & 1 & 1 \end{bmatrix}$

(c) $\begin{bmatrix} 1 & 2 & 4 \\ 2 & 3 & 5 \\ 3 & 4 & 6 \end{bmatrix}$ (d) $\begin{bmatrix} 2 & 2 & 4 \\ 4 & 3 & 5 \\ 6 & 4 & 6 \end{bmatrix}$

(e) $\begin{bmatrix} 4 & 2 & 2 \\ 5 & 3 & 4 \\ 6 & 4 & 6 \end{bmatrix}$ (f) $\begin{bmatrix} 4 & 5 & 6 \\ 2 & 3 & 4 \\ 2 & 4 & 6 \end{bmatrix}$

3. Use both cofactor expansions (1.2.2) and (1.2.3) to compute the determinant of the matrix

$$\begin{bmatrix} 1 & 2 & 3 & 4 \\ -2 & 2 & 2 & 4 \\ -3 & 4 & 3 & 5 \\ -4 & 6 & 4 & 6 \end{bmatrix}$$

4. Let

$$A = \begin{bmatrix} A_{11} & \cdots & A_{1p} \\ & \ddots & \vdots \\ & & A_{pp} \end{bmatrix}$$

be a block triangular matrix with square diagonal matrices A_{ii}, $i = 1, \ldots, p$. Use the cofactor expansion 1.2.6 to conclude that $\det A = \det A_{11} \det A_{22} \cdots \det A_{pp}$.

5. An $n \times n$ *Hadamard matrix* A has elements that are all ± 1 and satisfies $A^T A = nI$. Show that $|\det A| = n^{n/2}$.

1.3. Linear Equations and Inverses

One of the most common situations in which matrices arise is with systems of linear equations

$$\sum_{j=1}^{n} a_{ij} x_j = b_i, \qquad i = 1, \ldots, n \tag{1.3.1}$$

Here, the a_{ij} and b_i are assumed to be known coefficients, and we wish to find values of $x_1, \ldots, x_n$ for which (1.3.1) is satisfied. We will usually write (1.3.1) in matrix-vector form as

$$A\mathbf{x} = \mathbf{b} \tag{1.3.2}$$

We will now show that the system (1.3.2) has a unique solution $\mathbf{x}$ if $\det A \neq 0$. The method for doing this will be the Gaussian elimination process, which, with suitable modifications, forms the basis for reliable and efficient computer programs for solving such systems in practice.

Assume that $a_{11} \neq 0$. Then multiply the first equation of (1.3.1) by $a_{11}^{-1} a_{21}$ and subtract it from the second equation. Next, multiply the first equation by $a_{11}^{-1} a_{31}$ and subtract it from the third equation. Continue this process until the corresponding multiple of the first row has been subtracted from each of the other rows. The system then has the form

$$\begin{aligned} a_{11}x_1 + a_{12}x_2 + \cdots + a_{1n}x_n &= b_1 \\ a_{22}^1 x_2 + \cdots + a_{2n}^1 x_n &= b_2^1 \\ \vdots \qquad\qquad\quad \vdots \quad &\quad \vdots \\ a_{n2}^1 x_2 + \cdots + a_{nn}^1 x_n &= b_n^1 \end{aligned} \tag{1.3.3}$$

where the 1's on the coefficients indicate that they will, in general, have been changed from their original values. The effect of these modifications, of course, is to eliminate the unknown x_1 from the last $n - 1$ equations.

The modified system (1.3.3) has the same solutions as the original system (1.3.1), because if $x_1, \ldots, x_n$ are a solution of (1.3.1), they also satisfy (1.3.3). Conversely, by doing the analogous modifications of (1.3.3) (i.e., add $a_{11}^{-1} a_{21}$ times the first row to the second row, etc.) to return to the original system (1.3.1), it follows that if $x_1, \ldots, x_n$ are a solution of (1.3.3), they are also a solution of (1.3.1).

Now assume that $a_{22}^1 \neq 0$ and repeat the process for the last $n-1$ equations to eliminate the unknown x_2 from the last $n-2$ equations. Continuing in this fashion, we arrive at the triangular system of equations

$$\begin{aligned} a_{11}x_1 + a_{12}x_2 + \cdots + a_{1n}x_n &= b_1 \\ a_{22}^1 x_2 + \cdots + a_{2n}^1 x_n &= b_2^1 \\ \ddots \qquad \vdots \quad &\quad \vdots \\ a_{n-1,n-1}^{(n-2)} x_{n-1} + a_{n-1,n}^{(n-2)} x_n &= b_{n-1}^{(n-2)} \\ a_{nn}^{(n-1)} x_n &= b_n^{(n-1)} \end{aligned} \tag{1.3.4}$$

where the superscript indicates the number of times the coefficient may have been changed. By the same arguments as above, this system of equations has exactly the same solutions as the original system.

Next, assume that $a_{nn}^{(n-1)} \neq 0$. Then we obtain x_n uniquely from the last equation. With x_n known we obtain x_{n-1} uniquely from the next to last equation, and working our way back up the triangular system we obtain, in turn, $x_{n-2}, \ldots, x_1$. Thus, under the assumptions we have made, we have shown that the system (1.3.4), and hence the system (1.3.1), has a unique solution.

We now show that the assumption $\det A \neq 0$ implies that we can ensure that all the divisor elements $a_{11}, a_{22}^1, a_{33}^{(2)}, \ldots, a_{nn}^{(n-1)}$, usually called the *pivot* elements, can be made nonzero. First suppose that $a_{11} = 0$. We search the first column for a nonzero coefficient, say a_{k1}, and then interchange the first and kth rows of the system. Clearly, this does not change the solution of the equations (1.3.1) since we can write these equations in any order. We are guaranteed that at least one element in the first column of A is nonzero, since otherwise, by 1.2.1, $\det A = 0$, which contradicts our assumption. Next, consider the reduced system (1.3.3). If $a_{22}^1 = 0$, choose a nonzero element, say a_{k2}^1, with $k > 2$, and interchange the kth row with the second row. Again, the assumption $\det A \neq 0$ will ensure that at least one of these elements in the second column is nonzero. For suppose otherwise. Then the determinant of the $(n-1) \times (n-1)$ submatrix of (1.3.3) consisting of the last $n-1$ rows and columns is zero, since the first column of the matrix is zero. Thus, by the cofactor expansion 1.2.6 the determinant of the whole matrix of (1.3.3) is zero. But by 1.2.2 and 1.2.5 the determinant is unchanged by the operations that transformed (1.3.1) to (1.3.3), or, at most, it changes sign if two rows were interchanged. Hence, this is a contradiction. Continuing in this way, we conclude that all of the elements $a_{i,i}^{(i-1)}$, $i = 3, \ldots, n$, are either nonzero or can be made nonzero by interchanges of rows.

The above arguments are equally valid for either real or complex numbers. Hence, we have proved the following basic theorem.

1.3.1. SOLUTION OF LINEAR SYSTEMS. *Let A be an $n \times n$ real or complex matrix with* $\det A \neq 0$ *and* $\mathbf{b}$ *a given n-vector. Then the system* $A\mathbf{x} = \mathbf{b}$ *has a unique solution.*

We now give an example of the Gaussian elimination process. Let

$$A = \begin{bmatrix} 2 & 2 & 1 & 1 \\ 2 & 2 & 2 & 2 \\ 4 & 3 & 1 & 1 \\ 2 & 1 & 2 & 1 \end{bmatrix}, \qquad \mathbf{b} = \begin{bmatrix} 1 \\ 2 \\ 1 \\ 2 \end{bmatrix}$$

The first step, (1.3.3), produces the reduced system

$$\begin{aligned} 2x_1 + 2x_2 + x_3 + x_4 &= 1 \\ x_3 + x_4 &= 1 \\ -x_2 - x_3 - x_4 &= -1 \\ -x_2 + x_3 &= 1 \end{aligned}$$

The element a_{22}^1 is zero, so we interchange the second and third rows and proceed to the triangular system (1.3.4):

$$\begin{aligned} 2x_1 + 2x_2 + x_3 + x_4 &= 1 \\ -x_2 - x_3 - x_4 &= -1 \\ x_3 + x_4 &= 1 \\ -x_4 &= 0 \end{aligned}$$

Solving this system gives $x_4 = 0$, $x_3 = 1$, $x_2 = 0$, and $x_1 = 0$. This is the unique solution.

Inverses

It is sometimes useful to write the solution of $A\mathbf{x} = \mathbf{b}$ in terms of another matrix, called the *inverse* of A and denoted by A^{-1}. We can define A^{-1} in different (but equivalent) ways, and we shall return to this point in Chapter 2. Here, we define it computationally as follows. Assume that $\det A \neq 0$, and let $\mathbf{e}_i$ again be the vector with a 1 in the ith position and 0's elsewhere. Then by 1.3.1 each of the systems

$$A\mathbf{x}_i = \mathbf{e}_i, \qquad i = 1, \ldots, n \tag{1.3.5}$$

has a unique solution, and we let $X = (\mathbf{x}_1, \ldots, \mathbf{x}_n)$ be the matrix with these solutions as its columns. Then (1.3.5) is equivalent to

$$AX = I \tag{1.3.6}$$

where, as usual, I is the identity matrix. As an example, consider

$$A = \begin{bmatrix} 2 & 1 \\ 2 & 2 \end{bmatrix}, \qquad A\mathbf{x}_1 = \begin{bmatrix} 1 \\ 0 \end{bmatrix}, \qquad A\mathbf{x}_2 = \begin{bmatrix} 0 \\ 1 \end{bmatrix}$$

The solutions of the two systems are $\mathbf{x}_1^T = (1, -1)$ and $\mathbf{x}_2^T = \frac{1}{2}(-1, 2)$, so that

$$X = \frac{1}{2}\begin{bmatrix} 2 & -1 \\ -2 & 2 \end{bmatrix}$$

Now, by 1.2.7, $\det I = 1$, and by the product rule 1.2.8 applied to (1.3.6) we obtain

$$\det A \det X = 1$$

This shows, in particular, that $\det X \neq 0$. Hence, as above, we can find a matrix Y such that

$$XY = I$$

Multiplying by A and using (1.3.6), we conclude that $Y = A$, so that $XA = I$. We now define $A^{-1} = X$. Then the relations $XA = I$ and $AX = I$ are equivalent to

$$A^{-1}A = I = AA^{-1} \tag{1.3.7}$$

which are the basic relations that the inverse A^{-1} must satisfy. We can again apply the product rule 1.2.8 to (1.3.7) to conclude that

$$\det A \det A^{-1} = 1 \qquad \text{or} \qquad \det A^{-1} = (\det A)^{-1} \tag{1.3.8}$$

We can also multiply the equation $A\mathbf{x} = \mathbf{b}$ by A^{-1} write the solution as

$$\mathbf{x} = A^{-1}\mathbf{b} \tag{1.3.9}$$

which is sometimes useful theoretically (but, we stress, not in practice; that is, do *not* solve $A\mathbf{x} = \mathbf{b}$ by finding A^{-1} and then $A^{-1}\mathbf{b}$).

The above shows that if $\det A \neq 0$, then a matrix A^{-1} that satisfies (1.3.7) exists. Conversely, if such a matrix exists, then (1.3.8) shows that $\det A \neq 0$. We shall say that A is *nonsingular* or *invertible* if $\det A \neq 0$ or, equivalently, if A^{-1} exists; otherwise, A is *singular*. Note that by 1.2.8 the determinant of a product is the product of the determinants. We thus arrive at the following useful result, the proof of which is left as an exercise (1.3-9).

1.3.2. If A and B are $n \times n$ matrices, then AB is nonsingular if and only if both A and B are nonsingular. Moreover, $(AB)^{-1} = B^{-1}A^{-1}$.

We note that we can compute the inverse of diagonal matrices trivially, and we have already seen another case where this is true. If A is a real orthogonal matrix, then, by (1.1.12), $A^{-1} = A^T$, whereas $A^{-1} = A^*$ if A is unitary.

Matrix Formulation of Gaussian Elimination

It is useful to view the Gaussian elimination process in a matrix-theoretic way. We assume first that all the divisors in the process are nonzero, so that no interchanges of rows are necessary. Then the computation that led to the reduced system (1.3.3) can be represented by the matrix multiplication $L_1 A\mathbf{x} = L_1\mathbf{b}$, where

$$L_1 = \begin{bmatrix} 1 & & & & \\ -l_{21} & 1 & & & \\ 0 & 0 & \ddots & & \\ \vdots & \vdots & & \ddots & \\ -l_{n1} & 0 & \cdots & 0 & 1 \end{bmatrix}, \qquad l_{i1} = \frac{a_{i1}}{a_{11}}, \quad i = 2, \ldots, n \tag{1.3.10}$$

Similarly, the kth step is effected by multiplication by the matrix

$$L_k = \begin{bmatrix} 1 & & & & & \\ 0 & \ddots & & & & \\ \cdot & 0 & & & & \\ \cdot & \cdot & 1 & & & \\ \cdot & \cdot & -l_{k+1k} & \ddots & & \\ & & \vdots & & \ddots & \\ 0 & 0 & -l_{nk} & 0 & 0 & 1 \end{bmatrix}, \qquad l_{ik} = \frac{a_{ik}^{(k-1)}}{a_{kk}^{(k-1)}}, \quad i = k+1, \ldots, n \tag{1.3.11}$$

Thus the triangular system (1.3.4) is obtained by

$$\hat{L}A\mathbf{x} = \hat{L}\mathbf{b}, \qquad \hat{L} = L_{n-1} \cdots L_2 L_1 \tag{1.3.12}$$

The matrix $\hat{L}$ is lower triangular (Exercise 1.1-10), and it is easy to see that its diagonal elements are equal to 1. The detailed verification of the above steps is left as an exercise (1.3-4).

Denote by U the upper triangular matrix of the system (1.3.4). Then, from (1.3.12),

$$\hat{L}A = U \tag{1.3.13}$$

Since $\hat{L}$ is lower triangular with 1's on the main diagonal, $det\ \hat{L} = 1$, and the inverse of $\hat{L}$ exists. Multiplying (1.3.13) by $\hat{L}^{-1}$ and denoting $\hat{L}^{-1}$ by L give

$$A = LU \tag{1.3.14}$$

It is easy to see (Exercise 1.3-5) that L is also lower triangular (and also has 1's on the main diagonal). Hence, (1.3.14) is the factorization of A into the product of lower and upper triangular matrices; this is known as the *LU factorization* or *LU decomposition* of A. A very simple example is

$$\begin{bmatrix} 2 & 1 \\ 2 & 2 \end{bmatrix} = \begin{bmatrix} 1 & 0 \\ 1 & 1 \end{bmatrix} \begin{bmatrix} 2 & 1 \\ 0 & 1 \end{bmatrix}$$

The factorization (1.3.14) was predicated on the assumption that no divisor of the Gaussian elimination process was zero. If a divisor is zero, we saw before that we can circumvent the problem by interchanging rows of A, and we wish to represent these interchanges by matrix multiplications. A *permutation matrix* is an $n \times n$ matrix with exactly one 1 in each row and column and 0's elsewhere. Examples are

$$\begin{bmatrix} 0 & 0 & 1 \\ 0 & 1 & 0 \\ 1 & 0 & 0 \end{bmatrix}, \qquad \begin{bmatrix} 1 & 0 & 0 \\ 0 & 0 & 1 \\ 0 & 1 & 0 \end{bmatrix} \tag{1.3.15}$$

The identity matrix is the trivial example of a permutation matrix, but any permutation matrix can be obtained from the identity matrix by rearranging rows or columns. If P is a permutation matrix, the effect of the multiplications PA and AP is to rearrange rows or columns of A, respectively, in exactly the same pattern that P itself was obtained from the identity matrix. For example, multiplication of a 3×3 matrix on the left by the first matrix of (1.3.15) interchanges the first and third rows, and multiplication on the right interchanges the first and third columns. It is easy to show (Exercise 1.3-6) that a permutation matrix is orthogonal and, hence, nonsingular.

Now suppose that $a_{11} = 0$, so that to carry out Gaussian elimination we must interchange the first row with, say, the kth. This can be represented by P_1A, where P_1 is a permutation matrix. Then the matrix L_1 of (1.3.10) (with the elements of A now replaced by the elements of P_1A) is applied: L_1P_1A. For the second step, a permutation of rows is again necessary if $a_{22}^1 = 0$, and we represent this by $P_2L_1P_1A$ for a suitable permutation matrix P_2. Continuing in this way, we see that (1.3.12) is replaced by

$$\hat{L}A\mathbf{x} = \hat{L}\mathbf{b}, \qquad \hat{L} = L_{n-1}P_{n-1}\cdots L_2P_2L_1P_1 \tag{1.3.16}$$

Note that the permutation matrix P_i is just the identity if no interchange is needed at the ith stage. The matrices L_i of (1.3.16) are, again, all lower triangular with 1's on the main diagonal, but, because of the permutation matrices, $\hat{L}$ itself need not be lower triangular. $\hat{L}^{-1}$ still exists, however, by 1.3.2, since a permutation matrix is nonsingular (Exercise 1.3-6), and thus $\hat{L}$ is a product of nonsingular matrices. Therefore, we can write $A = \hat{L}^{-1}U$, which corresponds to the representation (1.3.14), although $\hat{L}^{-1}$ is now not necessarily lower triangular. An example is given by

$$A = \begin{bmatrix} 2 & 2 & 1 \\ 2 & 2 & 2 \\ 2 & 3 & 1 \end{bmatrix}, \quad \hat{L} = \begin{bmatrix} 1 & 0 & 0 \\ 0 & 0 & 1 \\ 0 & 1 & 0 \end{bmatrix}\begin{bmatrix} 1 & 0 & 0 \\ -1 & 1 & 0 \\ -1 & 0 & 1 \end{bmatrix} = \begin{bmatrix} 1 & 0 & 0 \\ -1 & 0 & 1 \\ -1 & 1 & 0 \end{bmatrix}$$

$$\hat{L}^{-1} = \begin{bmatrix} 1 & 0 & 0 \\ 1 & 1 & 0 \\ 1 & 0 & 1 \end{bmatrix}\begin{bmatrix} 1 & 0 & 0 \\ 0 & 0 & 1 \\ 0 & 1 & 0 \end{bmatrix} = \begin{bmatrix} 1 & 0 & 0 \\ 1 & 0 & 1 \\ 1 & 1 & 0 \end{bmatrix}, \qquad U = \begin{bmatrix} 2 & 2 & 1 \\ 0 & 1 & 0 \\ 0 & 0 & 1 \end{bmatrix}$$

where $\hat{L}^{-1}$ is obtained as $L^{-1}P^{-1}$.

The decomposition process when permutations are involved can be viewed in a somewhat different way, which gives the following useful result.

1.3.3. If A is a nonsingular $n \times n$ matrix, then there is a permutation matrix P such that

$$PA = LU \tag{1.3.17}$$

where L is lower triangular with 1's on the main diagonal, and U is upper triangular.

We could prove (1.3.17) by formal matrix manipulations, but it is easier to argue as follows. Suppose that on the ith step of the elimination process there is a zero in the ith diagonal position, so that an interchange with, say, the kth row is necessary. Now imagine that we had interchanged the ith and kth rows of the original matrix A before the elimination began. Then when we arrive at the ith stage, we will find in the ith diagonal position exactly the nonzero element that we would have interchanged into that position in the original process. Thus, if we carry out the original process and discover what rows need to be interchanged and then make exactly the same row interchanges on the original matrix A to obtain the matrix PA, the elimination process can be carried out on PA with no interchanges.

If we use 1.3.3, the previous example becomes

$$PA = \begin{bmatrix} 1 & 0 & 0 \\ 0 & 0 & 1 \\ 0 & 1 & 0 \end{bmatrix} \begin{bmatrix} 2 & 2 & 1 \\ 2 & 2 & 2 \\ 2 & 3 & 1 \end{bmatrix} = \begin{bmatrix} 2 & 2 & 1 \\ 2 & 3 & 1 \\ 2 & 2 & 2 \end{bmatrix}$$

$$L = \begin{bmatrix} 1 & 0 & 0 \\ 1 & 1 & 0 \\ 1 & 0 & 1 \end{bmatrix}, \qquad U = \begin{bmatrix} 2 & 2 & 1 \\ 0 & 1 & 0 \\ 0 & 0 & 1 \end{bmatrix}$$

We note that the matrices L of (1.3.17) and $\hat{L}$ of (1.3.16) are related by $\hat{L}^{-1} = P^{-1}L$, and that the matrix U of (1.3.17) is exactly that produced by the Gaussian elimination process with interchanges. We also note that (1.3.17) shows that $\det P \det A = \det L \det U$. Since P is obtained from the identity matrix by interchange of rows, 1.2.2 shows that $\det P = +1$ if an even number of interchanges are made, and $\det P = -1$ otherwise. Hence, since $\det L = 1$ and $\det U$ is the product of its diagonal elements u_{ii}, we have

$$\det A = \pm u_{11} \cdots u_{nn} \tag{1.3.18}$$

where the minus sign is taken if an odd number of row interchanges are made in the elimination process. Thus, $\det A$ is easily calculated as a by-product of Gaussian elimination by just forming the product of the diagonal elements of the final upper triangular matrix and keeping track of the number of row interchanges made. This is, in general, the most effective way to compute determinants.

Other Criteria for Nonsingularity

We now interpret the singularity or nonsingularity of A in another useful way, which we will expand upon in Chapter 2. A special case of Theorem 1.3.1 is when $\mathbf{b} = 0$; in this case, if A is nonsingular, $\mathbf{x} = 0$ is the only solution of $A\mathbf{x} = 0$. By the representation (1.1.11) of $A\mathbf{x}$ as the linear combination

$$A\mathbf{x} = \sum_{i=1}^{n} x_i \mathbf{a}_i$$

of the columns $\mathbf{a}_1, \ldots, \mathbf{a}_n$ of A, the statement that $A\mathbf{x} = 0$ has only the solution $\mathbf{x} = 0$ is equivalent to saying that the columns of A are linearly independent vectors. Thus, A is nonsingular if and only if its columns are linearly independent.

It is now useful to collect the various results on nonsingularity of this section into the following theorem.

1.3.4. Criteria for Nonsingularity. *Let A be an $n \times n$ real or complex matrix. Then the following are equivalent:*

(a) *$\det A \neq 0$;*
(b) *There is an $n \times n$ matrix A^{-1} such that $AA^{-1} = A^{-1}A = I$;*
(c) *$A\mathbf{x} = \mathbf{b}$ has a unique solution for any $\mathbf{b}$;*
(d) *$A\mathbf{x} = 0$ has only the solution $\mathbf{x} = 0$;*
(e) *The columns (or rows) of A are linearly independent.*

We have already proved that (a) and (b) are equivalent and that (a) implies (c). (d) is a special case of (c), and we just saw that (e) and (d) are equivalent for the columns of A. The statement in (e) about the rows follows immediately from Theorem 1.2.4 ($\det A = \det A^T$), so that linear independence of the columns of A^T is equivalent to that of the rows of A. Finally, we will show that (d) implies (a), and this will complete the proof.

Suppose that (d) holds, but that $\det A = 0$. We will show that this leads to a contradiction. If $\det A = 0$, the Gaussian elimination process must lead to a point in the elimination process where the reduced matrix has the form

$$\hat{A} = \begin{bmatrix} * & \cdots & * & \cdots & * \\ & \ddots & \vdots & & \vdots \\ & * & * & \cdots & * \\ & & 0 & * \; * \cdots & * \\ & & \vdots & \ddots & \vdots \\ & & 0 & * \cdots & * \end{bmatrix} = \begin{bmatrix} A_1 & \mathbf{a} & B \\ 0 & \mathbf{0} & C \end{bmatrix} \tag{1.3.19}$$

Here the asterisks indicate elements that are not necessarily zero, and in the kth column the diagonal element and all elements below it are zero. If this did not happen for some column, then the elimination process could be successfully completed with every element on the main diagonal nonzero, and we would have, from (1.3.18), det $A \neq 0$. Suppose, first, that $k = 1$, that is, the first column of A is all zero. Then $A\mathbf{e}_1 = 0$, where, again, $\mathbf{e}_1$ is the vector with a 1 in the first position and 0's elsewhere. Thus (d) is violated. Assume then that $k \geq 2$, and partition the matrix as indicated in (1.3.19), where A_1 is a $(k-1) \times (k-1)$ nonsingular triangular matrix, $\mathbf{a}$ is a $(k-1)$-long column vector, and B and C are $(k-1) \times (n-k)$ and $(n-k+1) \times (n-k)$ matrices, respectively. Since A_1 is nonsingular, the system $A_1\mathbf{x}_1 = \mathbf{a}$ has a solution $\mathbf{x}_1$. Then

$$\begin{bmatrix} A_1 & \mathbf{a} & B \\ 0 & \mathbf{0} & C \end{bmatrix} \begin{bmatrix} \mathbf{x}_1 \\ -1 \\ \mathbf{0} \end{bmatrix} = \begin{bmatrix} A_1\mathbf{x}_1 - \mathbf{a} \\ \mathbf{0} \end{bmatrix} = 0$$

Thus there is a non-zero vector $\hat{\mathbf{x}}$ such that $\hat{A}\hat{\mathbf{x}} = 0$. Since

$$\hat{A} = L_{k-1}P_{k-1} \cdots L_1P_1A$$

where the L_i and P_i are the lower triangular and permutation matrices of the elimination process, we have

$$\hat{A}\hat{\mathbf{x}} = L_{k-1}P_{k-1} \cdots L_1P_1A\hat{\mathbf{x}} = 0$$

Thus, since the L_i and P_i are all nonsingular, we must have $A\hat{\mathbf{x}} = 0$ for $\hat{\mathbf{x}} \neq 0$. This is a contradiction, and therefore det $A \neq 0$.

We will see in the next section a much easier way to prove this last result by means of eigenvalues, but the above proof provides useful practice in the topics of this section.

Echelon Form and Rank

We now assume that the matrix A is singular or, more generally, an $m \times n$ matrix. We can apply the Gaussian elimination process as before, using row interchanges if necessary; but it may happen that when we encounter a zero pivot element, every other element below it in that column is also zero. In this case we simply move to the next column, in the same row, and continue the process. For example, let

$$A = \begin{bmatrix} 2 & 1 & 2 & 1 & 2 \\ 2 & 1 & 3 & 2 & 1 \\ 2 & 1 & 3 & 2 & 1 \\ 2 & 1 & 3 & 2 & 1 \end{bmatrix}$$

The first stage of Gaussian elimination produces the reduced matrix

$$\begin{bmatrix} 2 & 1 & 2 & 1 & 2 \\ 0 & 0 & 1 & 1 & -1 \\ 0 & 0 & 1 & 1 & -1 \\ 0 & 0 & 1 & 1 & -1 \end{bmatrix}$$

Since the second column is zero from the second position down, we move to the third column, and the (2, 3) element becomes the pivot element for subsequent elimination. The next stage produces

$$\begin{bmatrix} 2 & 1 & 2 & 1 & 2 \\ 0 & 0 & 1 & 1 & -1 \\ 0 & 0 & 0 & 0 & 0 \\ 0 & 0 & 0 & 0 & 0 \end{bmatrix} \tag{1.3.20}$$

and the process is complete for this matrix. In general, the reduced form may have an appearance like

$$\begin{bmatrix} * & * & * & \cdot & \cdot & \cdot & \cdots & * \\ 0 & 0 & 0 & * & \cdot & \cdot & \cdots & * \\ \cdot & & & 0 & * & \cdot & \cdots & * \\ \cdot & & & & 0 & * & \cdots & * \\ \cdot & & & & & & \cdots & 0 \\ \cdot & & & & & & & \vdots \\ 0 & \cdot & \cdot & \cdot & & & & 0 \end{bmatrix} \tag{1.3.21}$$

where the asterisks indicate nonzero elements. This is the *echelon* or *row echelon* form of the matrix. In general, the first nonzero element in each row appears at least one position to the right of the first nonzero element in the previous row.

As with Gaussian elimination for nonsingular matrices, we can express this process as

$$PA = LU \tag{1.3.22}$$

where P is an $m \times m$ permutation matrix, L is an $m \times m$ nonsingular lower triangular matrix, and U is an $m \times n$ matrix of the form (1.3.21). If A is $n \times n$ and nonsingular, then, as before, U is $n \times n$ with nonzero diagonal elements, and thus the row echelon form of a nonsingular matrix is a nonsingular upper triangular matrix.

If we consider the columns of A as vectors, the number of linearly independent columns is the *column rank* of A. Similarly, the number of linearly independent rows is the *row rank* of A. The row rank of the matrix U of (1.3.21) is just the number of nonzero rows, since each nonzero row will have leading nonzero elements that do not appear in any subsequent rows; for example, the first two rows of (1.3.20) are clearly linearly independent, and the row rank of that matrix is 2. Next, consider the columns of U that contain the pivot elements. These columns are linearly independent because each has a nonzero element in a position that does not appear in any previous column; for example, the pivot elements of the matrix (1.3.20) are in columns 1 and 3, and, clearly, these columns are linearly independent. Moreover, every other column is a linear combination of these columns with the pivot elements, since finding the coefficients $c_1, \ldots, c_r$ of the linear combination is equivalent to solving a triangular system of equations

$$\begin{bmatrix} * & & \cdots & * \\ & * & & \vdots \\ & & \ddots & \vdots \\ & & & * \end{bmatrix} \begin{bmatrix} c_1 \\ \vdots \\ c_r \end{bmatrix} = \mathbf{b}$$

where the columns of this system are the columns of U with pivot elements, shortened to length r. Since the pivot elements are nonzero, this matrix is nonsingular. Each row of U contains exactly one pivot element, and thus we have shown that the number of linearly independent rows of U is the number of linearly independent columns; that is, row rank(U) = column rank(U). We next wish to conclude the same for A itself.

The operations that transformed the rows of A to those of U—interchanges and additions of multiples of one row to another row—cannot change the number of linearly independent rows, just as the same operations applied to a nonsingular matrix do not change the nonsingularity. More precisely, suppose that A has r linearly independent rows. Since, clearly, interchanges do not affect linear independence, we may assume that these are the first r rows $\mathbf{a}_1, \ldots, \mathbf{a}_r$ of A and that the corresponding rows of U are given by

$$\mathbf{u}_i = \mathbf{a}_i - \sum_{j=1}^{i-1} \alpha_{ij}\mathbf{a}_j, \qquad i = 1, \ldots, r$$

If $\sum_{i=1}^{r} c_i\mathbf{u}_i = 0$, then collecting coefficients of the $\mathbf{a}_i$ shows that the c_i satisfy the $r \times r$ linear system

$$\begin{bmatrix} 1 & -\alpha_{21} & \cdots & -\alpha_{r1} \\ & 1 & \ddots & \vdots \\ & & \ddots & -\alpha_{r,r-1} \\ & & & 1 \end{bmatrix} \begin{bmatrix} c_1 \\ \vdots \\ c_r \end{bmatrix} = 0$$

Since this triangular coefficient matrix is nonsingular, the c_i's must all be zero, so that $\mathbf{u}_1, \ldots, \mathbf{u}_r$ are linearly independent. In a similar way one shows that if $\mathbf{u}_1, \ldots, \mathbf{u}_r$ are linearly independent, then the corresponding rows of A are linearly independent. Hence, row rank(A) = row rank(U).

For the columns we can write (1.3.22) in the form $BA = U$, where $B = L^{-1}P$ is nonsingular. If $\mathbf{a}_1, \ldots, \mathbf{a}_r$ are linearly independent columns of A, then so are $B\mathbf{a}_1, \ldots, B\mathbf{a}_r$. This follows from assuming that $\sum_{i=1}^r c_i B\mathbf{a}_i = 0$ for nonzero coefficients $c_1, \ldots, c_r$. Then

$$B\left(\sum_{i=1}^{r} c_i \mathbf{a}_i\right) = 0$$

which would show that $B\mathbf{x} = 0$ has a nonzero solution in violation of Theorem 1.3.1. The same argument can be applied to $B^{-1}U = A$, and we conclude that column rank(A) = column rank(U). Thus we have proved the following result.

1.3.5. The row rank and column rank of any $m \times n$ matrix A are equal.

On the basis of 1.3.5 we define the *rank* of an $m \times n$ matrix A, denoted by rank(A), as the number of linearly independent rows of A or, equivalently, the number of linearly independent columns. We will see in Chapter 2 that rank(A) plays a key role in ascertaining the number of solutions of a system of equations $A\mathbf{x} = \mathbf{b}$ when A is singular.

Exercises 1.3

1. Solve the system

$$\begin{bmatrix} 1 & 2 & 2 \\ 2 & 3 & 2 \\ 3 & 4 & 3 \end{bmatrix} \begin{bmatrix} x_1 \\ x_2 \\ x_3 \end{bmatrix} = \begin{bmatrix} 1 \\ 2 \\ 3 \end{bmatrix}$$

by Gaussian elimination and also write the solution process in terms of the matrices L_i of (1.3.12).

2. If A is the block matrix diag($A_1, \ldots, A_p$) and each A_i is nonsingular, show that $A^{-1} = \text{diag}(A_1^{-1}, \ldots, A_p^{-1})$. Apply this result to find the inverse of

$$\begin{bmatrix} 1 & 1 & 1 & 0 & 0 & 0 \\ 0 & 2 & 2 & 0 & 0 & 0 \\ 0 & 0 & 3 & 0 & 0 & 0 \\ 0 & 0 & 0 & 1 & 1 & 1 \\ 0 & 0 & 0 & 0 & 2 & 2 \\ 0 & 0 & 0 & 0 & 0 & 3 \end{bmatrix}$$

3. Ascertain whether the following matrices are singular or nonsingular:

(a) $\begin{bmatrix} 2 & 1 \\ 1 & 2 \end{bmatrix}$

(b) $\begin{bmatrix} 1 & 1 & 1 \\ 2 & 2 & 2 \\ 3 & 3 & 3 \end{bmatrix}$

(c) $\begin{bmatrix} 4 & 4 & 4 & 4 \\ 0 & 3 & 3 & 3 \\ 0 & 0 & 2 & 2 \\ 0 & 0 & 0 & 1 \end{bmatrix}$

4. Show that $L_1 A\mathbf{x} = L_1 \mathbf{b}$ is the reduced equation (1.3.3), where L_1 is given by (1.3.10). More generally, verify that (1.3.4) is the same as (1.3.12), where the L_k are given by (1.3.11). Show also that the diagonal elements of the matrix of $\hat{L}$ of (1.3.12) are all 1's.
5. Show that the inverse of a lower (upper) triangular matrix is lower (upper) triangular and that it has all 1's on the main diagonal if the original matrix does.
6. Show that a permutation matrix is orthogonal and, hence, nonsingular.
7. Show that if A is an orthogonal matrix, then A^{-1} is an orthogonal matrix.
8. Show that m n-vectors cannot be linearly independent if $m > n$.
9. Let A and B be nonsingular $n \times n$ matrices. Show that $(AB)^{-1} = B^{-1}A^{-1}$.
10. Verify that if $ad - bc \neq 0$, then

$$\begin{bmatrix} a & b \\ c & d \end{bmatrix}^{-1} = \frac{1}{ad - bc}\begin{bmatrix} d & -b \\ -c & a \end{bmatrix}$$

11. Compute the inverses of

(a) $\begin{bmatrix} 2 & 1 \\ 1 & 2 \end{bmatrix}$

(b) $\begin{bmatrix} 1 & 3 \\ 2 & 2 \end{bmatrix}$

(c) $\begin{bmatrix} 4 & 2 \\ 4 & 1 \end{bmatrix}$

by solving the systems (1.3.5). Check your results by using Exercise 10.

12. Let

$$A = \begin{bmatrix} A_{11} & A_{12} \\ A_{21} & A_{22} \end{bmatrix}$$

be a partitioning of A with A_{11} nonsingular. Show that

$$A = \begin{bmatrix} I & 0 \\ A_{21}A_{11}^{-1} & I \end{bmatrix} \begin{bmatrix} A_{11} & A_{12} \\ 0 & A_{22} - A_{21}A_{11}^{-1}A_{12} \end{bmatrix}$$

Use this decomposition to show that

$$\det A = \det A_{11} \det(A_{22} - A_{21}A_{11}^{-1}A_{12})$$

Also show that if A is nonsingular then

$$A^{-1} = \begin{bmatrix} B_{11} & B_{12} \\ B_{21} & B_{22} \end{bmatrix}$$

with

$$B_{22} = (A_{22} - A_{21}A_{11}^{-1}A_{12})^{-1}, \qquad B_{12} = -A_{11}^{-1}A_{12}B_{22},$$

$$B_{21} = -B_{22}A_{21}A_{11}^{-1}, \qquad B_{11} = A_{11}^{-1}(I - A_{12}B_{21})$$

13. An $n \times n$ matrix A is *idempotent* if $A^2 = A$. Show that an idempotent matrix is singular unless it is the identity matrix.
14. Let A be a real nonsingular $n \times n$ matrix, and let $\mathbf{u}$ and $\mathbf{v}$ be real column vectors such that $\mathbf{v}^T A^{-1}\mathbf{u} \neq -1$. Show that $A + \mathbf{u}\mathbf{v}^T$ is nonsingular and verify the *Sherman-Morrison* formula

$$(A + \mathbf{u}\mathbf{v}^T)^{-1} = A^{-1} - \frac{A^{-1}\mathbf{u}\mathbf{v}^T A^{-1}}{1 + \mathbf{v}^T A^{-1}\mathbf{u}}$$

(*Hint*: Multiply $A + \mathbf{u}\mathbf{v}^T$ by its alleged inverse.) More generally, if U and V are real $n \times m$ matrices such that $I + V^T A^{-1} U$ is nonsingular, show that $A + UV^T$ is nonsingular, and verify the *Sherman-Morrison-Woodbury* formula

$$(A + UV^T)^{-1} = A^{-1} - A^{-1}U(I + V^T A^{-1} U)^{-1} V^T A^{-1}$$

Formulate and prove the corresponding results for complex matrices.

1.4. Eigenvalues and Eigenvectors

A large portion of this book will deal with various aspects of eigenvalues and eigenvectors of a matrix, and we will cover only the most basic background in this section.

An *eigenvalue* (also called a proper value, characteristic value, or latent root) of a $n \times n$ matrix A is a real or complex scalar λ satisfying the equation

$$A\mathbf{x} = \lambda\mathbf{x} \tag{1.4.1}$$

for some nonzero vector $\mathbf{x}$, called an *eigenvector*. Note that an eigenvector is only determined up to a scalar multiple, since if $\mathbf{x}$ is an eigenvector and α is a nonzero scalar, then $\alpha\mathbf{x}$ is also an eigenvector. We stress that an eigenvector is nonzero, since (1.4.1) is trivially satisfied for the zero vector for any scalar λ.

We can rewrite (1.4.1) in the form

$$(A - \lambda I)\mathbf{x} = 0 \tag{1.4.2}$$

Thus, from Theorem 1.3.1, (1.4.2) can have a nonzero solution $\mathbf{x}$ only if $A - \lambda I$ is singular. Therefore, any eigenvalue λ must satisfy the equation

$$\det(A - \lambda I) = 0 \tag{1.4.3}$$

By the cofactor expansion 1.2.6 it is easy to see that (1.4.3) is simply a polynomial of degree n in λ. For example, for $n = 3$,

$$\det\begin{bmatrix} a_{11}-\lambda & a_{12} & a_{13} \\ a_{21} & a_{22}-\lambda & a_{23} \\ a_{31} & a_{32} & a_{33}-\lambda \end{bmatrix}$$

$$= (a_{11}-\lambda)\det\begin{bmatrix} a_{22}-\lambda & a_{23} \\ a_{32} & a_{33}-\lambda \end{bmatrix}$$

$$- a_{21}\det\begin{bmatrix} a_{12} & a_{13} \\ a_{32} & a_{33}-\lambda \end{bmatrix} + a_{31}\det\begin{bmatrix} a_{12} & a_{13} \\ a_{22}-\lambda & a_{23} \end{bmatrix} \tag{1.4.4}$$

Expanding these 2×2 determinants, we see that the first term gives rise to cubic and squared terms in λ, while the last two terms give only linear and constant terms. Thus, it is a polynomial of degree 3. It is left to the reader (Exercise 1.4-1) to verify in detail that (1.4.3), is, indeed, a polynomial of degree n.

The polynomial (1.4.3) is known as the *characteristic polynomial* of A, and its roots are eigenvalues of A. The set of eigenvalues is called the *spectrum* of A. The collection of both eigenvalues and eigenvectors is called the *eigensystem* of A. By the Fundamental Theorem of Algebra a polynomial of degree n has exactly n real or complex roots, counting multiplicities. Hence, an $n \times n$ matrix has exactly n eigenvalues, although they are not necessarily distinct. For example, if $A = I$ is the identity matrix, then (1.4.3) reduces to $(1-\lambda)^n = 0$, which has the root 1 with multiplicity n. Thus, the eigenvalues of the identity matrix are $1, 1, \ldots, 1$ (n times).

If A is a triangular matrix, then, by 1.2.7

$$\det(A - \lambda I) = (a_{11}-\lambda)(a_{22}-\lambda)\cdots(a_{nn}-\lambda)$$

Thus, in this case, the eigenvalues are simply the diagonal elements of A. The same also holds, of course, for diagonal matrices. Except for triangular matrices and a few other special cases, eigenvalues are relatively difficult to compute, but a number of well-founded algorithms for their computation are now known. [We stress that one does *not* compute the characteristic polynomial (1.4.3) and then find its roots.] However, it is easy to compute at least some of the eigenvalues and eigenvectors of certain functions of a matrix A if the eigenvalues and eigenvectors of A are known. As the simplest example, let λ and $\mathbf{x}$ be any eigenvalue and corresponding eigenvector of A. Then

$$A^2\mathbf{x} = A(A\mathbf{x}) = A(\lambda\mathbf{x}) = \lambda A\mathbf{x} = \lambda^2\mathbf{x}$$

so that λ^2 is an eigenvalue of A^2 and $\mathbf{x}$ is a corresponding eigenvector. Proceeding in the same way for any integer power of A, we conclude the following:

1.4.1. If the $n \times n$ matrix A has eigenvalues $\lambda_1, \ldots, \lambda_n$, then for any positive integer m, $\lambda_1^m, \ldots, \lambda_n^m$ are eigenvalues of A^m. Moreover, any eigenvector of A is an eigenvector of A^m.

We note that the converse of 1.4.1 is not true. For example, if

$$A = \begin{bmatrix} 1 & 0 \\ 0 & -1 \end{bmatrix}, \qquad A^2 = \begin{bmatrix} 1 & 0 \\ 0 & 1 \end{bmatrix}$$

then every vector is an eigenvector of A^2, but the only eigenvectors of A are multiples of $(1, 0)^T$ and $(0, -1)^T$. Similarly,

$$A = \begin{bmatrix} 0 & 1 \\ 0 & 0 \end{bmatrix}$$

has only one linearly independent eigenvector, but, again, every vector is an eigenvector of A^2. Moreover, all three matrices

$$A = \begin{bmatrix} 1 & 0 \\ 0 & -1 \end{bmatrix}, \qquad A = \begin{bmatrix} -1 & 0 \\ 0 & -1 \end{bmatrix}, \qquad A = \begin{bmatrix} 1 & 0 \\ 0 & 1 \end{bmatrix}$$

give $A^2 = I$, so that knowing the eigenvalues of A^2 does not allow a precise determination of the eigenvalues of A.

By taking linear combinations of powers of a matrix, we can form a *polynomial* in A defined by

$$p(A) \equiv \alpha_0 I + \alpha_1 A + \alpha_2 A^2 + \cdots + \alpha_m A^m \tag{1.4.5}$$

for given scalars $\alpha_0, \ldots, \alpha_m$. By means of 1.4.1 it is immediate that if λ and $\mathbf{x}$ are any eigenvalue and corresponding eigenvector of A, then

$$p(A)\mathbf{x} = \alpha_0\mathbf{x} + \alpha_1 A\mathbf{x} + \cdots + \alpha_m A^m\mathbf{x} = (\alpha_0 + \alpha_1\lambda + \cdots + \alpha_m\lambda^m)\mathbf{x}$$

We can summarize this as follows by using the somewhat ambiguous (but standard) notation of p for both the matrix polynomial (1.4.5) as well as the scalar polynomial with the same coefficients.

1.4.2. If the $n \times n$ matrix A has eigenvalues $\lambda_1, \ldots, \lambda_n$, then $p(\lambda_1), \ldots, p(\lambda_n)$ are eigenvalues of the matrix $p(A)$ of (1.4.5). *Moreover, any eigenvector of A is an eigenvector of $p(A)$.*

A particularly useful special case of this last result is when $p(A) = \alpha I + A$. Thus, the eigenvalues of $\alpha I + A$ are just $\alpha + \lambda_i$, $i = 1, \ldots, n$, where the λ_i are the eigenvalues of A. An example is

$$A = \begin{bmatrix} 0 & 1 \\ 1 & 0 \end{bmatrix}, \qquad B = \begin{bmatrix} \alpha & 1 \\ 1 & \alpha \end{bmatrix}$$

Here A has eigenvalues ± 1, so that B has eigenvalues $\alpha + 1$ and $\alpha - 1$.

We can also immediately compute the eigenvalues of A^{-1}, if A is nonsingular, by multiplying the basic relation $A\mathbf{x} = \lambda\mathbf{x}$ by A^{-1}. Thus,

$$\mathbf{x} = \lambda A^{-1}\mathbf{x} \qquad \text{or} \qquad A^{-1}\mathbf{x} = \lambda^{-1}\mathbf{x}$$

so that the eigenvalues of A^{-1} are the inverses of the eigenvalues of A. But what if A has a zero eigenvalue? This cannot happen if A is nonsingular, for then we would have a nonzero vector $\mathbf{x}$ satisfying $A\mathbf{x} = 0$, which would contradict Theorem 1.3.4. Hence, we can conclude that

1.4.3. If A is an $n \times n$ nonsingular matrix with eigenvalues $\lambda_1, \ldots, \lambda_n$, then $\lambda_1^{-1}, \ldots, \lambda_n^{-1}$ are eigenvalues of A^{-1}. Moreover, any eigenvector of A is an eigenvector of A^{-1}.

The eigenvalues of A^T or A^* are also immediately obtained from the eigenvalues of A. Indeed, by 1.2.4,

$$\det(A^* - \lambda I) = \overline{\det(A - \bar{\lambda} I)}$$

which shows that the eigenvalues of A^* are the complex conjugates of the eigenvalues of A. In particular, if A is real, its characteristic polynomial has real coefficients, and thus its roots are either real or occur in complex conjugate pairs. Therefore, we have the following result.

1.4.4. *If A is a real $n \times n$ matrix, A and A^T have the same eigenvalues. If A is complex, then the eigenvalues of A^* are the complex conjugates of the eigenvalues of A.*

A and A^T do not necessarily have the same eigenvectors, however (see Exercise 1.4-3). If $A^T\mathbf{x} = \lambda\mathbf{x}$, then $\mathbf{x}^T A = \lambda\mathbf{x}^T$, and the row vector $\mathbf{x}^T$ is called a *left eigenvector* of A.

Eigenvalue calculations also simplify considerably for block triangular matrices. If

$$A = \begin{bmatrix} A_{11} & & \cdots & A_{1p} \\ & A_{22} & & \vdots \\ & & \ddots & \\ & & & A_{pp} \end{bmatrix}$$

where each A_{ii} is a square matrix, the cofactor expansion 1.2.6 allows us to conclude (Exercise 1.4-6) that

$$\det(A - \lambda I) = \det(A_{11} - \lambda I)\det(A_{22} - \lambda I)\cdots\det(A_{pp} - \lambda I) \quad (1.4.6)$$

Thus, the characteristic polynomial of A factors into the product of the characteristic polynomials of the matrices A_{ii}, and to compute the eigenvalues of A we need only compute the eigenvalues of the smaller matrices A_{ii}.

We next relate the determinant of A to its eigenvalues. We can write the characteristic polynomial of A in terms of the eigenvalues $\lambda_1, \ldots, \lambda_n$ of A as

$$\det(A - \lambda I) = (\lambda_1 - \lambda)(\lambda_2 - \lambda)\cdots(\lambda_n - \lambda) \quad (1.4.7)$$

This is just the factorization of a polynomial in terms of its roots. By evaluating (1.4.7) for $\lambda = 0$, we conclude that

$$\det A = \lambda_1 \cdots \lambda_n \quad (1.4.8)$$

so that the determinant of A is just the product of its n eigenvalues. Thus we have another characterization of the nonsingularity of a matrix to complement Theorem 1.3.4.

1.4.5. *An $n \times n$ real or complex matrix A is nonsingular if and only if all of its eigenvalues are nonzero.*

We note that 1.4.5 allows a much simpler proof of the assertion of Theorem 1.3.4 that the linear independence of the columns of A implies that $\det A \neq 0$. The linear independence ensures that there is no zero eigenvalue since $A\mathbf{x} = 0$ for a nonzero $\mathbf{x}$ would violate the linear independence of the columns of A.

We also note that (1.4.7) can be written in the form

$$\det(A - \lambda I) = (\lambda_1 - \lambda)^{r_1}(\lambda_2 - \lambda)^{r_2} \cdots (\lambda_p - \lambda)^{r_p}$$

where the λ_i are distinct and $\sum_{i=1}^{p} r_i = n$. Then r_i is the multiplicity of the eigenvalue λ_i. An eigenvalue of multiplicity 1 is called *simple.*

Symmetric and Hermitian Matrices

As noted previously, the eigenvalues of a matrix A may be either real or complex, even if A itself is real. This is simply a manifestation of the fact that a polynomial with real coefficients may have complex roots. There is an important case, however, in which the eigenvalues must be real. Recall that A is Hermitian if $A^* = A$. Suppose that λ is an eigenvalue of A and $\mathbf{x}$ is a corresponding eigenvector, that is, $A\mathbf{x} = \lambda\mathbf{x}$. Multiplying both sides by $\mathbf{x}^*$ gives

$$\mathbf{x}^*A\mathbf{x} = \lambda\mathbf{x}^*\mathbf{x} \tag{1.4.9}$$

Now take the complex conjugate of $\mathbf{x}^*A\mathbf{x}$:

$$\overline{\mathbf{x}^*A\mathbf{x}} = (\mathbf{x}^*A\mathbf{x})^* = \mathbf{x}^*A^*\mathbf{x} = \mathbf{x}^*A\mathbf{x}$$

The first step is valid since $\mathbf{x}^*A\mathbf{x}$ is a scalar; the second step is just the product rule for transposes; and the third step uses the fact that A is Hermitian. Thus, $\mathbf{x}^*A\mathbf{x}$ is real, and since $\mathbf{x}^*\mathbf{x}$ is real and nonzero (Section 1.1), it follows from (1.4.9) that λ must be real. Since a real symmetric matrix is a special case of a Hermitian matrix, we have proved the following important result.

1.4.6. If the $n \times n$ matrix A is complex and Hermitian or real and symmetric, its eigenvalues are all real.

This is only one of the many important facts about Hermitian and symmetric matrices that we will develop at various points in this book.

A real $n \times n$ matrix A is *skew-symmetric* if $A^T = -A$, and *skew-Hermitian* if A is complex and $A^* = -A$. As we shall see later, skew matrices have many of the same properties as symmetric or Hermitian matrices. An important difference is the following:

1.4.7. If the $n \times n$ matrix A is complex and skew-Hermitian or real and skew-symmetric, its eigenvalues are all pure imaginary.

The proof parallels that given for Hermitian matrices, but now

$$(\mathbf{x}^*A\mathbf{x})^* = -\mathbf{x}^*A\mathbf{x}$$

so that $\mathbf{x}^*A\mathbf{x}$ is pure imaginary. Hence, from (1.4.9), it follows that λ is pure imaginary.

The quantity $\mathbf{x}^*A\mathbf{x}$, which was used in the above proofs, is important in many applications, as we shall see later. It also used to define the concept of definiteness of a matrix.

1.4.8. DEFINITION. An $n \times n$ complex Hermitian (or real symmetric) matrix A is

(a) *positive definite* if $\mathbf{x}^*A\mathbf{x} > 0$ for all $\mathbf{x} \neq 0$;
(b) *positive semidefinite* if $\mathbf{x}^*A\mathbf{x} \geq 0$ for all $\mathbf{x}$;
(c) *negative definite* if $\mathbf{x}^*A\mathbf{x} < 0$ for all $\mathbf{x} \neq 0$;
(d) *negative semidefinite* if $\mathbf{x}^*A\mathbf{x} \leq 0$ for all $\mathbf{x}$;
(e) *indefinite* if none of the above hold.

An important type of positive definite (or semidefinite) matrix arises as a product $A = B^*B$, where B is an $m \times n$ matrix. For a given $\mathbf{x}$ set $\mathbf{y} = B\mathbf{x}$. Then $\mathbf{x}^*A\mathbf{x} = \mathbf{y}^*\mathbf{y} > 0$ unless $\mathbf{y} = 0$. If B is $n \times n$ and nonsingular, then $\mathbf{y}$ cannot be zero for nonzero $\mathbf{x}$; hence, A is positive definite. If B is singular, then A is positive semidefinite since $\mathbf{y}^*\mathbf{y} \geq 0$.

We note that if A is a positive definite matrix and λ is an eigenvalue with corresponding eigenvector $\mathbf{x}$, then $\lambda = \mathbf{x}^*A\mathbf{x}/\mathbf{x}^*\mathbf{x} > 0$, which shows that all eigenvalues of a positive definite matrix are positive. In a similar way one may prove (Exercise 1.4-13) the following statements.

1.4.9. If A is an $n \times n$ Hermitian (or real and symmetric) matrix, then all of its eigenvalues are

(a) positive if A is positive definite;
(b) nonnegative if A is positive semidefinite;
(c) negative if A is negative definite;
(d) nonpositive if A is negative semidefinite.

We will see in Chapter 3 that the converses of the above statements also hold, so that the signs of the eigenvalues of a Hermitian matrix characterize its definiteness properties.

Submatrices and Choleski Decomposition

A simple but useful property of positive definite matrices concerns certain submatrices. A *principal* submatrix of an $n \times n$ matrix A (not necessarily symmetric) is any $m \times m$ submatrix obtained by deleting $n - m$ rows of A and the corresponding columns. A *leading principal submatrix* of A of order m is obtained by deleting the last $n - m$ rows and columns of A. In Figure 1.2a,a 2×2 leading principal submatrix is obtained by deleting the last row and column, and in Fig. 1.2b, a 2×2 principal submatrix is obtained by deleting the second row and column.

$$\text{(a)} \begin{bmatrix} 1 & 2 & 4 \\ 3 & 2 & 1 \\ 4 & 5 & 6 \end{bmatrix} \qquad \text{(b)} \begin{bmatrix} 1 & 2 & 4 \\ 3 & 2 & 1 \\ 4 & 5 & 6 \end{bmatrix}$$

Figure 1.2. (a) Leading principal submatrix. (b) Principal submatrix.

1.4.10. If A is an $n \times n$ positive definite Hermitian matrix, then every principal submatrix of A is also Hermitian and positive definite. In particular, the diagonal elements of A are positive.

PROOF. Let A_p be any $p \times p$ principal submatrix. It is clear that A_p is Hermitian, because by deleting corresponding rows and columns of A to obtain A_p we maintain symmetry in the elements of A_p. Now let $\mathbf{x}_p$ be any nonzero p-vector, and let $\mathbf{x}$ be the n-vector obtained from $\mathbf{x}_p$ by inserting zeros in those positions corresponding to the rows deleted from A. Then a direct calculation shows that

$$\mathbf{x}_p^* A_p \mathbf{x}_p = \mathbf{x}^* A \mathbf{x} > 0$$

Thus, A_p is positive definite. □

An immediate consequence of (1.4.8), 1.4.9 and 1.4.10 is the next result.

1.4.11. The determinant of any principal submatrix of a Hermitian positive definite matrix A is positive. In particular, $\det A > 0$.

We end this section with an important variation of the LU decomposition discussed in Section 1.3.

1.4.12. CHOLESKI DECOMPOSITION. *If A is an $n \times n$ positive definite Hermitian matrix, there is a nonsingular lower triangular matrix L with positive diagonal elements such that $A = LL^*$.*

PROOF. The proof is by induction. Clearly, the result is true for $n = 1$, and the induction hypothesis is that it is true for any $(n-1) \times (n-1)$ positive definite Hermitian matrix. Let

$$A = \begin{bmatrix} B & \mathbf{b} \\ \mathbf{b}^* & a_{nn} \end{bmatrix}$$

where B is $(n-1)\times(n-1)$. By 1.4.10 the principal submatrix B is positive definite and thus, by the induction hypothesis, there is a nonsingular lower triangular matrix L_{n-1} with positive diagonal elements such that $B = L_{n-1}L_{n-1}^*$. Let

$$L = \begin{bmatrix} L_{n-1} & 0 \\ \mathbf{c}^* & \alpha \end{bmatrix}$$

where $\mathbf{c}$ and α are to be obtained by equating LL^* to A:

$$\begin{bmatrix} L_{n-1} & 0 \\ \mathbf{c}^* & \alpha \end{bmatrix}\begin{bmatrix} L_{n-1}^* & \mathbf{c} \\ 0 & \bar{\alpha} \end{bmatrix} = \begin{bmatrix} B & \mathbf{b} \\ \mathbf{b}^* & a_{nn} \end{bmatrix} \tag{1.4.10}$$

This implies that $\mathbf{c}$ and α must satisfy

$$L_{n-1}\mathbf{c} = \mathbf{b}, \qquad \mathbf{c}^*\mathbf{c} + |\alpha|^2 = a_{nn} \tag{1.4.11}$$

Since L_{n-1} is nonsingular, the first equation of (1.4.11) determines $\mathbf{c} = L_{n-1}^{-1}\mathbf{b}$. The second equation allows a positive α, provided that $a_{nn} - \mathbf{c}^*\mathbf{c} > 0$. From Exercise 1.3-12 and (1.4.11) we have

$$0 < \det A = \det\begin{bmatrix} B & \mathbf{b} \\ \mathbf{b}^* & a_{nn} \end{bmatrix} = \det B(a_{nn} - \mathbf{b}^*B^{-1}\mathbf{b})$$

$$= \det B(a_{nn} - \mathbf{c}^*L_{n-1}^*(L_{n-1}L_{n-1}^*)^{-1}L_{n-1}\mathbf{c}) = \det B(a_{nn} - \mathbf{c}^*\mathbf{c})$$

Thus, since $\det B > 0$ by 1.4.11, it follows that $a_{nn} - \mathbf{c}^*\mathbf{c} > 0$, and we may take $\alpha = (a_{nn} - \mathbf{c}^*\mathbf{c})^{1/2}$. □

A simple example of a Choleski decomposition is

$$\begin{bmatrix} 4 & 2 \\ 2 & 4 \end{bmatrix} = \begin{bmatrix} 2 & 0 \\ 1 & \sqrt{3} \end{bmatrix}\begin{bmatrix} 2 & 1 \\ 0 & \sqrt{3} \end{bmatrix}$$

Additional examples are given in Exercise 1.4-12.

Exercises 1.4

1. Complete the expansion of (1.4.4) and exhibit the cubic polynomial. More generally, use the cofactor expansion 1.2.6 for general n to conclude that (1.4.3) is a polynomial of degree n in λ.

2. Show that the eigenvalues of the matrix $[\begin{smallmatrix} a & b \\ c & d \end{smallmatrix}]$ are $\{a + d \pm [(a-d)^2 + 4bc]^{1/2}\}/2$. Then compute the eigenvalues of

(a) $\begin{bmatrix} 1 & 2 \\ 2 & 2 \end{bmatrix}$

(b) $\begin{bmatrix} 1 & 4 \\ 1 & 2 \end{bmatrix}$

(c) $\begin{bmatrix} 0 & 1 \\ -1 & 0 \end{bmatrix}$

3. Compute the eigenvalues and eigenvectors of A and A^T, where

$$A = \begin{bmatrix} 2 & 2 \\ 1 & 1 \end{bmatrix}$$

and conclude that A and A^T do not have the same eigenvectors.

4. Compute the eigenvalues and eigenvectors of

$$A = \begin{bmatrix} 2 & 1 \\ 1 & 2 \end{bmatrix}$$

Then use 1.4.2 to compute the eigenvalues and eigenvectors of

(a) $\begin{bmatrix} 1 & 1 \\ 1 & 1 \end{bmatrix}$

(b) $\begin{bmatrix} 0 & 1 \\ 1 & 0 \end{bmatrix}$

(c) $\begin{bmatrix} -1 & 1 \\ 1 & -1 \end{bmatrix}$

5. For the matrix A of Exercise 4, find $\det A^4$ without computing A^4. In addition, find the eigenvalues and eigenvectors of A^{-1}, A^m, $m = 2, 3, 4$, and $I + 2A + 4A^2$, without computing these matrices.
6. Use Exercise 1.2-4 to verify Equation (1.4.6).
7. Use (1.4.6) to compute the eigenvalues of the matrix

$$A = \begin{bmatrix} 1 & 2 & 2 & 4 \\ 2 & 2 & 1 & 5 \\ 0 & 0 & 1 & 4 \\ 0 & 0 & 1 & 2 \end{bmatrix}$$

8. Given the polynomial $p(\lambda) = a_0 + a_1\lambda + \cdots + a_{n-1}\lambda^{n-1} + \lambda^n$, the matrix

$$A = \begin{bmatrix} 0 & 1 & & & \\ & & 1 & & \\ & & & \ddots & \\ & & & & 1 \\ -a_0 & -a_1 & & \cdots & -a_{n-1} \end{bmatrix}$$

is called the *companion matrix* (or *Frobenius matrix*) of p. Use the cofactor expansion 1.2.6 to show that $\det(\lambda I - A) = p(\lambda)$.

9. Ascertain if the matrices (a) and (b) of Exercise 4 are positive definite. If not, what are they?
10. Show that the determinant of a negative definite $n \times n$ Hermitian matrix is positive if n is even and negative if n is odd.
11. List all leading principle submatrices and all principle submatrices of the matrix of Exercise 7.
12. Find the Choleski decomposition of the matrix A of Exercise 4. Try to find the Choleski decomposition of the matrix $\left[\begin{smallmatrix} 1 & 1 \\ 1 & 0 \end{smallmatrix}\right]$. What goes wrong?
13. Prove 1.4.9 in detail.

Review Questions—Chapter 1

Answer whether the following statements are true or false, in general, and justify your assertions. Note that although many of the statements are true for special cases, they may not be true in general under the conditions given. Justification can include citing a particular theorem, giving a counterexample, and so on.

EXAMPLE. If A is an $n \times n$ matrix, then A is nonsingular.

Although this is true for some $n \times n$ matrices, it is false in general.

1. If A is nonsingular, it can be written in the form $A = LU$, where L is lower triangular and U is upper triangular.
2. If $A = \text{diag}(A_1, \ldots, A_p)$, where each A_i is square, then A is nonsingular if and only if all A_i are nonsingular.
3. If $\mathbf{x}$ is a real or complex vector, then $\mathbf{x}^T\mathbf{x}$ is always nonnegative.
4. The Schur product of A and B is the same as the usual product when A and B have the same dimensions.
5. If $\mathbf{x}$ and $\mathbf{y}$ are linearly independent column n-vectors, then the rank of $\mathbf{x}\mathbf{y}^T$ is 2.
6. A matrix A is orthogonal if its columns are orthogonal.
7. An $n \times n$ matrix A is nonsingular if and only if $\text{rank}(A) = n$.
8. If A and B are $n \times n$ matrices, then $\det(AB) = 0$ if and only if A is zero or B is zero.

9. If A is nonsingular, then A^{-1} may be either singular or nonsingular.
10. If A is a real $n \times n$ matrix, then all its eigenvalues must be real.
11. An $n \times n$ matrix A may have more than n eigenvalues if all the multiplicities are counted.
12. If A^2 has positive eigenvalues $\lambda_1^2, \ldots, \lambda_n^2$, then $|\lambda_1|, \ldots, |\lambda_n|$ are the eigenvalues of A.
13. A skew-Hermitian matrix has complex eigenvalues of absolute value 1.
14. If A is a Hermitian negative definite matrix, then $\det A < 0$.
15. The inverse of an orthogonal matrix is orthogonal.
16. If A is an indefinite real symmetric matrix, then $\det A < 0$.
17. If $\mathbf{u}$ and $\mathbf{v}$ are nonzero column n-vectors, then $\mathbf{u}$ is an eigenvector of $\mathbf{uv}^*$.

References and Extensions: Chapter 1

1. Proofs for the results on determinants in Section 1.2 can be found in almost any introductory book on linear algebra.
2. An alternative, more modern, definition of the determinant is the following. Let $A = (\mathbf{a}_1, \ldots, \mathbf{a}_n)$ be the column representation of A. Then $\det A$ is a scalar-valued function that satisfies the following axioms:

   (a) $\det(\mathbf{a}_1, \ldots, \mathbf{a}_{j-1}, \alpha\mathbf{a}_j, \mathbf{a}_{j+1}, \ldots, \mathbf{a}_n) = \alpha \det A$ for any scalar α.
   (b) $\det(\mathbf{a}_1, \ldots, , \mathbf{a}_{i-1}, \mathbf{a}_i + \mathbf{a}_j, \mathbf{a}_{i+1}, \ldots, \mathbf{a}_n) = \det A$ for any i, j.
   (c) $\det I = 1$.

   The formula (1.1) as well as all the other properties of determinants then follows from these three axioms. For this approach to the determinant, see, for example, Samelson [1974]. There are other sets of equivalent axioms; see, for example, Mostow and Sampson [1969].
3. The product rule $\det AB = \det A \det B$ is a special case of the Binet-Cauchy formula for the determinant of an $n \times n$ product of rectangular matrices. See, for example, Gantmacher [1959].
4. The Gaussian elimination algorithm presented in this chapter can be numerically unstable in practice. The row interchange portion should be modified to select the element of maximum absolute value in the column to be zeroed, and then a row with this maximum value is interchanged. Although not foolproof, this interchange strategy works well in practice. The Choleski decomposition provides a useful alternative to Gaussian elimination when the coefficient matrix is positive definite. It is numerically stable without any interchanges, as is Gaussian elimination in this case. See, for example, Ortega and Poole [1981], or almost any text on numerical analysis, for further discussion.
5. The number of multiplications necessary to carry out Gaussian elimination is $n^3/3 +$ lower-order terms in n. Although other methods are known that have a lower operation count for large n, they are more complicated and have not come into general use.

6. Another method for solving linear systems of equations is *Cramer's rule*, in which the components of the solution vector are given by

$$x_i = \frac{\det A_i}{\det A}, \qquad i = 1, \ldots, n$$

where A_i is the matrix A with its ith column replaced by the right-hand side $\mathbf{b}$. Although this is not recommended as a numerical approach, it is sometimes a useful theoretical tool.

7. Cramer's rule is really based on the *adjoint* or *adjugate* matrix defined by

$$\mathrm{adj}(A) = (A_{ij})^T$$

where A_{ij} is the cofactor of a_{ij}. It can be shown that

$$A\,\mathrm{adj}(A) = (\det A)I \tag{1}$$

so that if A is nonsingular then

$$A^{-1} = \frac{\mathrm{adj}(A)}{\det A}$$

For $n = 2$ this is the formula of Exercise 1.3-10. The solution $A^{-1}\mathbf{b}$ of a linear system is then $\mathrm{adj}(A)\mathbf{b}/\det A$, which can be shown to be Cramer's rule.

8. The eigenvalues of a matrix are continuous functions of the elements of the matrix; see, for example, Ortega [1972] for a proof. The same is true of eigenvectors, suitably normalized, that correspond to a simple eigenvalue. However, at multiple eigenvalues the eigenvectors need not be continuous. A simple example is the following. Let

$$A = \begin{bmatrix} 1 + t\cos(2/t) & -t\sin(2/t) \\ -t\sin(2/t) & 1 - t\cos(2/t) \end{bmatrix}$$

which has eigenvalues $1 \pm t$ and corresponding eigenvectors $(\sin 1/t, \cos 1/t)^T$, $(\sin 1/t, -\cos 1/t)^T$. As $t \to 0$, A converges to the identity matrix, and the eigenvalues converge to 1. However, the eigenvectors do not converge.

2

Linear Spaces and Operators

Matrix theory can be studied with no mention of linear spaces, and most of the results in this book are of such a nature. However, the introduction of linear spaces and the role of matrices in defining or representing linear transformations on such spaces add considerably to our insight. Most importantly, perhaps, the notions of linear spaces and linear transformations give a geometrical basis to matrix theory, which aids both in understanding as well as in suggesting proofs and new results.

The organization of this chapter is as follows. In Section 2.1 we introduce linear spaces and some of their basic properties. In Section 2.2 we study linear transformations on linear spaces and relate them to matrices. In Section 2.3 we study linear equations, inverses, rank, eigenvalues, and other aspects of linear operators, relating these results to matrices. In Section 2.4 we add the idea of an inner product, which allows orthogonality of vectors. Finally, in Section 2.5 we add distances, which lead to normed linear spaces.

2.1. Linear Spaces

We begin with the formal definition of a linear space.

2.1.1 DEFINITION. A *linear space* (also called a *vector space*) R is a collection of objects, called *vectors*, for which the following axioms hold for all elements $\mathbf{x}, \mathbf{y}, \mathbf{z} \in R$ and scalars α, β:

(a) $\mathbf{x} + \mathbf{y} \in R$, $\alpha\mathbf{x} \in R$;
(b) $\mathbf{x} + \mathbf{y} = \mathbf{y} + \mathbf{x}$;
(c) $(\mathbf{x} + \mathbf{y}) + \mathbf{z} = \mathbf{x} + (\mathbf{y} + \mathbf{z})$;
(d) There is an element $\mathbf{0} \in R$ such that $0\mathbf{x} = \mathbf{0}$;

(e) $1\mathbf{x} = \mathbf{x}$;
(f) $\alpha(\beta\mathbf{x}) = (\alpha\beta)\mathbf{x}$;
(g) $(\alpha + \beta)\mathbf{x} = \alpha\mathbf{x} + \beta\mathbf{x}$;
(h) $\alpha(\mathbf{x} + \mathbf{y}) = \alpha\mathbf{x} + \alpha\mathbf{y}$.

If the underlying scalar field is the real numbers, we will say that R is a *real* linear space; if the scalars are the complex numbers, we call R a *complex* linear space.

Note that axiom (a), the *closure axiom*, states that sums and scalar multiples remain in the space. The remaining axioms simply require that the familiar operations in R^n or C^n hold. Indeed, it is quite easy (Exercise 2.1-1) to verify that the spaces R^n and C^n of real or complex n-vectors satisfy the axioms of 2.1.1. But a wide variety of other collections of objects do also, as the following examples indicate.

2.1.2. EXAMPLES

(a) The collection of all polynomials of degree less than or equal to n with real or complex coefficients is a real or complex linear space.
(b) The collection of polynomials of all degrees with real or complex coefficients is a real or complex linear space.
(c) If $\phi_1, \ldots, \phi_n$ are given real- or complex-valued functions on an interval $[a, b]$, then the set of all linear combinations $\sum_{i=1}^n c_i\phi_i$ for real or complex coefficients $c_1, \ldots, c_n$ is a real or complex linear space.
(d) The collection of all continuous functions on an interval $[a, b]$ is a linear space.
(e) The collection of all (Riemann)-integrable functions on an interval $[a, b]$ is a linear space.
(f) The collection of all infinite sequences $a_1, a_2, \ldots$ of real or complex numbers is a linear space.

Again, it is easy to verify that these examples all satisfy the axioms of 2.1.1 (Exercise 2.1-2).

Linear Independence

As in R^n, we can define linearly independent elements in a linear space R as follows: $\mathbf{x}_1, \ldots, \mathbf{x}_n \in R$ are *linearly dependent* if

$$\sum_{i=1}^{n} c_i\mathbf{x}_i = 0$$

and the scalars $c_1, \ldots, c_n$ are not all zero. If elements in R are not linearly dependent, then they are *linearly independent*. For example, the polynomials t and t^2 in 2.1.2(a) are linearly independent because no linear combination $c_1 t + c_2 t^2$ can give the identically zero polynomial unless $c_1 = c_2 = 0$. On the other hand, $t + 2t^2$, t, t^2 are linearly dependent since the first polynomial can be written as a linear combination of the second two.

Subspaces and Spanning Sets

If R is a linear space and R_1 is a subset of R, which is itself a linear space, then R_1 is a *subspace* of R. As examples: a line through the origin is a subspace of R^2; a plane through the origin is a subspace of R^3; the collection of all polynomials of degree 3 or less is a subspace of the linear space of polynomials of degree n. However, subsets such as the unit circle in R^2 or a line in R^2 not through the origin are not subspaces.

The *span* of m vectors in R, denoted by $\text{span}(\mathbf{x}_1, \ldots, \mathbf{x}_m)$, is the set of all linear combinations of $\mathbf{x}_1, \ldots, \mathbf{x}_m$. It is easy to see (Exercise 2.1-4) that $\text{span}(\mathbf{x}_1, \ldots, \mathbf{x}_m)$ is a linear space and, hence, a subspace of R. The vectors $\mathbf{x}_1, \ldots, \mathbf{x}_m$ constitute a *spanning set* for this subspace. For example, the span of the polynomials 1, t, t^2 is the subspace of all polynomials of degree 2 or less.

Bases

If $\mathbf{e}_1, \ldots, \mathbf{e}_n$ are, again, the coordinate vectors of R^n ($\mathbf{e}_i$ has 1 in the ith position and 0's elsewhere), then, clearly, any vector $\mathbf{x} \in R^n$ can be written as $\sum_{i=1}^n x_i \mathbf{e}_i$. This extends to the following definition for linear spaces.

2.1.3. DEFINITION. A set of linearly independent elements $\mathbf{x}_1, \ldots, \mathbf{x}_n$ in a linear space R is a *basis* for R if any element $\mathbf{x} \in R$ can be written as a linear combination of $\mathbf{x}_1, \ldots, \mathbf{x}_n$.

Consider again the examples of 2.1.2. A basis for the space of polynomials of degree n is the set 1, t, $t^2, \ldots, t^n$ because any polynomial of degree n can be written as

$$p(t) = a_0 + a_1 t + \cdots + a_n t^n$$

and the polynomials 1, $t, \ldots, t^n$ are linearly independent (Exercise 2.1-3). However, for the space of polynomials of all degrees, a basis would require *all* powers 1, t, $t^2, \ldots$. This leads to an important dichotomy in linear spaces: The linear space R is *finite dimensional* if it has a basis of a finite number of elements, and *infinite dimensional* otherwise. Thus, 2.1.2(a) and

2.1.2(c), as well as R^n and C^n, are examples of finite-dimensional linear spaces. On the other hand, 2.1.2(b) is an infinite-dimensional space as 2.1.2(d), (e), and (f).

Does a linear space always have a basis? First, we note that the trivial linear space, which consists of only the zero element, is a special case, as we shall see more clearly in a moment. Let R be any other linear space, let $\mathbf{x}_1$ be a nonzero element in R, and let R_1 be all scalar multiples of $\mathbf{x}_1$. If $R_1 \neq R$, then there must be a nonzero element $\mathbf{x}_2 \in R$ such that $\mathbf{x}_1$ and $\mathbf{x}_2$ are linearly independent. Let $R_2 = \text{span}(\mathbf{x}_1, \mathbf{x}_2)$. If $R_2 \neq R$, then there must be an $\mathbf{x}_3$ such that $\mathbf{x}_1$, $\mathbf{x}_2$, $\mathbf{x}_3$ are linearly independent. Continuing in this way, we either reach an integer n such that $\text{span}(\mathbf{x}_1, \ldots, \mathbf{x}_n) = R$, in which case $\mathbf{x}_1, \ldots, \mathbf{x}_n$ is a basis for R, or the process does not terminate, in which case R must be infinite dimensional. In the former case, R is called *finitely generated*, and we have proved the following result.

2.1.4. Any finitely generated linear space has a basis.

A given finitely generated linear space may have several different bases. For example, $\mathbf{e}_1$ and $\mathbf{e}_2$ form a basis for R^2 but so do $\mathbf{e}_1 + \mathbf{e}_2$, and $\mathbf{e}_1 - \mathbf{e}_2$. The important thing is that the number of elements in any two bases must be the same, as we now prove.

2.1.5. Let $\mathbf{x}_1, \ldots, \mathbf{x}_n$ and $\mathbf{y}_1, \ldots, \mathbf{y}_m$ be two bases for a linear space R. Then $n = m$.

Proof. Suppose that $m > n$; if $n > m$, we would just reverse the roles of the $\mathbf{x}_i$ and $\mathbf{y}_i$ in the following. Since $\mathbf{x}_1, \ldots, \mathbf{x}_n$ is a basis, we can write

$$\mathbf{y}_i = \sum_{j=1}^{n} a_{ji}\mathbf{x}_j, \qquad i = 1, \ldots, m$$

for suitable coefficients a_{ij}. Now consider the system of linear equations

$$\begin{bmatrix} a_{11} & \cdots & a_{1m} \\ a_{21} & \cdots & a_{2m} \\ \vdots & & \vdots \\ a_{n1} & \cdots & a_{nm} \\ 0 & & 0 \\ \vdots & & \vdots \\ 0 & \cdots & 0 \end{bmatrix} \begin{bmatrix} \alpha_1 \\ \\ \\ \vdots \\ \\ \\ \alpha_m \end{bmatrix} = 0$$

The $m \times m$ coefficient matrix is singular because of the zero rows. Hence, by Theorem 1.3.4(d), there must be a nonzero solution $\boldsymbol{\alpha}$. Thus,

$$\sum_{k=1}^{m} \alpha_k \mathbf{y}_k = \sum_{k=1}^{m} \alpha_k \sum_{j=1}^{n} a_{jk}\mathbf{x}_j = \sum_{j=1}^{n} \left(\sum_{k=1}^{m} a_{jk}\alpha_k \right) \mathbf{x}_j = 0$$

which shows that $\mathbf{y}_1, \ldots, \mathbf{y}_m$ are linearly dependent. This contradiction proves the result. □

An example is given by R^n. As mentioned before, $\mathbf{e}_1, \ldots, \mathbf{e}_n$ is a basis for R^n, and we will call this the *natural basis.* Let A be any nonsingular $n \times n$ matrix with columns $\mathbf{a}_1, \ldots, \mathbf{a}_n$. Then, by Theorem 1.3.4, the columns of A are linearly independent, and the system $A\mathbf{x} = \mathbf{b}$ has a unique solution for any n-vector $\mathbf{b}$. Thus, $\mathbf{b} = \sum_{i=1}^{n} x_i \mathbf{a}_i$ and $\mathbf{a}_1, \ldots, \mathbf{a}_n$ is a basis for R^n. Any $n + 1$ vectors in R^n are linearly dependent (Exercise 2.1-5) and thus do not form a basis. On the other hand, by 2.1.5 any collection of less than n vectors cannot form a basis for R^n. The same considerations hold for C^n.

Dimension

The number, n, of elements in a basis for a finitely generated linear space R is the *dimension* of the space, and we write $\dim R = n$. By 2.1.5 the dimension is well defined because any two bases must have the same number of elements. We define the dimension of the linear space consisting of only the zero element to be zero. We shall see the utility of this zero-dimensional space in a moment. However, in the sequel, we shall tacitly assume that a linear space under consideration is not the zero space unless explicitly stated. Clearly, n is the dimension of R^n and C^n, and the dimension of the space of all polynomials of degree n or less is $n + 1$ since $1, t, \ldots, t^n$ is a basis of $n + 1$ elements.

If R_1 and R_2 are two subspaces of R, we can define the *intersection* of R_1 and R_2 as the set of all elements of R that are in both R_1 and R_2:

$$R_1 \cap R_2 = \{\mathbf{z}: \mathbf{z} \in R_1, \mathbf{z} \in R_2\}$$

It is easy to verify (Exercise 2.1-6) that $R_1 \cap R_2$ is also a linear space and, hence, a subspace whose dimension may range from 0 to $\min(\dim R_1, \dim R_2)$. For example, if R_1 and R_2 are two distinct planes through the origin in R^3, then $R_1 \cap R_2$ will be a line in R^3. If R_1 is a plane and R_2 is a line not in R_1, then $R_1 \cap R_2$ is the origin, the zero-dimensional subspace consisting only of the zero vector.

We can also define the union of two subspaces, but this is not, in general, a linear space; for example, the union of two lines is only the two lines. A more useful concept is the *sum* of two subspaces, defined by

$$R_1 + R_2 = \{\mathbf{x} + \mathbf{y}: \mathbf{x} \in R_1, \mathbf{y} \in R_2\}$$

Thus, the sum of R_1 and R_2, denoted by $R_1 + R_2$, is the collection of all possible sums of elements in R_1 and R_2. It is easy to see (Exercise 2.1-7) that $R_1 + R_2$ is the linear space $\text{span}(\mathbf{x}_1, \ldots, \mathbf{x}_m, \mathbf{y}_1, \ldots, \mathbf{y}_p)$, where

$\mathbf{x}_1, \ldots, \mathbf{x}_m$ and $\mathbf{y}_1, \ldots, \mathbf{y}_p$ are bases for R_1 and R_2, respectively. For example, if R_1 and R_2 are distinct lines through the origin in R^3, then $R_1 + R_2$ will be the plane generated by these two lines. If R_1 and R_2 are two-dimensional subspaces of R^n such that $\dim(R_1 \cap R_2) = 1$, then $R_1 + R_2$ will be a three-dimensional subspace. This last result is a special case of the following general counting theorem for two subspaces R_1 and R_2 of any finite-dimensional linear space:

$$\dim(R_1 + R_2) = \dim R_1 + \dim R_2 - \dim(R_1 \cap R_2) \qquad (2.1.1)$$

The proof of (2.1.1) is left to Exercise 2.1-7.

An important special case of $R_1 + R_2$ is when $R_1 \cap R_2 = \{0\}$. Then $R_1 + R_2$ is called the *direct sum* and denoted by $R_1 \dot{+} R_2$. In this case the subspaces R_1 and R_2 are called *complements*, and any element $\mathbf{x} \in R_1 \dot{+} R_2$ has a unique representation $\mathbf{x} = \mathbf{x}_1 + \mathbf{x}_2$, where $\mathbf{x}_i \in R_i$. This follows from the fact that if $\mathbf{x} = \mathbf{y}_1 + \mathbf{y}_2$, $\mathbf{y}_i \in R_i$, is another representation, then $\mathbf{y}_1 + \mathbf{y}_2 = \mathbf{x}_1 + \mathbf{x}_2$ or $\mathbf{y}_1 - \mathbf{x}_1 = \mathbf{x}_2 - \mathbf{y}_2$. But $\mathbf{y}_1 - \mathbf{x}_1 \in R_1$, $\mathbf{x}_2 - \mathbf{y}_2 \in R_2$, and since R_1 and R_2 have only the zero element in common, we must have $\mathbf{y}_1 - \mathbf{x}_1 = 0$ and $\mathbf{x}_2 - \mathbf{y}_2 = 0$.

Closely related to the idea of a direct sum is the following.

2.1.6. EXTENSION OF BASIS. *Let R be an n-dimensional linear space, and let R_1 be a linear subspace of R with basis $r_1, \ldots, r_m$, $m < n$. Then there is a basis $\hat{r}_i$, $i = 1, \ldots, n$, for R such that $\hat{r}_i = r_i$, $i = 1, \ldots, m$.*

Proof. We use the same argument as in the proof of 2.1.4. Since $R_1 \neq R$, there must be a nonzero element $\hat{r}_{m+1} \notin R_1$. Then, $\dim \operatorname{span}(r_1, \ldots, r_m, \hat{r}_{m+1}) = m + 1$. If $m + 1 \neq n$, there must be a nonzero $\hat{r}_{m+2}$ such that $\dim \operatorname{span}(r_1, \ldots, r_m, \hat{r}_{m+1}, \hat{r}_{m+2}) = m + 2$. If $m + 2 \neq n$, we continue this process until we have $\dim \operatorname{span}(r_1, \ldots, r_m, \hat{r}_{m+1}, \ldots, \hat{r}_n) = n$. □

Sums or direct sums may be extended to any number of subspaces. Thus, the direct sum

$$R = R_1 \dot{+} \cdots \dot{+} R_p$$

indicates that R is the sum of p subspaces $R_1, \ldots, R_p$, any two of which have only the zero element in common. An example is

$$R^n = \operatorname{span}(\mathbf{e}_1) \dot{+} \operatorname{span}(\mathbf{e}_2) \dot{+} \cdots \dot{+} \operatorname{span}(\mathbf{e}_n)$$

in which R^n is viewed as the direct sum of the n one-dimensional subspaces spanned by the unit vectors $\mathbf{e}_1, \ldots, \mathbf{e}_n$. As we shall see in the next chapter, direct sums play an important role in the study of eigenvalues.

Coordinates

Let R be a linear space with basis $\mathbf{x}_1, \ldots, \mathbf{x}_n$. Then any element $\mathbf{x} \in R$ may be written as a linear combination

$$\mathbf{x} = \sum_{i=1}^{n} x_i \mathbf{x}_i$$

The scalars $x_1, \ldots, x_n$ are called the *coordinates* of $\mathbf{x}$ in the basis $\mathbf{x}_1, \ldots, \mathbf{x}_n$, and $\mathbf{x}_c = (x_1, \ldots, x_n)^T$ is the *coordinate vector* for $\mathbf{x}$. These coordinates are unique, because if

$$\mathbf{x} = \sum_{i=1}^{n} \hat{x}_i \mathbf{x}_i$$

then subtracting gives

$$0 = \sum_{i=1}^{n} (x_i - \hat{x}_i)\mathbf{x}_i$$

Thus, since $\mathbf{x}_1, \ldots, \mathbf{x}_n$ are linearly independent, we must have that $x_i - \hat{x}_i = 0$, $i = 1, \ldots, n$.

Consider R^n with the natural basis $\mathbf{e}_1, \ldots, \mathbf{e}_n$. If $\mathbf{x}$ is a column vector in R^n, then

$$\mathbf{x} = \sum_{i=1}^{n} x_i \mathbf{e}_i$$

so that the components of $\mathbf{x}$ are just the coordinates of $\mathbf{x}$ in the basis $\mathbf{e}_1, \ldots, \mathbf{e}_n$. On the other hand, if $\mathbf{x}_1, \ldots, \mathbf{x}_n$ is another basis in R^n, then the components of a column vector $\mathbf{x}$ are not, in general, the coordinates of $\mathbf{x}$ in this basis. For example, in R^2, let

$$\mathbf{x}_1 = \begin{pmatrix} 1 \\ -1 \end{pmatrix}, \quad \mathbf{x}_2 = \begin{pmatrix} 1 \\ 1 \end{pmatrix}, \quad \mathbf{x} = \begin{pmatrix} 1 \\ -3 \end{pmatrix} \tag{2.1.2}$$

Then $\mathbf{x} = 2\mathbf{x}_1 - \mathbf{x}_2$, so that the coordinates of $\mathbf{x}$ in the basis of (2.1.2) are 2 and -1, not 1 and -3. We will see in the next section the general relationship of coordinates in two different bases.

As another example, consider the space of real polynomials of degree n with basis $1, t, \ldots, t^n$. Then the coordinate vector in this basis for a polynomial $a_0 + a_1 t + \cdots + a_n t^n$ is just $(a_0, \ldots, a_n)^T$, a vector in R^{n+1}. This illustrates the general principle that any n-dimensional linear space can be related to R^n (or C^n if the space is complex) by means of coordinates relative to a particular basis.

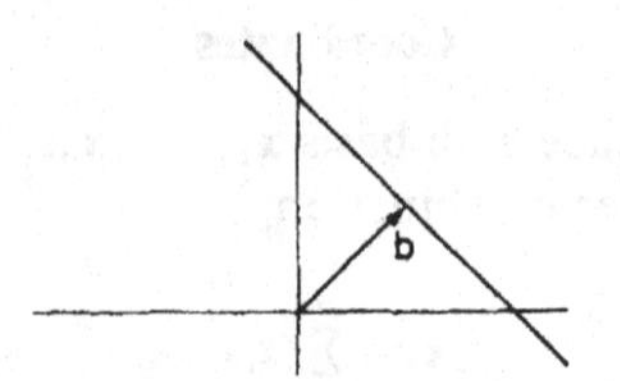

Figure 2.1. An affine subspace.

Affine Spaces

Many times we need to consider, for example, lines in R^n that do not go through the origin, as illustrated in Figure 2.1. Such a line is not a linear subspace because it does not contain the zero vector. Rather, it is an example of an *affine subspace* (also called a *linear manifold*), defined as follows. If R_1 is a linear subspace of a linear space R and $\mathbf{b} \in R$, then

$$R_{\mathbf{b}} = \{\mathbf{x} + \mathbf{b}: \mathbf{x} \in R_1\} \tag{2.1.3}$$

An affine subspace is just a linear subspace "translated" by a fixed element $\mathbf{b}$.

Summary

This section has developed several basic properties of abstract linear spaces that parallel familiar properties of the spaces R^n and C^n. The most important concepts of this section are

- Linear independence of vectors;
- Subspaces;
- Spanning sets and bases;
- Dimension;
- Coordinates.

We note, in particular, that the use of coordinates essentially equates any linear space of dimension n to R^n or C^n, depending on whether the space is real or complex.

Exercises 2.1

1. Verify that R^n and C^n are linear spaces.
2. Verify that the examples of 2.1.2 are all linear spaces.
3. Show that the polynomials $1, t, \ldots, t^n$ are linearly independent in the linear space 2.1.2(a).

4. If $\mathbf{x}_1, \ldots, \mathbf{x}_m$ are m vectors in a linear space R, verify that $\operatorname{span}(\mathbf{x}_1, \ldots, \mathbf{x}_m)$ is a linear space.
5. Let $\mathbf{x}_1, \ldots, \mathbf{x}_{n+1}$ be any $n+1$ vectors in R^n. Show that they must be linearly dependent.
6. If R_1 and R_2 are linear subspaces of a linear space R, show that $R_1 \cap R_2$ is a linear subspace.
7. Show that $R_1 + R_2$ is the linear space $\operatorname{span}(\mathbf{x}_1, \ldots, \mathbf{x}_m, \mathbf{y}_1, \ldots, \mathbf{y}_p)$, where $\mathbf{x}_1, \ldots, \mathbf{x}_m$ and $\mathbf{y}_1, \ldots, \mathbf{y}_p$ are any bases for R_1 and R_2. Show also that $\dim(R_1 + R_2) = m + p - \dim(R_1 \cap R_2)$.
8. Ascertain whether the following are linear spaces:
   (a) The set of all twice continuously differentiable functions on the interval $(0, 1)$;
   (b) The set of functions $\sin k\pi x$, $\cos k\pi x$, $k = 0, 1, \ldots$, on the interval $[0, 1]$;
   (c) The set of vectors of the form (a, b, c) when a, b and c are real and positive.
9. Let $R^{n,m}$ be the collection of all real $m \times n$ matrices. Show that $R^{n,m}$ is a linear space of dimension mn. Exhibit a basis for $R^{n,m}$.
10. Ascertain whether the following vectors are a basis for R^3;

$$\begin{bmatrix} 1 \\ 2 \\ 3 \end{bmatrix}, \quad \begin{bmatrix} 4 \\ 5 \\ 6 \end{bmatrix}, \quad \begin{bmatrix} 7 \\ 8 \\ 9 \end{bmatrix}$$

11. If R_1 is a subspace of a vector space R and $0 < \dim R_1 < \dim R$, show that R_1 has infinitely many complements R_2. (*Hint*: Consider first R_1 as a line through the origin in R^2.)
12. Let $\mathbf{x}_1^T = (1, 2)$, $\mathbf{x}_2^T = (2, 1)$. Find the coordinates of the vector $\mathbf{x} = (3, 2)^T$ in the basis $\mathbf{e}_1, \mathbf{e}_2$ and in the basis $\mathbf{x}_1, \mathbf{x}_2$.
13. Show that the polynomials 2, $1 + t$, $t + t^2$ form a basis for the space of polynomials of degree 2. Find the coordinates of the polynomial $3 + 4t + 2t^2$ in this basis.
14. Let $R_1 = \operatorname{span}(\mathbf{x})$, where $\mathbf{x}^T = (1, 2)$, and let $\mathbf{b}^T = (1, 1)$. Sketch the affine subspace of R^2 defined by (2.1.3).

2.2. Linear Operators

We begin now the study of linear operators. If R and S are linear spaces, an *operator* $\mathbf{A}: R \to S$ is a rule for assigning elements $\mathbf{x}$ of R to elements of S. We also call $\mathbf{A}$ a *mapping, function,* or *transformation.* R is called the *domain space* of $\mathbf{A}$, and S is the *range space.* The operator $\mathbf{A}$ is *linear* if

$$\mathbf{A}(\alpha \mathbf{x} + \beta \mathbf{y}) = \alpha \mathbf{A}\mathbf{x} + \beta \mathbf{A}\mathbf{y} \tag{2.2.1}$$

for all $\mathbf{x}, \mathbf{y} \in R$ and all scalars α, β.

If $R = R^n$ and $S = R^m$, then a real $m \times n$ matrix A defines a linear operator $\mathbf{A}$ by the matrix-vector multiplication $A\mathbf{x}$, where $\mathbf{x}$ is a column vector in R^n. Similarly, a real or complex $m \times n$ matrix defines a linear operator from C^n to C^m. Other examples are listed below.

2.2.1. Examples of Linear Operators

(a) $R = S =$ the collection of all polynomials of degree n or less. $\mathbf{A}p = p'$ (the derivative).
(b) $R =$ the collection of all continuous functions on an interval $[a, b]$. $S = R^1$. $\mathbf{A}f = \int_a^b f(s)\, ds$.
(c) $R = S =$ the collection of all continuous functions on an interval $[a, b]$. $K(s, t)$ is a given continuous function of two variables $a \leq s$, $t \leq b$. $(\mathbf{A}f)(s) = \int_a^b K(s, t) f(t)\, dt$.
(d) $R = S =$ the collection of twice continuously differentiable functions on (a, b). $\mathbf{A}f = f''$.

It is easy to verify (Exercise 2.2-1) that the operators $\mathbf{A}$ defined in the above examples are, indeed, linear. In the first and last examples, $\mathbf{A}$ is a differential operator. The other two examples give integral operators.

We note that the collection of all linear operators from R to S is itself a linear space (see Exercise 2.2-10).

Matrix Representation of Linear Operators

We noted previously that an $m \times n$ matrix defines a linear transformation from C^n to C^m. Conversely, any linear transformation between finite-dimensional spaces can be represented by a matrix that depends on a choice of bases for the two spaces. Let R and S be finite-dimensional linear spaces with bases $\mathbf{r}_1, \ldots, \mathbf{r}_n$ and $\mathbf{s}_1, \ldots, \mathbf{s}_m$, respectively, and let $\mathbf{A}: R \to S$ be a linear operator. Then we can represent $\mathbf{A}$ in terms of the bases of R and S by an $m \times n$ matrix in the following way. If we write an arbitrary element $\mathbf{r} \in R$ in terms of the basis as

$$\mathbf{r} = \sum_{j=1}^{n} x_j \mathbf{r}_j$$

then successive applications of the linearity property (2.2.1) allows writing $\mathbf{A}\mathbf{r}$ as a linear combination of the $\mathbf{A}\mathbf{r}_j$:

$$\mathbf{A}\mathbf{r} = \sum_{j=1}^{n} x_j \mathbf{A}\mathbf{r}_j \tag{2.2.2}$$

The elements $\mathbf{Ar}_j$ can be expressed in terms of the basis for S by

$$\mathbf{Ar}_j = \sum_{i=1}^{m} a_{ij}\mathbf{s}_i, \qquad j = 1, \ldots, n, \tag{2.2.3}$$

for suitable coefficients a_{ij}. If we put these representations of $A\mathbf{r}_j$ into (2.2.2), interchange summation, and collect coefficients of the $\mathbf{s}_i$, we obtain

$$\mathbf{Ar} = \sum_{j=1}^{n} x_j \sum_{i=1}^{m} a_{ij}\mathbf{s}_i = \sum_{i=1}^{m}\sum_{j=1}^{n} a_{ij}x_j\mathbf{s}_i = \sum_{i=1}^{m} y_i\mathbf{s}_i$$

where

$$y_i = \sum_{j=1}^{n} a_{ij}x_j, \qquad i = 1, \ldots, m \tag{2.2.4}$$

Therefore, if A is the $m \times n$ matrix with elements a_{ij}, and $\mathbf{x}$, $\mathbf{y}$ are the n- and m-vectors with components $x_1, \ldots, x_n$ and $y_1, \ldots, y_m$, then (2.2.4) is equivalent to

$$\mathbf{y} = A\mathbf{x} \tag{2.2.5}$$

Thus, the effect of the linear operator $\mathbf{A}$ on elements $\mathbf{r}$ of R to give elements $\mathbf{Ar}$ of S is represented by the matrix-vector multiplication (2.2.5), and A is called the *matrix representation* of the operator $\mathbf{A}$ with respect to the particular bases. This is an underlying reason that matrix-vector multiplication is defined the way it is. Note that the matrix A is completely defined by (2.2.3).

We give a simple example. Let R and S be the linear spaces of all real polynomials of degree n or less and $n-1$ or less, respectively, and let $\mathbf{A}$ be the first-derivative operator: $\mathbf{A}p = p'$. Let $1, t, \ldots, t^n$ be a basis for R, and similarly for S with powers to t^{n-1}. Then

$$p(t) = \sum_{i=0}^{n} x_i t^i$$

is the representation of a polynomial in R in terms of the basis, and

$$\mathbf{A}p(t) = p'(t) = \sum_{i=1}^{n} i x_i t^{i-1}$$

is the representation of $\mathbf{A}p$ in terms of the basis for S. In particular, $\mathbf{A}(t^i) = it^{i-1}$, and by (2.2.3), with $\mathbf{r}_i = t^{i-1}$ $(i = 1, \ldots, n+1)$ and $\mathbf{s}_i = t^{i-1}$ $(i = 1, \ldots, n)$, we have $a_{ij} = 0$ $(j \neq i+1)$ and $a_{ii+1} = i$ $(i = 1, \ldots, n)$. Thus, the matrix representation of $\mathbf{A}$ in these bases is the $n \times n+1$ matrix

$$A = \begin{bmatrix} 0 & 1 & 0 & \cdots & & 0 \\ 0 & 0 & 2 & 0 & \cdots & 0 \\ & & & \cdots & & \vdots \\ & & & \cdots & & 0 \\ 0 & & & \cdots & 0 & n \end{bmatrix}$$

Composition and Products

We next consider two linear operators $\mathbf{A}: R \to S$ and $\mathbf{B}: S \to T$, where R, S, and T are linear spaces. Then the *composition* or *product* of $\mathbf{A}$ and $\mathbf{B}$ is the operator $\mathbf{BA}: R \to T$, defined by $(\mathbf{BA})\mathbf{r} = \mathbf{B}(\mathbf{A}\mathbf{r})$ for all $\mathbf{r} \in R$. It is sometimes useful to view this process geometrically as indicated in Figure 2.2.

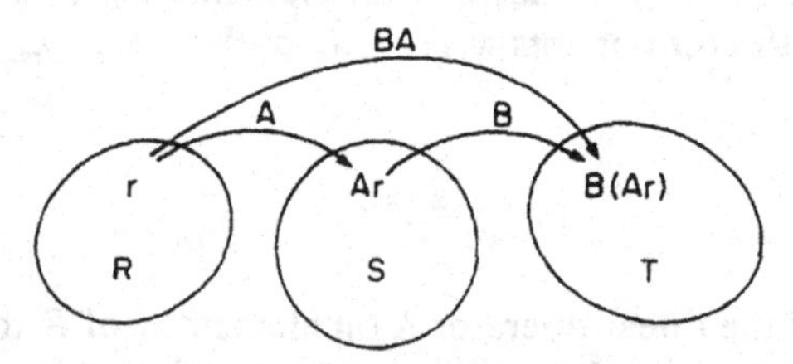

Figure 2.2. Composition of operators.

Now suppose that $\mathbf{r}_1, \ldots, \mathbf{r}_n$, $\mathbf{s}_1, \ldots, \mathbf{s}_m$, and $\mathbf{t}_1, \ldots, \mathbf{t}_p$ are bases for R, S, and T, respectively, and let A and B be the matrix representations of $\mathbf{A}$ and $\mathbf{B}$ relative to these bases. By (2.2.5) the coordinates of a vector $\mathbf{r} = \sum_{i=1}^n x_i\mathbf{r}_i$ are transformed to the coordinates of $\mathbf{s} = \sum_{i=1}^m y_i\mathbf{s}_i$, and similarly the coordinates of $\mathbf{s}$ will be transformed to the coordinates of $\mathbf{t} = \sum_{i=1}^p z_i\mathbf{t}_i$ by

$$\mathbf{z} = B\mathbf{y} \tag{2.2.6}$$

where B is the matrix representation of $\mathbf{B}$ defined by

$$\mathbf{B}\mathbf{s}_j = \sum_{i=1}^{p} b_{ij}\mathbf{t}_i, \qquad j = 1, \ldots, m$$

Hence, the matrix representation of the composition $\mathbf{BA}$ is obtained from (2.2.6) and (2.2.5) by

$$\mathbf{z} = B\mathbf{y} = B(A\mathbf{x}) = (BA)\mathbf{x}$$

That is, the matrix product BA gives the representation of **BA** relative to the given bases. It is for this reason that matrix multiplication is defined as it is. Note that the product BA is indeed well defined since B is a $p \times m$ matrix and A is an $m \times n$ matrix; thus BA is $p \times n$.

Change of Basis; Equivalence Transformations

The above matrix representations are relative to given bases. As noted before, a linear space may have many different bases. How does the matrix representation change as the basis changes? Let R be a linear space with bases $\mathbf{r}_1, \ldots, \mathbf{r}_n$ and $\hat{\mathbf{r}}_1, \ldots, \hat{\mathbf{r}}_n$. Since each $\hat{\mathbf{r}}_j$ must be a linear combination of $\mathbf{r}_1, \ldots, \mathbf{r}_n$, we can express one basis in terms of the other by

$$\hat{\mathbf{r}}_j = \sum_{i=1}^{n} q_{ij}\mathbf{r}_i, \qquad j = 1, \ldots, n \tag{2.2.7}$$

for suitable coefficients q_{ij}. Let $\mathbf{r}$ be an element of R with coordinates $x_1, \ldots, x_n$ and $\hat{x}_1, \ldots, \hat{x}_n$ in the two bases. Then substituting (2.2.7) into the representation of $\mathbf{r}$ in terms of the basis $\hat{\mathbf{r}}_1, \ldots, \hat{\mathbf{r}}_n$, interchanging summations, and collecting coefficients of the $\mathbf{r}_i$, we obtain

$$\mathbf{r} = \sum_{j=1}^{n} \hat{x}_j\hat{\mathbf{r}}_j = \sum_{j=1}^{n} \hat{x}_j \sum_{i=1}^{n} q_{ij}\mathbf{r}_i = \sum_{i=1}^{n}\left(\sum_{j=1}^{n} q_{ij}\hat{x}_j\right)\mathbf{r}_i \tag{2.2.8}$$

Thus, $\sum_{j=1}^{n} q_{ij}\hat{x}_j$, $i = 1, \ldots, n$, are the coordinates of $\mathbf{r}$ in the basis $\mathbf{r}_1, \ldots, \mathbf{r}_n$. But so are $x_1, \ldots, x_n$, and since the coordinates in a given basis are unique, we must have

$$x_i = \sum_{j=1}^{n} q_{ij}\hat{x}_j, \qquad i = 1, \ldots, n$$

Thus,

$$\mathbf{x} = Q\hat{\mathbf{x}} \tag{2.2.9}$$

where $\mathbf{x}$ and $\hat{\mathbf{x}}$ are vectors with components $x_1, \ldots, x_n$ and $\hat{x}_1, \ldots, \hat{x}_n$, respectively, and Q is the $n \times n$ matrix $Q = (q_{ij})$. Q is nonsingular, because if there were a nonzero vector $\hat{\mathbf{x}}$ such that $Q\hat{\mathbf{x}} = 0$, then (2.2.8) shows that $\hat{\mathbf{r}}_1, \ldots, \hat{\mathbf{r}}_n$ would be linearly dependent. Hence, we can also write

$$\hat{\mathbf{x}} = Q^{-1}\mathbf{x} \tag{2.2.10}$$

The relations (2.2.9) and (2.2.10) describe how the coordinates of a given element $\mathbf{r} \in R$ relate to one another in two different bases. We also note that given a basis $\mathbf{r}_1, \ldots, \mathbf{r}_n$ and a nonsingular $n \times n$ matrix Q, (2.2.7) *defines* a new basis $\hat{\mathbf{r}}_1, \ldots, \hat{\mathbf{r}}_n$ (Exercise 2.2-8).

As a simple example, let R be the space of polynomials of degree n or less, and let $1, t, \ldots, t^n$ and $1, 1 + t, 1 + t + t^2, \ldots, 1 + t + \cdots + t^n$ be two bases. The representation of the second basis in terms of the first, (2.2.7), gives the upper triangular matrix

$$Q = \begin{bmatrix} 1 & 1 & \cdots & 1 \\ & 1 & & \\ & & 1 & \vdots \\ & & \ddots & \\ & & & 1 \end{bmatrix}$$

Now suppose that $\mathbf{A}: R \to S$ is a linear operator, where S is an m-dimensional linear space. If $\mathbf{s}_1, \ldots, \mathbf{s}_m$ and $\hat{\mathbf{s}}_1, \ldots, \hat{\mathbf{s}}_m$ are two bases for S, then, as above, there will be a nonsingular $m \times m$ matrix T so that coordinates of an element $\mathbf{s} \in S$ are related, as in (2.2.9) and (2.2.10), by

$$\mathbf{y} = T\hat{\mathbf{y}}, \qquad \hat{\mathbf{y}} = T^{-1}\mathbf{y} \tag{2.2.11}$$

If A is the matrix representation of $\mathbf{A}$ in the bases $\mathbf{r}_1, \ldots, \mathbf{r}_n$ and $\mathbf{s}_1, \ldots, \mathbf{s}_m$, then

$$\mathbf{y} = A\mathbf{x} = AQ\hat{\mathbf{x}}$$

so that

$$\hat{\mathbf{y}} = T^{-1}\mathbf{y} = T^{-1}AQ\hat{\mathbf{x}} \tag{2.2.12}$$

This shows that the matrix representation of $\mathbf{A}$ in the bases $\hat{\mathbf{r}}_1, \ldots, \hat{\mathbf{r}}_n$ and $\hat{\mathbf{s}}_1, \ldots, \hat{\mathbf{s}}_m$ is $T^{-1}AQ$. We summarize the above in the following basic theorem.

2.2.2. CHANGE OF BASIS THEOREM. *Let R be a linear space with bases $\mathbf{r}_1, \ldots, \mathbf{r}_n$ and $\hat{\mathbf{r}}_1, \ldots, \hat{\mathbf{r}}_n$, and S a linear space with bases $\mathbf{s}_1, \ldots, \mathbf{s}_m$ and $\hat{\mathbf{s}}_1, \ldots, \hat{\mathbf{s}}_m$. Let Q and T be $n \times n$ and $m \times m$ nonsingular matrices that give the change in coordinates between the bases in R and in S, respectively* [*as defined by* (2.2.9) *and* (2.2.11)]. *Let $\mathbf{A}: R \to S$ be a linear operator whose matrix representation in the bases $\mathbf{r}_1, \ldots, \mathbf{r}_n$ and $\mathbf{s}_1, \ldots, \mathbf{s}_m$ is A. Then with $P = T^{-1}$, the matrix representation of $\mathbf{A}$ in the bases $\hat{\mathbf{r}}_1, \ldots, \hat{\mathbf{r}}_n$ and $\hat{\mathbf{s}}_1, \ldots, \hat{\mathbf{s}}_m$ is*

$$B = PAQ \tag{2.2.13}$$

The relationship (2.2.13) is of fundamental importance, and we isolate the key aspect in the following definition.

2.2.3. DEFINITION: *Equivalence.* If A and B are $m \times n$ matrices and P and Q are nonsingular $m \times m$ and $n \times n$ matrices, respectively, then A and B are *equivalent* if (2.2.13) holds, and (2.2.13) is an *equivalence transformation* of A.

If two matrices are equivalent, they represent the same linear transformation with respect to different bases. This leads to the question of the simplest matrix representation of a linear transformation, which will be addressed in the next chapter.

Similarity Transformations

A very important special case of 2.2.3 is when $m = n$ and $P = Q^{-1}$.

2.2.4. DEFINITION: *Similarity.* If A and B are $n \times n$ matrices and Q is a nonsingular $n \times n$ matrix such that

$$B = Q^{-1}AQ \tag{2.2.14}$$

then A and B are *similar*, and (2.2.14) is a *similarity transformation.*

Similarity transformations arise when we make a change of basis for a linear operator $\mathbf{A}: R \to R$ whose range space is the same as its domain space. In this case R also plays the role of S in the previous development. Thus, if A is the representation of $\mathbf{A}$ in a basis $\mathbf{r}_1, \ldots, \mathbf{r}_n$ for R, and $\hat{\mathbf{r}}_1, \ldots, \hat{\mathbf{r}}_n$ is another basis, where the two bases are related by (2.2.7), then (2.2.9) still holds, and in (2.2.11) the matrix T is Q also. Thus, (2.2.12) becomes

$$\hat{\mathbf{y}} = Q^{-1}AQ\hat{\mathbf{x}} \tag{2.2.15}$$

so that the matrix representation of A in the second basis is $B = Q^{-1}AQ$, which is (2.2.14).

We summarize this in the following basic theorem.

2.2.5. SIMILARITY THEOREM. *If* $\mathbf{A}: R \to R$ *is a linear operator from a finite-dimensional linear space R into itself, and if A and* $\hat{A}$ *are any two matrix representations of* $\mathbf{A}$ *with respect to different bases for R, then A and* $\hat{A}$ *are similar.*

Similarity transformations play an extremely important role in dealing with eigenvalues, as we shall see later.

Application to R^n and C^n

Let us apply the above results to the concrete case of R^n. The same discussion applies immediately to C^n if we consider complex scalars.

Let $\mathbf{e}_1, \ldots, \mathbf{e}_n$ be the natural basis of unit vectors for R^n. Assuming the corresponding basis for R^m, a given $m \times n$ matrix A defines a linear transformation $\mathbf{A}: R^n \to R^m$, where the matrix representation in the aforementioned bases is just the matrix A. Now let $\mathbf{q}_1, \ldots, \mathbf{q}_n$ and $\mathbf{t}_1, \ldots, \mathbf{t}_m$ be linearly independent sets of vectors in R^n and R^m, respectively. Then these sets are bases for R^n and R^m. In (2.2.7) we take $\mathbf{r}_i = \mathbf{e}_i$, $i = 1, \ldots, n$, as the original basis vectors, and $\hat{\mathbf{r}}_i = \mathbf{q}_i$, $i = 1, \ldots, n$, as the new basis vectors. Then q_{ij} in (2.2.7) is just the ith component of $\mathbf{q}_j$. Hence, the matrix Q of (2.2.9) is just the matrix with columns $\mathbf{q}_1, \ldots, \mathbf{q}_n$. Thus, if $\mathbf{x}$ is a given n-vector, then its components, which are its coordinates in the $\mathbf{e}_1, \ldots, \mathbf{e}_n$ basis, are related to the coordinates $\hat{\mathbf{x}}$ of $\mathbf{x}$ in the basis $\mathbf{q}_1, \ldots, \mathbf{q}_n$ by (2.2.9): $\mathbf{x} = Q\hat{\mathbf{x}}$. This is just the usual linear "change of variable" formula in R^n. Similarly, $\mathbf{y} = T\hat{\mathbf{y}}$ for vectors in R^m, where T is the matrix with columns $\mathbf{t}_i$. Hence, with $P = T^{-1}$, $\hat{A} = PAQ$ is the matrix representation of $\mathbf{A}$ in the new bases.

Summary

The most important results and concepts of this section are

- Representation of linear operators in terms of matrices;
- Equivalence transformation of a matrix, which expresses how the matrix representation changes under a change of basis;
- Similarity transformation as the special case of an equivalence transformation when there is only one linear space.

In particular, the results of this section show that, for finite-dimensional linear spaces, linear operators and matrices are equivalent in the following sense. If R and S are linear spaces with bases $\mathbf{r}_1, \ldots, \mathbf{r}_n$ and $\mathbf{s}_1, \ldots, \mathbf{s}_m$, then a given $m \times n$ matrix A defines a linear operator $\mathbf{A}: R \to S$ by (2.2.4), where $x_1, \ldots, x_n$ and $y_1, \ldots, y_m$ are the coordinates of elements $\mathbf{r} \in R$ and $\mathbf{s} \in S$. Conversely, given a linear operator $\mathbf{A}: R \to S$, we can represent it in terms of bases $\mathbf{r}_1, \ldots, \mathbf{r}_n$ and $\mathbf{s}_1, \ldots, \mathbf{s}_m$ by a matrix A determined by (2.2.3).

Exercises 2.2

1. Verify that the operators $\mathbf{A}$ defined in 2.2.1 are linear.
2. Let $\mathbf{A}$ be the operator defined on the space of polynomials of degree n by $\mathbf{A}p = p^{(k)}$, the kth derivative of p. Show that $\mathbf{A}$ is linear.

3. Let $\mathbf{A}$ be the operator defined on the space of polynomials of all degrees by $\mathbf{A}p = p^2$, the square of the polynomial p. Show that $\mathbf{A}$ is not linear.

4. Let $\mathbf{A}$ be the operator from the set of real continuous functions on the interval $[0, 1]$ to R^1, defined by $\mathbf{A}f = f(\frac{1}{2})$; that is, $\mathbf{A}f$ is the value of f at $\frac{1}{2}$. Is $\mathbf{A}$ linear?

5. Let R and S be linear spaces with bases $\mathbf{r}_1, \ldots, \mathbf{r}_n$ and $\mathbf{s}_1, \ldots, \mathbf{s}_n$, respectively. Let $\mathbf{A}: R \to S$ be a linear operator such that

$$\mathbf{A}\mathbf{r}_1 = \mathbf{s}_1, \qquad \mathbf{A}\mathbf{r}_i = \mathbf{s}_i + \mathbf{s}_{i-1}, \qquad i = 2, \ldots, n$$

Find the matrix representation of $\mathbf{A}$ with respect to these bases.

6. Let $\mathbf{r}_1, \mathbf{r}_2, \mathbf{r}_3$ be a basis for a linear space R, and let $\mathbf{A}: R \to R$ be a linear operator such that

$$\mathbf{A}\mathbf{r}_1 = \mathbf{r}_1 + \mathbf{r}_2, \qquad \mathbf{A}\mathbf{r}_2 = \mathbf{r}_3, \qquad \mathbf{A}\mathbf{r}_3 = \mathbf{r}_1 - \mathbf{r}_2$$

Find the matrix representation of $\mathbf{A}$ with respect to this basis.

7. Let $\mathbf{y}_1, \mathbf{y}_2, \mathbf{y}_3$ be another basis for R related to the basis of Exercise 6 by

$$\mathbf{y}_1 = \mathbf{r}_1 - \mathbf{r}_2, \qquad \mathbf{y}_2 = \mathbf{r}_1 + \mathbf{r}_2, \qquad \mathbf{y}_3 = \mathbf{r}_1 - \mathbf{r}_3$$

Obtain the matrix representation of the linear operator of Exercise 6 relative to the basis $\mathbf{y}_1, \mathbf{y}_2, \mathbf{y}_3$.

8. Let $\mathbf{r}_1, \ldots, \mathbf{r}_n$ be a basis for a linear space R, and let $Q = (q_{ij})$ be a nonsingular $n \times n$ matrix. Show that the elements $\hat{\mathbf{r}}_1, \ldots, \hat{\mathbf{r}}_n$ of (2.2.7) are also a basis for R.

9. Let

$$A = \begin{bmatrix} 2 & 1 & 2 \\ 3 & 2 & 2 \end{bmatrix},$$

and let $\mathbf{A}: R^3 \to R^2$ be the linear operator whose matrix representation in the natural bases for R^3 and R^2 is A. Let

$$\begin{bmatrix} 1 \\ 0 \\ 1 \end{bmatrix}, \begin{bmatrix} 1 \\ 1 \\ 0 \end{bmatrix}, \begin{bmatrix} 0 \\ 1 \\ 1 \end{bmatrix}, \quad \text{and} \quad \begin{bmatrix} 1 \\ 1 \end{bmatrix}, \begin{bmatrix} 1 \\ -1 \end{bmatrix}$$

be two other bases for R^3 and R^2. Find the matrix representation of $\mathbf{A}$ in these bases.

10. Let R and S be linear spaces, and let $[R, S]$ denote the collection of all linear operators from R to S. Show that $[R, S]$ is a linear space and $\dim [R, S] = nm$ if $\dim R = n$ and $\dim S = m$. Compare this result with that of Exercise 2.1-9.

11. Let R_1 be a subspace of a linear space R, and let $\mathbf{A}: R \to S$ be a linear operator. Show that $\{\mathbf{A}\mathbf{r}: \mathbf{r} \in R_1\}$ is a subspace of S.

12. If $\mathbf{A}: R \to S$ is a linear operator, then the operator $\mathbf{B}: R \to S$ defined by $\mathbf{B}\mathbf{r} = \mathbf{A}\mathbf{r} + \mathbf{b}$, where $\mathbf{b}$ is a fixed element of S, is called an *affine operator*. Show that if R_1 is a subspace of R, then $\{\mathbf{B}\mathbf{r}: \mathbf{r} \in R_1\}$ is an affine subspace of S.

2.3. Linear Equations, Rank, Inverses, and Eigenvalues

In this section we use the framework of the previous two sections to rephrase and expand many of the results in Chapter 1 having to do with eigenvalues and solving linear equations.

Range and Rank

If R and S are finite-dimensional linear spaces and $\mathbf{A}: R \to S$ is a linear operator from R to S, the *range* of $\mathbf{A}$, denoted by range($\mathbf{A}$) or $\mathbf{A}(R)$ (also sometimes called the *image* of $\mathbf{A}$), is $\{\mathbf{Ar}: \mathbf{r} \in R\}$. It is easy to verify that range($\mathbf{A}$) is a linear subspace of S (Exercise 2.3-1) and that range($\mathbf{A}$) = span($\mathbf{Ar}_1, \ldots, \mathbf{Ar}_n$) if $\mathbf{r}_1, \ldots, \mathbf{r}_n$ is any basis for R. The *rank* of $\mathbf{A}$ is the dimension of range($\mathbf{A}$).

In Section 1.3 we defined the rank of a matrix A to be the number of linearly independent columns (or rows) of A. We next show the connection between these two concepts of rank.

2.3.1. If $\mathbf{A}: R \to S$ is a linear operator and R and S are finite-dimensional linear spaces, then rank($\mathbf{A}$) is equal to the rank of any matrix representation of $\mathbf{A}$.

Proof. Let $\mathbf{r}_1, \ldots, \mathbf{r}_n$ and $\mathbf{s}_1, \ldots, \mathbf{s}_m$ be any bases for R and S, respectively, and let A be the matrix representation of $\mathbf{A}$ in these bases. If r = rank($\mathbf{A}$), then r is the number of linearly independent elements in the set $\mathbf{Ar}_1, \ldots, \mathbf{Ar}_n$. Equivalently, r is the number of linearly independent elements in the set

$$\sum_{i=1}^{m} a_{ij}\mathbf{s}_i, \qquad j = 1, \ldots, n \tag{2.3.1}$$

where we have used (2.2.3) and the matrix representation to express the $\mathbf{Ar}_i$ in terms of the $\mathbf{s}_i$. We may assume, without loss of generality, that the linearly independent elements of the set (2.3.1) are the first r, for otherwise we could simply renumber the basis elements $\mathbf{r}_1, \ldots, \mathbf{r}_n$. Now suppose that a linear combination of the first r columns of A is zero:

$$\sum_{j=1}^{r} a_{ij}c_j = 0, \qquad i = 1, \ldots, m \tag{2.3.2}$$

But then

$$0 = \sum_{i=1}^{m}\left(\sum_{j=1}^{r} a_{ij}c_j\right)\mathbf{s}_i = \sum_{j=1}^{r} c_j \sum_{i=1}^{m} a_{ij}\mathbf{s}_i \tag{2.3.3}$$

and since the first r elements of the set (2.3.1) are linearly independent, we must have that $c_1 = \cdots = c_r = 0$. Therefore, the first r columns of A are linearly independent, so that rank(A) $\geq r$. On the other hand, if more than r columns of A were linearly independent, say the first $r + 1$, then there is no nonzero vector $\mathbf{c}$ so that (2.3.2) holds, with the summation now going to $r + 1$. Consequently, since the $\mathbf{s}_i$ are linearly independent, there is no nonzero $\mathbf{c}$ for which (2.3.3) holds, again with the sum going to $r + 1$. Hence, the first $r + 1$ elements of the set (2.3.1) are linearly independent, which would contradict that rank($\mathbf{A}$) $= r$. Thus, rank(A) $= r$, which was to be proved. □

In the special case that $R = R^n$ and $S = R^m$ with $\mathbf{A}$ defined by a given real $m \times n$ matrix A, then range($\mathbf{A}$) is just the set of column vectors $\{A\mathbf{x}: \mathbf{x} \in R^n\}$. The same is true if R^n and R^m are replaced by C^n and C^m and A is allowed to be complex. Hence, we define the *range of a matrix A* as the set of all linear combinations of the columns of A:

$$\text{range}(A) = \text{span}(\mathbf{a}_1, \ldots, \mathbf{a}_n)$$

where $\mathbf{a}_1, \ldots, \mathbf{a}_n$ are the columns of A. Thus, range(A) is sometimes called the *column space* of A. Theorem 2.3.1 shows that

$$\dim \text{range}(A) = \text{rank}(A) = \text{number of linearly independent columns of } A$$

As a corollary of 2.3.1, we obtain the important result that multiplication of a matrix on the left or right by nonsingular matrices does not change its rank.

2.3.2. Rank Preservation Theorem. *Let A be an m × n matrix and P and Q nonsingular m × m and n × n matrices, respectively. Then*

$$\text{rank}(A) = \text{rank}(PA) = \text{rank}(AQ) = \text{rank}(PAQ)$$

Proof. Let $\mathbf{A}: C^n \to C^m$ be the linear operator defined by A. Then the matrices PA, AQ, and PAQ are all equivalence transformations of A and, hence, are matrix representations of $\mathbf{A}$ in different bases in C^n and/or C^m. Theorem 2.3.1 then applies to give the conclusion. □

The Null Space

The *null space*, also called the *kernel* of $\mathbf{A}$ or ker($\mathbf{A}$), of a linear operator $\mathbf{A}: R \to S$ is the set of all elements $\mathbf{r} \in R$ such that $\mathbf{A}\mathbf{r} = 0$; it will be denoted by null($\mathbf{A}$). It is easy to show that null($\mathbf{A}$) is a linear subspace of R (Exercise 2.3-1). The dimension of null($\mathbf{A}$) is sometimes called the *nullity* of A. If A is an $m \times n$ matrix, null(A), the null space of the matrix, is the set of all solutions of $A\mathbf{x} = 0$.

An extremely basic and useful result is that the rank of **A** and the dimension of null(**A**) must always sum to the dimension of the domain space of **A**.

2.3.3. Let **A** *be a linear operator from an n-dimensional space to an m-dimensional space. Then dim range*(**A**) + *dim null*(**A**) = *n.*

Proof. Let $q = \dim \text{null}(\mathbf{A})$, and let $\mathbf{r}_1, \ldots, \mathbf{r}_q$ be a basis for null(**A**). By 2.1.6 this can be extended to a basis $\mathbf{r}_1, \ldots, \mathbf{r}_n$ for the whole space. Then $\mathbf{A}\mathbf{r}_i$, $i = q+1, \ldots, n$, are linearly independent; for if this were not the case, then there would be a linear combination

$$0 = \sum_{i=q+1}^{n} c_i \mathbf{A}\mathbf{r}_i = \mathbf{A} \sum_{i=q+1}^{n} c_i \mathbf{r}_i$$

with not all c_i equal to zero. Hence, $\sum_{i=q+1}^{n} c_i \mathbf{r}_i \in \text{null}(\mathbf{A})$ and is a linear combination of $\mathbf{r}_1, \ldots, \mathbf{r}_q$, which would contradict $\mathbf{r}_1, \ldots, \mathbf{r}_n$ being a basis for R unless $q = n$. In this case range (**A**) = {0}, and we are done. Otherwise, we still need to show that $\mathbf{A}\mathbf{r}_i$, $i = q+1, \ldots, n$, is a basis for range(**A**). But this is clear, since for any element $\mathbf{r} \in R$ with coordinates $x_1, \ldots, x_n$ in the basis $\mathbf{r}_1, \ldots, \mathbf{r}_n$,

$$\mathbf{A}\mathbf{r} = \mathbf{A}\left(\sum_{i=1}^{n} x_i \mathbf{r}_i\right) = \sum_{i=q+1}^{n} x_i \mathbf{A}\mathbf{r}_i$$

This completes the proof. □

As a simple example of 2.3.3, let P_n and P_{n-1} be the spaces of polynomials of degrees n and $n-1$, and let $\mathbf{A}: P_n \to P_{n-1}$ be the first-derivative operator $\mathbf{A}p = p'$. Clearly, range(**A**) = P_{n-1}, and thus dim null(**A**) = $n + 1 - n = 1$. Null(**A**) is just the set of constant polynomials, which are mapped into zero by the first-derivative operator. Similarly, if $\mathbf{A}p = p'''$, the third derivative, then dim null(**A**) = 3, and dim range(**A**) = $n - 2$.

The null and range spaces in the previous example were obvious. However, for an $m \times n$ matrix, where m and n are at all large, it is no easy task to determine these spaces. We can, however, determine them in principle from the row echelon form of Section 1.3. We illustrate this by an example. From (1.3.22), $A = QU$, where Q is an $m \times m$ nonsingular matrix, and U is the $m \times n$ row echelon form of A. Now let

$$U = \begin{bmatrix} 1 & 2 & 3 & 4 & 5 \\ 0 & 0 & 1 & 0 & 0 \\ 0 & 0 & 0 & 0 & 1 \\ 0 & 0 & 0 & 0 & 0 \end{bmatrix}$$

be the row echelon form of a 4×5 matrix A. Clearly, $\text{rank}(U) = 3$, and, as noted in Section 1.3, $\text{rank}(A) = \text{rank}(U)$. Thus, $\dim \text{range}(A) = 3$ and, by 2.3.3, $\dim \text{null}(A) = 2$. The pivot columns 1, 3, and 5 of U are linearly independent. Therefore, by 2.3.2, the corresponding columns of A must be linearly independent because U is obtained from A by multiplication by a nonsingular matrix. Hence, columns 1, 3, and 5 of A are a basis for range(A). [Note that these same columns of U are *not* a basis for range(A)]. In general, linearly independent columns of U correspond to linearly independent columns of A and, if $\text{rank}(A) = r$, a basis for range(A) is obtained by choosing any r linearly independent columns of U and taking the corresponding columns of A. For null(A), we note, again since U is obtained from A by multiplication by a nonsingular matrix, that $\text{null}(A) = \text{null}(U)$. In the above example it is easy to calculate that a basis for null(U) is $(-2, 1, 0, 0, 0)^T$ and $(-4, 0, 0, 1, 0)^T$. In general, one can interchange the r linearly independent columns of U into the first r columns to obtain a system of the form

$$r\left\{\begin{bmatrix} * & \cdots & * & * & \cdots & * \\ & * & \vdots & \vdots & & \vdots \\ & & * & * & & * \\ 0 & & \cdot & \cdot & & 0 \\ \vdots & & & & & \vdots \\ 0 & & \cdot & \cdot & & 0 \end{bmatrix}\right. \begin{bmatrix} \hat{x}_1 \\ \vdots \\ \cdot \\ \cdot \\ \vdots \\ \hat{x}_n \end{bmatrix} = 0$$

where the indicated $r \times r$ principal submatrix is triangular and nonsingular. Hence, one can solve for $n - r$ sets of $\hat{x}_1, \ldots, \hat{x}_r$, given $\hat{x}_n = 1$, $\hat{x}_i = 0$ $(i = r + 1, \ldots, n - 1)$ for the first set and then $\hat{x}_n = 0$, $\hat{x}_{n-1} = 1$, $\hat{x}_i = 0$ $(i = r + 1, \ldots, n - 2)$, and so on to $\hat{x}_{r+1} = 1$, $\hat{x}_i = 0$ $(i = r + 2, \ldots, n)$. After permutation back to the correct positions, this gives $n - r$ basis vectors for null(U) and, hence, for null(A).

We next give some examples of the use of 2.3.3 to give information about the rank of products of matrices. The first concerns the matrix A^*A, where A^* is the conjugate transpose of A. As we will see in Chapter 4, A^*A is of fundamental importance in least squares problems.

2.3.4. Let A be a $m \times n$ matrix. Then $\text{null}(A^*A) = \text{null}(A)$, and $\text{rank}(A^*A) = \text{rank}(A)$.

Proof. We show first that $\text{null}(A) = \text{null}(A^*A)$. If $A\mathbf{x} = 0$, then $A^*A\mathbf{x} = 0$, so $\text{null}(A) \subset \text{null}(A^*A)$. If $A^*A\mathbf{x} = 0$, then $0 = \mathbf{x}^*A^*A\mathbf{x} = (A\mathbf{x})^*(A\mathbf{x})$, so that $A\mathbf{x} = 0$. Therefore, $\text{null}(A^*A) = \text{null}(A)$. Thus, by 2.3.3,

$$\dim \text{range}(A^*A) = n - \dim \text{null}(A^*A) = n - \dim \text{null}(A) = \dim \text{range}(A).$$
□

The previous result, which will be important later, is rather special. In general, the rank of a product will be less than the ranks of the individual matrices. For example, if $A = \begin{bmatrix} 0 & 1 \\ 0 & 0 \end{bmatrix}$, then $A^2 = 0$ and has rank 0. The general result on the rank of matrix products is given next.

2.3.5. RANK OF PRODUCTS. *Let B and C be $m \times n$ and $n \times p$ matrices of ranks b and c, respectively. Then*

$$\text{rank}(BC) \leq \min(b, c) \tag{2.3.4}$$

and equality holds in (2.3.4) *if and only if $d = \max(0, c - b)$, where*

$$S = \text{null}(B) \cap \text{range}(C), \qquad d = \dim S \tag{2.3.5}$$

Proof. If $\mathbf{x} \in \text{null}(C)$, then $BC\mathbf{x} = 0$, so that $\mathbf{x} \in \text{null}(BC)$. Thus, by 2.3.3,

$$\dim \text{null}(BC) \geq \dim \text{null}(C) = p - c$$

Therefore, again by 2.3.3,

$$\text{rank}(BC) = p - \dim \text{null}(BC) \leq p - (p - c) = c$$

On the other hand, since

$$\text{range}(BC) = B(\text{range}(C)) \subset \text{range}(B)$$

we have $\text{rank}(BC) \leq b$, which proves (2.3.4). Now consider the subspace S of (2.3.5), as illustrated in Figure 2.3 in the hatched area. The quantity

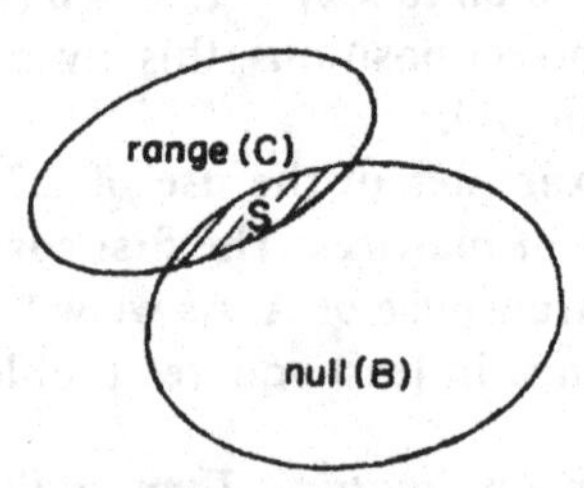

Figure 2.3. The set S.

d measures how much rank the product loses by part of range(C) lying in null(B):

$$\text{rank}(BC) = \text{rank}(C) - d$$

Thus, if $c < b$, equality holds in (2.3.4) if and only if $d = 0$. If $b \leq c$, equality holds if and only if $d = c - b$. □

We note that both 2.3.2 and 2.3.4 may be obtained as corollaries of 2.3.5, but because of their importance we have given independent proofs.

We next give a product theorem of a rather different type in which we decompose a given matrix of rank r into a product of matrices of rank r. This is sometimes called a *full-rank factorization* of A.

2.3.6. Let A be an $m \times n$ matrix of rank r. Then A may be written in the form $A = BC$, where B is $m \times r$, C is $r \times n$ and $\text{rank}(B) = \text{rank}(C) = r$.

Proof. Let $\mathbf{b}_1, \ldots, \mathbf{b}_r$ be a basis for range(A), and let B be the $m \times r$ matrix with these vectors as columns. Let $\mathbf{c}_i$ be the coordinate vector of the ith column, $\mathbf{a}_i$, of A in the basis $\mathbf{b}_1, \ldots, \mathbf{b}_r$; that is,

$$\mathbf{a}_i = \sum_{j=1}^{r} c_{ji}\mathbf{b}_j = B\mathbf{c}_i$$

Thus

$$A = (\mathbf{a}_1, \ldots, \mathbf{a}_n) = B(\mathbf{c}_1, \ldots, \mathbf{c}_n) = BC$$

The matrix B has rank r by construction, and C must also have rank r or else, by 2.3.5, we would have $\text{rank}(A) < r$. □

An important special case of 2.3.6 is when $\text{rank}(A) = 1$. In this case B is $m \times 1$ and C is $1 \times n$. Hence, we can interpret the product BC as the outer product of two column vectors $\mathbf{b}$ and $\mathbf{c}$:

$$A = \mathbf{b}\mathbf{c}^T$$

Rank one matrices appear in a variety of applications, as we shall see later. We only note here that the matrices L_k of the Gaussian elimination process [see (1.3.11)] can be written as

$$L_k = I - \mathbf{l}_k\mathbf{e}_k^T$$

where $\mathbf{l}_k^T = (0, \ldots, 0, l_{k+1,k}, \ldots, l_{nk})$ and $\mathbf{e}_k$ is the kth unit vector.

Linear Equations

Let $\mathbf{A}: R \to S$ and consider the linear equation

$$\mathbf{A}\mathbf{x} = \mathbf{b} \tag{2.3.6}$$

for given $\mathbf{b} \in S$. It is clear that this equation can have a solution if and only if $\mathbf{b} \in \text{range}(\mathbf{A})$. If this is the case, assume that $\hat{\mathbf{x}}$ is a solution. Then $\hat{\mathbf{x}} + \hat{\mathbf{y}}$ is also a solution for any $\hat{\mathbf{y}} \in \text{null}(\mathbf{A})$ since

$$\mathbf{A}(\hat{\mathbf{x}} + \hat{\mathbf{y}}) = \mathbf{A}\hat{\mathbf{x}} = \mathbf{b}$$

The vector $\hat{\mathbf{x}}$ is sometimes called a *particular solution* of (2.3.6). We summarize this as follows.

2.3.7. The equation (2.3.6) *has a solution if and only if* $\mathbf{b} \in$ *range*$(\mathbf{A})$. *If a solution exists, it is unique if and only if null*$(\mathbf{A}) = 0$. *Otherwise, if* $\hat{\mathbf{x}}$ *is any solution, then* $\hat{\mathbf{x}} + \hat{\mathbf{y}}$ *is also a solution for any* $\hat{\mathbf{y}} \in$ *null*$(\mathbf{A})$.

We interpret 2.3.7 in terms of the system of linear equations $A\mathbf{x} = \mathbf{b}$, where A is an $m \times n$ matrix. If $\mathbf{b} \in \text{range}(A)$, the system is said to be *consistent*; that is, there is a solution. Note that the condition $\mathbf{b} \in \text{range}(A)$ is equivalent to the statement that $\mathbf{b}$ is a linear combination of the columns of A. This, in turn, is equivalent (see Exercise 2.3-4) to $\text{rank}(A) = \text{rank}(A, \mathbf{b})$, where $(A, \mathbf{b})$ is the $m \times n + 1$ matrix with $\mathbf{b}$ added as the $(n + 1)$st column. If $m < n$, the system is sometimes said to be *underdetermined* since there are more unknowns than equations; in this case, if the system is consistent, there are necessarily infinitely many solutions since, by 2.3.3, $\dim \text{null}(A) = n - \dim \text{range}(A) \geq n - m > 0$. If $m > n$, so that there are more equations than unknowns, the system is sometimes said to be *overdetermined*; here, there can be a unique solution, no solution, or infinitely many solutions, depending on the particular A and $\mathbf{b}$. We give several examples.

Consider the system

$$\begin{bmatrix} 1 & 1 & 1 \\ 1 & 2 & 1 \\ 2 & 2 & 2 \end{bmatrix} \begin{bmatrix} x_1 \\ x_2 \\ x_3 \end{bmatrix} = \begin{bmatrix} 1 \\ 1 \\ 2 \end{bmatrix}$$

Clearly, A is singular and $\text{rank}(A) = 2$. Thus, $\dim \text{null}(A) = 1$ and $\text{null}(A) = \text{span}(\mathbf{y})$, $\mathbf{y}^T = (1, 0, -1)$. Since $\mathbf{e}_1$ is a solution, all solutions may be written as $\mathbf{e}_1 + c\mathbf{y}$ for arbitrary constants c. On the other hand, if $\mathbf{b} = \mathbf{e}_1$, then $\mathbf{b} \notin \text{range}(A)$, and there is no solution.

Consider next the system

$$\begin{bmatrix} 1 & 1 & 1 \\ 1 & 2 & 1 \end{bmatrix} \begin{bmatrix} x_1 \\ x_2 \\ x_3 \end{bmatrix} = \begin{bmatrix} 1 \\ 1 \end{bmatrix}$$

Again, rank$(A) = 2$ and dim null$(A) = 1$, so that the solutions are $\mathbf{e}_1 + c\mathbf{y}$, just as before. Finally, for the system

$$\begin{bmatrix} 1 & 1 \\ 1 & 2 \\ 2 & 2 \end{bmatrix} \begin{bmatrix} x_1 \\ x_2 \end{bmatrix} = \begin{bmatrix} 1 \\ 1 \\ 2 \end{bmatrix}$$

again rank$(A) = 2$; but now, since $n = 2$, dim null$(A) = 0$, and $\mathbf{e}_1$ is the unique solution. However, if $\mathbf{b}$ were changed to $\mathbf{e}_1$ there would be no solution. The range of A is a two-dimensional subspace of R^3, and only for right-hand sides in this subspace will there be a solution.

We consider next the interpretation of the condition that $\mathbf{b} \in$ range$(\mathbf{A})$ in terms of matrix representations. Let $\mathbf{r}_1, \ldots, \mathbf{r}_n$ and $\mathbf{s}_1, \ldots, \mathbf{s}_m$ be bases for R and S, respectively, and let A be the matrix representation of $\mathbf{A}$ in these bases. If $\mathbf{b} \in$ range(A), then

$$\mathbf{b} = \sum_{j=1}^{n} x_j \mathbf{A}\mathbf{r}_j \tag{2.3.7}$$

Hence, in terms of the basis $\mathbf{s}_1, \ldots, \mathbf{s}_m$, and using (2.2.3), (2.3.7) becomes

$$\mathbf{b} = \sum_{j=1}^{n} x_j \sum_{i=1}^{m} a_{ij}\mathbf{s}_i = \sum_{i=1}^{m} \left(\sum_{j=1}^{n} a_{ij} x_j \right) \mathbf{s}_i$$

Therefore, if $b_1, \ldots, b_m$ are the coordinates of $\mathbf{b}$ in the basis $\mathbf{s}_1, \ldots, \mathbf{s}_m$, we must have

$$b_i = \sum_{j=1}^{n} a_{ij} x_j, \qquad i = 1, \ldots, m \tag{2.3.8}$$

Equation (2.3.8) simply states that the coordinate vector with components $b_1, \ldots, b_m$ is a linear combination of the columns of the matrix A, as was noted previously for a linear system of equations.

Inverses

The linear operator $\mathbf{A}: R \to S$ is said to be *one-to-one* if $\mathbf{Ax} \neq \mathbf{Ay}$ whenever $\mathbf{x} \neq \mathbf{y}$. Clearly, $\mathbf{A}$ is one-to-one if and only if null$(\mathbf{A}) = 0$ (Exercise 2.3-2). Hence, an equivalent way of stating the existence and uniqueness portion of 2.3.7 is as follows:

2.3.8. The equation $\mathbf{Ax} = \mathbf{b}$ *has a unique solution if and only if* $\mathbf{b} \in$ *range*$(\mathbf{A})$, *and* $\mathbf{A}$ *is one-to-one.*

If **A** is one-to-one, what does this signify about its matrix representations? Since **A** is one-to-one if and only if null(**A**) = 0, by 2.3.3, **A** is one-to-one if and only if rank(**A**) = n = dim R. But, by 2.3.1, the rank of any matrix representation of **A** is equal to rank(**A**). Moreover, the rank of an $m \times n$ matrix A can be no greater than the maximum number of linearly independent rows. Hence, we can conclude that

2.3.9. The linear operator $\mathbf{A}: R \to S$, *where* dim $R = n$ *and* dim $S = m$, *is one-to-one if and only if* $m \geqslant n$ *and the rank of any matrix representation of* **A** *is equal to n. In particular, if* $m = n$, *then* **A** *is one-to-one if and only if any matrix representation of* **A** *is nonsingular; in this case* (2.3.6) *has a unique solution for any* $\mathbf{b} \in S$.

We note that if $m > n$ in 2.3.9, then necessarily range(**A**) is a proper subset of S. If $\mathbf{b} \in$ range(**A**), then (2.3.6) has a unique solution; otherwise, there is no solution.

If **A** is one-to-one, there is a linear operator $\mathbf{A}_L : S \to R$ that satisfies

$$\mathbf{A}_L \mathbf{A} \mathbf{x} = \mathbf{x} \qquad \text{for all } \mathbf{x} \text{ in } R \tag{2.3.9}$$

and $\mathbf{A}_L$ is called a *left inverse* of **A**. Note that, by Theorem 2.3.9, a necessary condition for $\mathbf{A}_L$ to exist is that $m \geqslant n$. In particular, if $n > m$, **A** cannot be one-to-one. However, if range(**A**) = S, there is a one-to-one linear operator $\mathbf{A}_R : S \to R$ that satisfies

$$\mathbf{A}\mathbf{A}_R \mathbf{y} = \mathbf{y} \qquad \text{for all } \mathbf{y} \text{ in } S$$

In this case $\mathbf{A}_R$ is a *right inverse* of **A**. Finally, if **A** has both left and right inverses, then it has an *inverse* $\mathbf{A}^{-1}$ that satisfies

$$\mathbf{A}^{-1} = \mathbf{A}_R = \mathbf{A}_L \qquad \text{and} \qquad \mathbf{A}^{-1}\mathbf{A} = \mathbf{A}\mathbf{A}^{-1} = \mathbf{I}$$

The above statements translate directly to matrices, and we summarize them in the following theorem:

2.3.10. Let A be an $m \times n$ *matrix.*

(*a*) *If* $m \geqslant n$ *and* rank(A) = n, *then there is an* $n \times m$ *matrix* A_L *of rank n, called a* left inverse *of A, such that*

$$A_L A = I_n$$

where I_n is the $n \times n$ identity matrix.

(*b*) *If $n \geq m$ and rank$(A) = m$, then there is an $n \times m$ matrix A_R of rank m, called a* right inverse *of A, such that*

$$AA_R = I_m$$

(*c*) *If $n = m$ and A has both left and right inverses, then A is nonsingular and $A^{-1} = A_L = A_R$.*

As examples, consider first the matrix

$$A = \begin{bmatrix} 1 & 1 \\ 1 & 2 \\ 1 & 1 \end{bmatrix}$$

As previously noted, rank$(A) = 2$, and so A has a left inverse A_L. One such A_L is given by

$$\begin{bmatrix} 2 & -1 & 0 \\ -1 & 1 & 0 \end{bmatrix}$$

as is easily verified. However, if we write A as

$$A = \begin{bmatrix} A_1 \\ A_2 \end{bmatrix}$$

where $A_2 = (1, 1)$, then A_1 is nonsingular, and

$$A_L = [A_1^{-1} - A_3A_2A_1^{-1}, A_3] \tag{2.3.10}$$

is also a left inverse of A for an arbitrary column vector A_3. This holds in general: If A_1 is $n \times n$ and nonsingular, then (2.3.10) gives a left inverse of A for any $n \times (m - n)$ matrix A_3. Similarly, if

$$A = \begin{bmatrix} 1 & 1 & 1 \\ 1 & 2 & 1 \end{bmatrix}$$

then

$$A_R = \begin{bmatrix} 2 & -1 \\ -1 & 1 \\ 0 & 0 \end{bmatrix}$$

is a right inverse. In general, if

$$A = [A_1, A_2]$$

where A_1 is $m \times m$ and nonsingular, then

$$A_R = \begin{bmatrix} A_1^{-1} - A_1^{-1}A_2A_3 \\ A_3 \end{bmatrix}$$

is a right inverse of A for any $(n - m) \times m$ matrix A_3.

In a later chapter we will give other formulas for left and right inverses in terms of certain canonical forms of A.

Eigenvalues

We end this section with a short discussion of eigenvalues and eigenvectors for linear operators. As for matrices, we can define eigenvalues and eigenvectors of a linear operator **A** by the basic equation

$$\mathbf{Ax} = \lambda \mathbf{x} \tag{2.3.11}$$

Here, since the element **x** must be in both the range and domain spaces of **A**, we are restricted to operators $\mathbf{A}: R \to R$, that is, operators that map a linear space R into itself. The scalar λ in (2.3.11) must then lie in the scalar field of R.

Now let $\mathbf{r}_1, \ldots, \mathbf{r}_n$ be a basis for R, and let A be the matrix representation of **A** in this basis. What is the relation of the eigenvalues of A to the eigenvalues of **A** defined by (2.3.11)?

2.3.11. Let $\mathbf{A}: R \to R$ *be a linear operator, where* R *is a complex linear space of dimension* n. *Then all matrix representations of* **A** *have the same eigenvalues* $\lambda_1, \ldots, \lambda_n$, *and these are precisely the eigenvalues of* **A**.

Proof. Let $\mathbf{r}_1, \ldots, \mathbf{r}_n$ be a basis for R, and let A be the matrix representation of **A** in this basis. Then A has eigenvalues $\lambda_1, \ldots, \lambda_n$. If λ is any eigenvalue of A and $\hat{\mathbf{x}}$ a corresponding eigenvector, let **x** be the element of R with coordinates $\hat{\mathbf{x}}$ in the basis $\mathbf{r}_1, \ldots, \mathbf{r}_n$. Then

$$\mathbf{Ax} = \sum_{j=1}^{n} \hat{x}_j \mathbf{Ar}_j = \sum_{j=1}^{n} \hat{x}_j \sum_{i=1}^{n} a_{ij}\mathbf{r}_i = \sum_{i=1}^{n} \left(\sum_{j=1}^{n} a_{ij}\hat{x}_j \right) \mathbf{r}_i = \sum_{i=1}^{n} \lambda \hat{x}_i \mathbf{r}_i = \lambda \mathbf{x}$$

so that λ is an eigenvalue of **A** with corresponding eigenvector **x**. Conversely, if λ and **x** are an eigenpair of **A**, and $\hat{\mathbf{x}}$ is the coordinate vector of **x** in the basis $\mathbf{r}_1, \ldots, \mathbf{r}_n$, then a similar calculation (Exercise 2.3-6) shows that λ and $\hat{\mathbf{x}}$ are an eigenvalue and an eigenvector of the matrix A. Finally, if $\hat{\mathbf{r}}_1, \ldots, \hat{\mathbf{r}}_n$ is another basis for R, and $\hat{A}$ is the matrix representation of **A** in this basis, then 2.2.5 shows that A and $\hat{A}$ are similar; that is, there is a nonsingular matrix P such that $\hat{A} = PAP^{-1}$. Hence,

$$\det(\hat{A} - \lambda I) = \det(PAP^{-1} - \lambda I) = \det P \det(A - \lambda I) \det P^{-1} = \det(A - \lambda I)$$

Therefore, A and $\hat{A}$ have the same characteristic polynomial and, thus, the same eigenvalues. □

The last part of the proof of 2.3.11 is of such importance that we restate the conclusion separately and also show how the eigenvectors of a matrix change under a similarity transformation.

*2.3.12. Similar matrices have the same characteristic polynomial and, hence, the same eigenvalues. Moreover, if $\hat{A} = PAP^{-1}$ and **x** is an eigenvector of A, then $P\mathbf{x}$ is an eigenvector of $\hat{A}$.*

To prove the last statement of 2.3.12, let $A\mathbf{x} = \lambda \mathbf{x}$. Then $PAP^{-1}P\mathbf{x} = \lambda P\mathbf{x}$.

We next discuss the assumption in 2.3.11 that R is a complex linear space. Suppose that $R = R^2$, and A is a matrix with characteristic polynomial $\lambda^2 + 1 = 0$. Then A has no real eigenvalues. Hence, if the defining relation (2.3.11) is restricted to real vectors **x** and real scalars λ, A has no eigenvalues or eigenvectors. The same is true more generally. If R is a real n-dimensional linear space, then a linear operator $\mathbf{A}: R \to R$ may have no real eigenvalues or any number between 1 and n (with the proviso that, since complex roots of a real polynomial occur in complex conjugate pairs, there are restrictions on the number of real eigenvalues). One can circumvent this problem by (usually tacitly) considering the corresponding complex space rather than the real space, and then **A** will have its full complement of n eigenvalues. Eigenvectors corresponding to complex eigenvalues will necessarily then be in the complex space, not the real one.

Summary

The main concepts and results of this section are as follows:

- The range and null space of a linear operator **A** or of a matrix;

- dim range(**A**) + dim null(**A**) = dim(domain space of **A**);
- A linear equation $\mathbf{Ax} = \mathbf{b}$ has a solution if and only if $\mathbf{b} \in$ range(**A**). If $\hat{\mathbf{x}}$ is any solution, then $\hat{\mathbf{x}} + \mathbf{y}$ is also a solution for any $\mathbf{y} \in$ null(**A**);
- Inverses, and left and right inverses, of a linear operator;
- The eigenvalues of a linear operator **A** are equal to the eigenvalues of any matrix representation of **A**.

It is also useful to state explicitly the main concrete results for an $m \times n$ matrix **A**

- range(A) = subspace of R^m (or C^m) spanned by the columns of A;
- null(A) = subspace of R^n (or C^n) of all solutions of $A\mathbf{x} = 0$;
- rank(A) + dim null(A) = n;
- rank(A) is preserved under multiplication by a nonsingular matrix;
- rank(A^*A) = rank(A);
- If $A = BC$, then rank(A) $\leq$ min(rank(B), rank(C));
- Similar $n \times n$ matrices have the same eigenvalues but different eigenvectors.

Exercises 2.3

1. Let R and S be linear spaces and $\mathbf{A}: R \to S$ a linear operator. Show that range(**A**) and null(**A**) are linear subspaces of S and R, respectively.
2. Show that a linear operator $\mathbf{A}: R \to S$ is one-to-one if and only if null(**A**) = 0.
3. Determine bases for the range and null spaces of the matrices

   (a) $A = \begin{bmatrix} 1 & 1 & 1 \\ 2 & 2 & 2 \end{bmatrix}$

   (b) $A = \begin{bmatrix} 1 & 1 & 1 & 1 \\ 1 & 2 & 3 & 4 \\ 2 & 4 & 6 & 8 \end{bmatrix}$

4. Let A be an $m \times n$ matrix, $\mathbf{b}$ an m-vector, and $(A, \mathbf{b})$ the $m \times (n+1)$ matrix with additional column $\mathbf{b}$. Show that $\mathbf{b} \in$ range(A) if and only if rank(A) = rank($A, \mathbf{b}$).
5. Let A and B be $m \times n$ matrices. Show that
   (a) rank($A + B$) $\leq$ rank(A) + rank(B)
   (b) rank($A - B$) $\geq$ |rank(A) − rank(B)|
6. Let λ and $\mathbf{x}$ be an eigenvalue and a corresponding eigenvector of the linear operator $\mathbf{A}: R \to R$. Let $\mathbf{r}_1, \ldots, \mathbf{r}_n$ be a basis for R, let $\hat{\mathbf{x}}$ be the coordinate vector of $\mathbf{x}$, and let A be the matrix representation of **A** in this basis. Show that λ and $\hat{\mathbf{x}}$ are an eigenvalue and an eigenvector of the matrix A.

7. Ascertain whether the following pairs of matrices are similar:

(a) $\begin{bmatrix} 1 & 1 \\ 0 & 1 \end{bmatrix}$ $\begin{bmatrix} 1 & 0 \\ 1 & 1 \end{bmatrix}$

(b) $\begin{bmatrix} 0 & 1 \\ 0 & 1 \end{bmatrix}$ $\begin{bmatrix} 0 & 1 \\ 0 & 0 \end{bmatrix}$

(c) $\begin{bmatrix} 1 & 2 & 3 \\ 4 & 5 & 6 \\ 7 & 8 & 9 \end{bmatrix}$ $\begin{bmatrix} 9 & 8 & 7 \\ 6 & 5 & 4 \\ 3 & 2 & 1 \end{bmatrix}$

8. Let A be an $m \times n$ matrix. Apply 2.3.3 to $\mathbf{A}^*$ to conclude that

$$\dim \operatorname{range}(A^*) + \dim \operatorname{null}(A^*) = m$$

9. Let A and P be $n \times n$ matrices with P nonsingular. Show that $\operatorname{Tr}(A) = \operatorname{Tr}(P^{-1}AP)$.

10. Let A and B be $m \times n$ and $n \times p$ matrices, respectively. Show that $\operatorname{range}(AB) \subset \operatorname{range}(A)$.

11. Let H be an $n \times n$ Hermitian positive definite matrix and A an $n \times m$ matrix. Show that $\operatorname{rank}(A) = \operatorname{rank}(A^*HA)$.

12. Let

$$A = \begin{bmatrix} 1 & 2 & 2 \\ 1 & 3 & 2 \\ 1 & 4 & 2 \end{bmatrix}$$

Find $\operatorname{range}(A)$ and $\operatorname{null}(A)$.

13. If A is the matrix of Exercise 12, and $r = \operatorname{rank}(A)$, write A as a product $A = BC$, where B is $3 \times r$ and C is $r \times 3$.

14. *If* A is an $m \times n$ matrix with a left inverse, show that A^* has a right inverse. Similarly, if A has a right inverse, show that A^* has a left inverse.

2.4. Inner Product Spaces

In Section 2.1 we abstracted many of the properties of R^n to give the notion of a linear space. Two fundamental properties that we did not discuss were orthogonality of vectors and distance between two points. We now add these basic properties, beginning in this section with orthogonality.

In Section 1.1 we defined the inner product of two vectors $\mathbf{x}$, $\mathbf{y}$ in R^n or C^n by $\mathbf{x}^T\mathbf{y}$ or $\mathbf{x}^*\mathbf{y}$, respectively. This inner product has four basic properties, which motivate the following definition.

2.4.1. DEFINITION: *Inner product.* Let R be a real or a complex linear space. Then an *inner product* on R is a real- or a complex-valued function of two vector variables, denoted by (,), that satisfies

(a) $(\mathbf{x}, \mathbf{x}) \geq 0$ for all $\mathbf{x} \in R$, and $(\mathbf{x}, \mathbf{x}) = 0$ only if $\mathbf{x} = 0$;
(b) $(\mathbf{x}, \mathbf{y}) = (\mathbf{y}, \mathbf{x})$ for all $\mathbf{x}, \mathbf{y} \in R$ (real);
$(\mathbf{x}, \mathbf{y}) = \overline{(\mathbf{y}, \mathbf{x})}$ for all $\mathbf{x}, \mathbf{y} \in R$ (complex);
(c) $(\alpha\mathbf{x}, \mathbf{y}) = \bar{\alpha}(\mathbf{x}, \mathbf{y})$ for all $\mathbf{x}, \mathbf{y} \in R$ and scalars α;
(d) $(\mathbf{x} + \mathbf{y}, \mathbf{z}) = (\mathbf{x}, \mathbf{z}) + (\mathbf{y}, \mathbf{z})$ for all $\mathbf{x}, \mathbf{y}, \mathbf{z} \in R$.

If R is a linear space with an inner product, then it is an *inner product space.* A real inner product space is sometimes called a *Euclidean space,* and a complex inner product space is called a *unitary space.*

One can immediately verify that the usual inner product $\mathbf{x}^T\mathbf{y}$ on R^n (or $\mathbf{x}^*\mathbf{y}$ on C^n) satisfies the properties of 2.4.1. More generally, if A is a real symmetric positive definite $n \times n$ matrix, then $\mathbf{x}^TA\mathbf{y}$ also defines an inner product on R^n, and $\mathbf{x}^*A\mathbf{y}$ defines an inner product of C^n if A is Hermitian and positive definite (Exercise 2.4-1).

Orthogonal Vectors and Bases

Orthogonal vectors in an inner product space as defined as follows.

2.4.2. DEFINITION. If R is an inner product space, the nonzero vectors $\mathbf{x}_1, \ldots, \mathbf{x}_m$ are *orthogonal* if

$$(\mathbf{x}_i, \mathbf{x}_j) = 0, \qquad i \neq j, \qquad j = 1, \ldots, m \tag{2.4.1}$$

and *orthonormal* if they are orthogonal and $(\mathbf{x}_i, \mathbf{x}_i) = 1$, $i = 1, \ldots, m$.

A set of orthogonal vectors is necessarily linearly independent. This can be easily shown directly (Exercise 2.4-2), but it is an immediate consequence of the following general criterion for linear independence in an inner product space. Let $\mathbf{x}_1, \ldots, \mathbf{x}_m$ be nonzero elements in the inner product space R and form the *Grammian matrix*

$$A = \begin{bmatrix} (\mathbf{x}_1, \mathbf{x}_1) & \cdots & (\mathbf{x}_1, \mathbf{x}_m) \\ \vdots & & \vdots \\ (\mathbf{x}_m, \mathbf{x}_1) & \cdots & (\mathbf{x}_m, \mathbf{x}_m) \end{bmatrix} \tag{2.4.2}$$

By 2.4.1(b), A is symmetric if R is a real space, and Hermitian if R is complex. The relation of A to the linear dependence of $\mathbf{x}_1, \ldots, \mathbf{x}_m$ is given by the following:

2.4.3. The Grammian matrix (2.4.2) is always positive semidefinite. It is positive definite if and only if the vectors $\mathbf{x}_1, \ldots, \mathbf{x}_m$ are linearly independent.

Proof. Let $\mathbf{c}$ be an arbitrary m-vector. Then

$$\mathbf{c}^*A\mathbf{c} = \sum_{i,j=1}^{m} \bar{c}_i c_j (\mathbf{x}_i, \mathbf{x}_j) = \sum_{i,j=1}^{m} (c_i\mathbf{x}_i, c_j\mathbf{x}_j) = \left(\sum_{i=1}^{m} c_i\mathbf{x}_i, \sum_{i=1}^{m} c_i\mathbf{x}_i\right) \geq 0 \quad (2.4.3)$$

Hence, A is positive semidefinite. If A is not positive definite, there is a nonzero vector $\mathbf{c}$ such that $\mathbf{c}^*A\mathbf{c} = 0$. Then (2.4.3) shows that $\sum c_i\mathbf{x}_i = 0$, so that $\mathbf{x}_1, \ldots, \mathbf{x}_m$ are linearly dependent. Therefore, if $\mathbf{x}_1, \ldots, \mathbf{x}_m$ are linearly independent, then A is positive definite. Conversely, if A is positive definite, then $\mathbf{c}^*A\mathbf{c} > 0$ for any $\mathbf{c} \neq 0$, and, hence, by (2.4.3), $\sum c_i\mathbf{x}_i \neq 0$. □

An important special case of 2.4.3 is for the Grammian matrix

$$\begin{bmatrix} (\mathbf{x}, \mathbf{x})(\mathbf{x}, \mathbf{y}) \\ (\mathbf{y}, \mathbf{x})(\mathbf{y}, \mathbf{y}) \end{bmatrix}$$

Since this matrix is positive semidefinite, its determinant is nonnegative, so that

$$|(\mathbf{x}, \mathbf{y})|^2 \leq (\mathbf{x}, \mathbf{x})(\mathbf{y}, \mathbf{y}) \quad (2.4.4)$$

This is the *Cauchy–Schwarz inequality* (also called the Cauchy–Schwarz–Buniakowsky inequality) for any two vectors in an inner product space.

Another important special case of a Grammian matrix is a matrix of the form B^*B, where B is an $m \times n$ matrix. If $\mathbf{b}_1, \ldots, \mathbf{b}_n$ are the columns of B, then

$$B^*B = \begin{bmatrix} \mathbf{b}_1^*\mathbf{b}_1 & \cdots & \mathbf{b}_1^*\mathbf{b}_n \\ \vdots & & \vdots \\ \mathbf{b}_n^*\mathbf{b}_1 & \cdots & \mathbf{b}_n^*\mathbf{b}_n \end{bmatrix} \quad (2.4.5)$$

so that B^*B is a Grammian matrix in the usual inner product on C^n. As we shall see in Chapter 4, such matrices are important in least squares problems. Recall that we proved in 2.3.4 that $\text{rank}(B) = \text{rank}(B^*B)$, a result that overlaps and complements 2.4.3. In particular, by 2.4.3, B^*B is positive definite if and only if $\text{rank}(B) = n$.

If $\mathbf{x}_1, \ldots, \mathbf{x}_m$ are orthogonal vectors in an inner product space R, then the Grammian matrix A is a diagonal matrix with nonzero diagonal elements. Hence, it is positive definite. Thus, by 2.4.3, $\mathbf{x}_1, \ldots, \mathbf{x}_m$ are linearly independent. If $m = \dim R = n$, then $\mathbf{x}_1, \ldots, \mathbf{x}_n$ form an *orthogonal* basis for R and an *orthonormal basis* if $\mathbf{x}_1, \ldots, \mathbf{x}_n$ are orthonormal. On the other hand, if $\mathbf{x}_1, \ldots, \mathbf{x}_n$ is a basis for R but not orthogonal, we can transform the $\mathbf{x}_i$ to an orthogonal basis by the *Gram-Schmidt process*:

$$\mathbf{y}_1 = \mathbf{x}_1, \qquad \mathbf{y}_k = \mathbf{x}_k - \sum_{j=1}^{k-1} \alpha_{jk}\mathbf{y}_j, \quad \alpha_{jk} = \frac{(\mathbf{y}_j, \mathbf{x}_k)}{(\mathbf{y}_j, \mathbf{y}_j)}, \; k = 2, \ldots, n \tag{2.4.6}$$

To see why these vectors are orthogonal, consider first $\mathbf{y}_2 = \mathbf{x}_2 - \alpha_{12}\mathbf{y}_1$. Then

$$(\mathbf{y}_1, \mathbf{y}_2) = (\mathbf{y}_1, \mathbf{x}_2) - \alpha_{12}(\mathbf{y}_1, \mathbf{y}_1) = 0$$

by the definition of α_{12}. Clearly, $\mathbf{y}_2 \neq 0$, or else $\mathbf{x}_1$ and $\mathbf{x}_2$ would be linearly dependent. Continuing in this way, it is easy to verify by induction that the $\mathbf{y}_i$ generated by (2.4.6) are all nonzero and orthogonal (Exercise 2.4-3).

As a simple example, consider the three vectors

$$\mathbf{x}_1 = \begin{bmatrix} 1 \\ 1 \\ 1 \end{bmatrix}, \qquad \mathbf{x}_2 = \begin{bmatrix} 1 \\ 2 \\ 2 \end{bmatrix}, \qquad \mathbf{x}_3 = \begin{bmatrix} -1 \\ 1 \\ -1 \end{bmatrix}$$

in R^3. The Gram-Schmidt process then produces the orthogonal vectors

$$\mathbf{y}_1 = \begin{bmatrix} 1 \\ 1 \\ 1 \end{bmatrix}, \qquad \mathbf{y}_2 = \begin{bmatrix} 1 \\ 2 \\ 2 \end{bmatrix} - \frac{5}{3}\begin{bmatrix} 1 \\ 1 \\ 1 \end{bmatrix} = \frac{1}{3}\begin{bmatrix} -2 \\ 1 \\ 1 \end{bmatrix}$$

$$\mathbf{y}_3 = \begin{bmatrix} -1 \\ 1 \\ -1 \end{bmatrix} - \frac{1}{3}\begin{bmatrix} -2 \\ 1 \\ 1 \end{bmatrix} + \frac{1}{3}\begin{bmatrix} 1 \\ 1 \\ 1 \end{bmatrix} = \begin{bmatrix} 0 \\ 1 \\ -1 \end{bmatrix}$$

Note that, as in the above example, the vectors produced by the Gram-Schmidt process are not necessarily orthonormal. However, if we like, we can always make the $\mathbf{y}_i$ orthonormal by the scaling

$$\hat{\mathbf{y}}_i = \frac{\mathbf{y}_i}{(\mathbf{y}_i, \mathbf{y}_i)^{1/2}}, \qquad i = 1, \ldots, n$$

By 2.1.4 every finite-dimensional space has a basis, and we can obtain from this basis an orthogonal or orthonormal basis by the Gram-Schmidt process. Moreover, if we have a set of m orthonormal vectors, by 2.1.5 we can extend these to n basis vectors and then orthogonalize the additional $n - m$ vectors by the Gram-Schmidt process. We summarize this in the following theorem.

2.4.4. A finite-dimensional inner product space R has an orthonormal basis. In particular, if $\mathbf{x}_1, \ldots, \mathbf{x}_m$ *are m orthonormal vectors, and dim* $R = n > m$, *then there is an orthonormal basis* $\mathbf{y}_1, \ldots, \mathbf{y}_n$ *for R with* $\mathbf{y}_i = \mathbf{x}_i$, $i = 1, \ldots, m$.

QR Orthogonalization

We consider next a different approach to the orthogonalization of a set of vectors in R^n (or C^n). Let $\mathbf{a}_1, \ldots, \mathbf{a}_n$ be linearly independent n-vectors, and let A be the matrix with these vectors as columns. Let Q_1 be the $n \times n$ matrix

$$Q_1 = \begin{bmatrix} \cos\theta_1 & \sin\theta_1 & \\ -\sin\theta_1 & \cos\theta_1 & \\ & & I_{n-2} \end{bmatrix}$$

where I_{n-2} is the $(n-2) \times (n-2)$ identity, and choose θ_1 so that

$$-a_{11}\sin\theta_1 + a_{21}\cos\theta_1 = 0, \qquad \theta_1 = \tan^{-1}\left(\frac{a_{21}}{a_{11}}\right).$$

Then the multiplication $Q_1 A$ produces a matrix whose (2, 1) element is zero. Next, let

$$Q_2 = \begin{bmatrix} \cos\theta_2 & 0 & \sin\theta_2 & \\ 0 & 1 & 0 & \\ -\sin\theta_2 & 0 & \cos\theta_2 & \\ & & & I_{n-3} \end{bmatrix}$$

where $-a_{11}^1\sin\theta_2 + a_{31}^1\cos\theta_2 = 0$ and a_{ij}^1 denotes the elements of $Q_1 A$. The multiplication $Q_2 Q_1 A$ produces a zero element in the (3, 1) position while preserving the previous zero in the (2, 1) position. Continuing in this fashion, we successively zero the elements of the first column by multiplication by matrices Q_i that have a cosine in positions (1, 1) and $(i+1, i+1)$ and sine and $-$sine in positions $(j, i+1)$ and $(i+1, j)$. At the end of this stage

$$Q_{n-1}Q_{n-2}\cdots Q_1 A = \begin{bmatrix} a_{11}^1 & a_{12}^1 & \cdots & a_{1n}^1 \\ 0 & a_{22}^1 & \cdots & a_{2n}^1 \\ \vdots & \vdots & & \vdots \\ 0 & a_{n2}^1 & \cdots & a_{nn}^1 \end{bmatrix}$$

where the a_{ij}^1 are the current elements and may have been changed several times from the original a_{ij}. We now repeat this process in order to zero the elements in the second column below the (2, 2) position. Q_n will have the form

$$Q_n = \begin{bmatrix} 1 & & & \\ & \cos\theta_n & \sin\theta_n & \\ & -\sin\theta_n & \cos\theta_n & \\ & & & I_{n-3} \end{bmatrix}$$

and the successive Q_i will have the cosine and sine terms in the appropriate places to effect the zeroing of the subdiagonal elements in successive columns. At the end of this process we will have multiplied by $p = n - 1 + n - 2 + \cdots + 1 = n(n-1)/2$ matrices Q_i to obtain

$$Q_p Q_{p-1} \cdots Q_1 A = R$$

where R is an upper triangular matrix.

The matrices Q_i used above are called *plane rotation* matrices (in numerical analysis they are also called *Givens transformations*). The Q_i are all orthogonal (Exercise 2.4-9); hence, their product is orthogonal (and therefore nonsingular). We then set $Q^{-1} = Q_p \cdots Q_1$ so that

$$A = QR \tag{2.4.7}$$

Since the inverse of an orthogonal matrix is orthogonal (Exercise 1.3-7), (2.4.7) shows that A may be written as a product of an orthogonal matrix and an upper triangular matrix. The columns of Q are orthonormal vectors, and they are the orthogonal set obtained from the columns of A.

To relate the QR process back to the Gram-Schmidt procedure, note that the linear independence of the columns of A ensures that A is nonsingular and, hence, R is nonsingular. Let $T = R^{-1}$; then R^{-1} is also upper triangular. If we equate the columns of $AT = Q$, we obtain

$$\mathbf{q}_i = \sum_{j=1}^{i} t_{ji}\mathbf{a}_j, \qquad i = 1, \ldots, n$$

and this is mathematically equivalent to the Gram-Schmidt process, using the inner product $\mathbf{x}^T\mathbf{y}$ and normalizing the orthogonal vectors at each step. However, computationally, the QR approach is preferred in practice.

Note that the QR process works just as above if there are $m < n$ vectors $\mathbf{a}_1, \ldots, \mathbf{a}_m$. In this case A and R are $n \times m$ matrices, and R has the form

$$R = \begin{bmatrix} R_1 \\ 0 \end{bmatrix}$$

where R_1 is a nonsingular $m \times m$ upper triangular matrix. This process is useful in least squares problems, as we shall see in Chapter 4.

Change of Basis

Let $\mathbf{r}_1, \ldots, \mathbf{r}_n$ and $\hat{\mathbf{r}}_1, \ldots, \hat{\mathbf{r}}_n$ be two orthonormal bases of an inner product space R related by

$$\hat{\mathbf{r}}_j = \sum_{k=1}^{n} q_{kj}\mathbf{r}_k, \qquad j = 1, \ldots, n$$

Since $(\mathbf{r}_i, \mathbf{r}_j) = (\hat{\mathbf{r}}_i, \hat{\mathbf{r}}_j) = \delta_{ij}$, where δ_{ij} is the Kronecker delta ($\delta_{ij} = 0$ if $i \neq j$ and $\delta_{ii} = 1$), we have

$$\delta_{ij} = (\hat{\mathbf{r}}_i, \hat{\mathbf{r}}_j) = \left(\sum_{k=1}^{n} q_{ki}\mathbf{r}_k, \sum_{k=1}^{n} q_{kj}\mathbf{r}_k \right) = \sum_{k=1}^{n} \bar{q}_{ki}q_{kj}, \qquad i, j = 1, \ldots, n \tag{2.4.8}$$

where the complex conjugate is needed if the q_{kj} are complex. The equations (2.4.8) are equivalent to

$$Q^*Q = I$$

for the matrix $Q = (q_{ij})$, so that Q is a unitary matrix (or orthogonal if Q is real). Then if A and $\hat{A}$ are matrix representations of a linear operator $\mathbf{A}: R \rightarrow R$, 2.2.5 shows that

$$\hat{A} = Q^{-1}AQ = Q^*AQ$$

We can summarize this as follows:

2.4.5. If R is a real (complex) finite-dimensional inner product space, then coordinate vectors in two orthonormal bases are related by an orthogonal (unitary) matrix, and matrix representations in the two bases are related by an orthogonal (unitary) similarity transformation.

Orthogonal Projection

The first step of the Gram–Schmidt process (2.4.6),

$$\mathbf{y}_2 = \mathbf{x}_2 - \alpha\mathbf{y}_1, \qquad \alpha = \frac{(\mathbf{y}_1, \mathbf{x}_2)}{(\mathbf{y}_1, \mathbf{y}_1)}$$

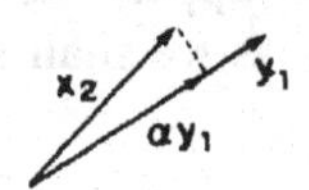

Figure 2.4. Orthogonal projection.

subtracts from $\mathbf{x}_2$ the orthogonal projection of $\mathbf{x}_2$ onto $\mathbf{y}_1$, as illustrated in Figure 2.4 for vectors in R^2 with the inner product $\mathbf{x}^T\mathbf{y}$.

More generally, if S is a subspace of an inner product space R, then the *orthogonal projection* of a vector $\mathbf{x} \in R$ onto S is a vector $\mathbf{x}_S$ in S such that $\mathbf{x} - \mathbf{x}_S$ is orthogonal to every vector in S. In the above Gram–Schmidt example, $S = \text{span}(\mathbf{y}_1)$, and $\mathbf{x}_S = \alpha \mathbf{y}_1$. Then $\mathbf{x}_2 - \mathbf{x}_S = \mathbf{y}_2$, which is orthogonal to every vector in S.

In general, we can represent $\mathbf{x}_S$ in the following way. Let $\mathbf{x}_1, \ldots, \mathbf{x}_m$ be a basis for S, and write

$$\mathbf{x}_S = \sum_{i=1}^{m} c_i \mathbf{x}_i$$

For $\mathbf{x} - \mathbf{x}_S$ to be orthogonal to every vector in S, it is necessary and sufficient that it be orthogonal to $\mathbf{x}_1, \ldots, \mathbf{x}_m$, so that

$$(\mathbf{x}_k, \mathbf{x} - \mathbf{x}_S) = 0, \qquad k = 1, \ldots, m$$

or

$$\left(\mathbf{x}_k, \sum_{i=1}^{m} c_i \mathbf{x}_i\right) = (\mathbf{x}_k, \mathbf{x}), \qquad k = 1, \ldots, m$$

This is a system of linear equations for the coefficients $c_1, \ldots, c_m$ and may be written as

$$A\mathbf{c} = \begin{bmatrix} (\mathbf{x}_1, \mathbf{x}) \\ \vdots \\ (\mathbf{x}_m, \mathbf{x}) \end{bmatrix} \tag{2.4.9}$$

where A is the Grammian matrix (2.4.2). Since $\mathbf{x}_1, \ldots, \mathbf{x}_m$ are linearly independent, Theorem 2.4.3 ensures that A is nonsingular, and thus (2.4.9) has a unique solution. In particular, if $\mathbf{x}_1, \ldots, \mathbf{x}_m$ is an orthonormal basis for S, then A is just the identity matrix. Therefore, $c_i = (\mathbf{x}_i, \mathbf{x})$, and the orthogonal projection is

$$\mathbf{x}_S = \sum_{i=1}^{m} (\mathbf{x}_i, \mathbf{x})\mathbf{x}_i \tag{2.4.10}$$

In this case, $\mathbf{x}_S$ is just the sum of the individual orthogonal projections of $\mathbf{x}$ onto the $\mathbf{x}_i$.

We apply the above to the following situation. Let $\mathbf{x}_1, \ldots, \mathbf{x}_m$ be m linearly independent vectors in C^n with $m < n$, let X be the $n \times m$ matrix whose columns are the $\mathbf{x}_i$, and let $S = \text{range}(X) = \text{span}(\mathbf{x}_1, \ldots, \mathbf{x}_m)$. Assume that C^n has the usual inner product $\mathbf{x}^*\mathbf{y}$. To find the orthogonal projection of a given vector $\mathbf{x}$ onto the subspace S, we note that, by 2.4.3 the Grammian matrix $A = X^*X$ is nonsingular, and the right-hand side of (2.4.9) is $X^*\mathbf{x}$; thus, from (2.4.9),

$$\mathbf{c} = (X^*X)^{-1}X^*\mathbf{x}$$

The vector $\mathbf{c}$ gives the coefficients of the orthogonal projection, $\mathbf{x}_S$, of $\mathbf{x}$ in terms of the vectors $\mathbf{x}_1, \ldots, \mathbf{x}_m$; therefore,

$$\mathbf{x}_S = \sum_{i=1}^{m} c_i \mathbf{x}_i = X\mathbf{c} = X(X^*X)^{-1}X^*\mathbf{x} \tag{2.4.11}$$

Before considering this formula further, we give a simple example. For the vectors

$$\mathbf{x}_1 = \begin{bmatrix} 1 \\ 0 \\ 1 \end{bmatrix}, \quad \mathbf{x}_2 = \begin{bmatrix} 1 \\ 2 \\ 0 \end{bmatrix}, \quad \mathbf{x} = \begin{bmatrix} 1 \\ 1 \\ 2 \end{bmatrix}$$

we wish to determine the orthogonal projection of $\mathbf{x}$ onto $\text{span}(\mathbf{x}_1, \mathbf{x}_2)$, where we assume that R^3 has the usual inner product. The matrices X and A and the system of equations (2.4.9) are

$$X = \begin{bmatrix} 1 & 1 \\ 0 & 2 \\ 1 & 0 \end{bmatrix}, \quad A = X^TX = \begin{bmatrix} 2 & 1 \\ 1 & 5 \end{bmatrix}, \quad A\mathbf{c} = \begin{bmatrix} 3 \\ 3 \end{bmatrix}$$

Thus $\mathbf{c} = \frac{1}{3}(4, 1)^T$, and the orthogonal projection is

$$\mathbf{x}_S = c_1\mathbf{x}_1 + c_2\mathbf{x}_2 = \frac{1}{3}\begin{bmatrix} 5 \\ 2 \\ 4 \end{bmatrix}$$

Returning to the general formula (2.4.11), we see that multiplication of $\mathbf{x}$ by the $n \times n$ matrix

$$P = X(X^*X)^{-1}X^* \tag{2.4.12}$$

gives the orthogonal projection of $\mathbf{x}$ onto range(X). Thus, P is called an *orthogonal projection* matrix or *orthogonal projector.* For the previous example,

$$P = \frac{1}{9}\begin{bmatrix} 1 & 1 \\ 0 & 2 \\ 1 & 0 \end{bmatrix}\begin{bmatrix} 5 & -1 \\ -1 & 2 \end{bmatrix}\begin{bmatrix} 1 & 0 & 1 \\ 1 & 2 & 0 \end{bmatrix} = \frac{1}{9}\begin{bmatrix} 5 & 2 & 4 \\ 2 & 8 & -2 \\ 4 & -2 & 5 \end{bmatrix}$$

In general, it is easy to see (Exercise 2.4-8) that P has the properties

$$P^2 = P, \qquad P^* = P \tag{2.4.13}$$

so that P is idempotent and Hermitian. P is also necessarily singular unless $m = n$ (see Exercise 1.3-13), in which case $P = I$ and the projection is trivial.

We next show that any Hermitian idempotent matrix must, in fact, be an orthogonal projection matrix.

2.4.6. *If P is an $n \times n$ matrix that satisfies* (2.4.13), *then P is an orthogonal projector onto range*(P).

Proof. It suffices to show that P can be written in the form (2.4.12). Let rank(P) = m. Then, by Theorem 2.3.6, P may be written as $P = XR$, where X and R are $n \times m$ and $m \times n$ matrices of rank m. Using (2.4.13), we have

$$XR = (XR)^2 = XRXR, \qquad R^*X^* = (XR)^* = XR$$

Multiplying the first equation on the right by R^* and replacing XR by R^*X^* gives

$$XRR^* = R^*X^*XRR^*$$

By 2.4.3 the Grammian matrices X^*X and RR^* are both nonsingular since the columns of X and R^* are linearly independent. Hence,

$$X = R^*X^*X \qquad \text{or} \qquad R^* = X(X^*X)^{-1}$$

Therefore, $R = (X^*X)^{-1}X^*$, so that $P = XR$ has the form (2.4.12). □

If P is an idempotent matrix but not Hermitian, it is still a projector onto range(P) but not an orthogonal projector. For example,

$$P = \begin{bmatrix} 1 & 1 \\ 0 & 0 \end{bmatrix}, \qquad P^2 = \begin{bmatrix} 1 & 1 \\ 0 & 0 \end{bmatrix}, \qquad P\mathbf{x} = \begin{bmatrix} x_1 + x_2 \\ 0 \end{bmatrix}$$

so that P projects a vector $\mathbf{x}$ onto range$(P) = \text{span}((1,0)^T)$.

We note another interpretation of the Cauchy-Schwarz inequality (2.4.4) in terms of orthogonal projections. If $\mathbf{x}$ and $\mathbf{y}$ are vectors in R^n, then the orthogonal projection of $\mathbf{y}$ onto span$(\mathbf{x})$ is $\alpha\mathbf{x}$, where $\alpha = \mathbf{x}^T\mathbf{y}/\mathbf{x}^T\mathbf{x}$. Thus, the cosine of the angle between $\mathbf{x}$ and $\mathbf{y}$ is given by

$$\cos\theta = \frac{\mathbf{x}^T\mathbf{y}}{(\mathbf{x}^T\mathbf{x})^{1/2}(\mathbf{y}^T\mathbf{y})^{1/2}}$$

as illustrated in Figure 2.5. The fact that $|\cos\theta| \leq 1$ is just the Cauchy-Schwarz inequality.

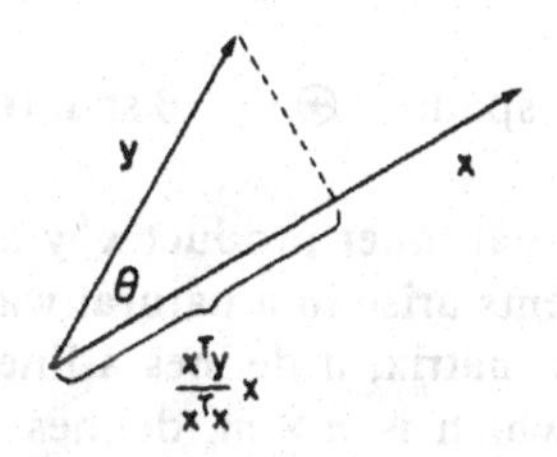

Figure 2.5. Angle between vectors.

Orthogonal Complements

In Section 2.1 we defined two subspaces S and T of a linear space R to be complements if $S \cap T = \{\mathbf{0}\}$ and $S + T = R$. We also used the notation $R = S \dotplus T$ to express R as the direct sum of S and T. If R is an inner product space, two subspaces S and T are *orthogonal* if $(\mathbf{x}, \mathbf{y}) = 0$ for all $\mathbf{x} \in S$, $\mathbf{y} \in T$; that is, every vector in S is orthogonal to every vector in T. Orthogonal subspaces necessarily satisfy $S \cap T = \{\mathbf{0}\}$. If, in addition, $R = S + T$, then S and T are *orthogonal complements* and we use the notation $R = S \oplus T$ to express the fact that R is the direct sum of orthogonal subspaces. We stress that two subspaces may be orthogonal without being orthogonal complements. For example, if S and T are perpendicular lines through the origin in R^3, they are orthogonal subspaces, but $R^3 \neq S + T$.

If S is a subspace of an inner product space R, then the *orthogonal complement* of S, denoted by $S^\perp$, is

$$S^\perp = \{\mathbf{y} \in R: (\mathbf{x}, \mathbf{y}) = 0 \text{ for all } \mathbf{x} \in S\}$$

That is, $S^\perp$ is the set of all vectors in R that are orthogonal to every vector in S. It is easy to verify (Exercise 2.4.10) that $S^\perp$ is a linear space and that S and $S^\perp$ are orthogonal complements. Thus,

$$R = S \oplus S^\perp \tag{2.4.14}$$

As an example, let $R = R^3$ with the inner product $\mathbf{x}^T\mathbf{y}$, and let S be a plane in R^3 through the origin. Then $S^\perp$ is the line through the origin perpendicular to the plane S.

As was done in Section 2.3 with the direct sums, (2.4.14) may be extended to more than two spaces. The subspaces $S_1, \ldots, S_m$ are *orthogonal* if they are all pairwise orthogonal. Then the notation

$$R = S_1 \oplus \cdots \oplus S_m$$

indicates that R is the direct sum of orthogonal subspaces. As in Section 2.3, an example is

$$R^n = \text{span}(\mathbf{e}_1) \oplus \cdots \oplus \text{span}(\mathbf{e}_n)$$

in which we assume the usual inner product $\mathbf{x}^T\mathbf{y}$ on R^n.

Orthogonal complements arise in a natural way in dealing with linear operators. If A is a $m \times n$ matrix, it defines a linear transformation from C^n to C^m, whereas A^*, which is $n \times m$, defines a linear transformation from C^m to C^n. Now suppose that $\mathbf{x} \in \text{range}(A^*)^\perp$. Then $\mathbf{x}$ is orthogonal to every column of A^*, so that $\mathbf{x}^*A^* = 0$ or $A\mathbf{x} = 0$. Thus $\mathbf{x} \in \text{null}(A)$. On the other hand, if $\mathbf{x} \in \text{null}(A)$, then $A\mathbf{x} = 0$ or $\mathbf{x}^*A^* = 0$, so that $\mathbf{x} \in \text{range}(A^*)^\perp$. Thus, we have shown the following result.

2.4.7. If A is an $m \times n$ matrix, then null(A) and range(A^*) are orthogonal complements in C^n, so that $C^n = \text{null}(A) \oplus \text{range}(A^*)$.

If $m = n$, and A is Hermitian, then 2.4.7 shows that null(A) and range(A) are orthogonal complements in C^n, and, hence,

$$C^n = \text{null}(A) \oplus \text{range}(A)$$

Note that, in general, this does not hold for matrices that are not Hermitian (Exercise 2.4.13).

As an example of the use of 2.4.7 we show that for any $m \times n$ matrix, $\text{range}(A^*A) = \text{range}(A^*)$; this extends 2.3.4, in which we proved that $\text{rank}(A^*A) = \text{rank}(A)$. In 2.3.4 we also showed that $\text{null}(A^*A) = \text{null}(A)$, so that $\text{null}(A^*A)^\perp = \text{null}(A)^\perp$. Then, by 2.4.7,

$$\text{range}(A^*A) = \text{null}(A^*A)^\perp = \text{null}(A)^\perp = \text{range}(A^*)$$

Next, we characterize orthogonal projectors in a somewhat different way by means of orthogonal complements.

2.4.8. An $n \times n$ matrix P is an orthogonal projector onto $S = range(P)$ if and only if

$$P\mathbf{x} = \mathbf{x} \quad \textit{for all } \mathbf{x} \in S \qquad \textit{and} \qquad P\mathbf{x} = 0 \quad \textit{for all } \mathbf{x} \in S^{\perp} \qquad (2.4.15)$$

Proof. If P is an orthogonal projector, it has the form (2.4.12), where the columns of X are a basis for S. If $\mathbf{x} \in S$, then $\mathbf{x} = X\mathbf{c}$ for some m-vector $\mathbf{c}$, and thus

$$P\mathbf{x} = X(X^*X)^{-1}X^*\mathbf{x} = X(X^*X)^{-1}X^*X\mathbf{c} = X\mathbf{c} = \mathbf{x}$$

If $\mathbf{x} \in S^{\perp}$, then $X^*\mathbf{x} = 0$, and thus $P\mathbf{x} = 0$. Conversely, assume that (2.4.15) holds, and write any $\mathbf{x}$ as $\mathbf{x} = \mathbf{u} + \mathbf{v}$, where $\mathbf{u} \in S$, $\mathbf{v} \in S^{\perp}$. Then for any $\mathbf{y} \in S$ we have

$$\mathbf{y}^*(\mathbf{x} - P\mathbf{x}) = \mathbf{y}^*(\mathbf{u} + \mathbf{v} - \mathbf{u}) = \mathbf{y}^*\mathbf{v} = 0$$

But this is just the definition of the orthogonal projection $\mathbf{p}$ of $\mathbf{x}$ onto S, and since this holds for all $\mathbf{x}$, P must be the orthogonal projection matrix. □

The characterization 2.4.8 provides a natural way to define orthogonal projector operators on inner product spaces. A linear operator $\mathbf{P}: R \to R$ on an inner product space R is an *orthogonal projector* onto a subspace S if

$$\mathbf{P}\mathbf{x} = \mathbf{x} \text{ for all } \mathbf{x} \in S \qquad \text{and} \qquad \mathbf{P}\mathbf{x} = 0 \quad \text{for all } \mathbf{x} \in S^{\perp}$$

Linear Equations

In the previous section we saw that if $\hat{\mathbf{x}}$ is a solution of the linear system $A\mathbf{x} = \mathbf{b}$, where A is an $m \times n$ matrix, then $\hat{\mathbf{x}} + \mathbf{y}$ is also a solution for any $\mathbf{y} \in \text{null}(A)$. In general, any of the infinitely many solutions of $A\mathbf{x} = \mathbf{b}$ could play the role of $\hat{\mathbf{x}}$. However, if $\hat{\mathbf{x}}$ is restricted to $\text{null}(A)^{\perp}$, then it is unique. To see this, let $\hat{\mathbf{x}}_1$ and $\hat{\mathbf{x}}_2$ be any two solutions, and let

$$\hat{\mathbf{x}}_i = \mathbf{u}_i + \mathbf{v}_i, \qquad i = 1, 2$$

be the decompositions of $\hat{\mathbf{x}}_i$ into the parts $\mathbf{u}_i$ in $\text{null}(A)$ and $\mathbf{v}_i$ in $\text{null}(A)^{\perp}$. Then

$$A\hat{\mathbf{x}}_i = A\mathbf{v}_i = \mathbf{b}$$

so that

$$A(\mathbf{v}_1 - \mathbf{v}_2) = 0$$

Thus, $\mathbf{v}_1 - \mathbf{v}_2$ is in both null(A) and null(A)$^\perp$, so it must be zero. Therefore, $\mathbf{v}_1 = \mathbf{v}_2$, which shows that the component in null(A)$^\perp$ of any solution is unique. We summarize the above as follows.

2.4.9. If A is an $m \times n$ matrix and $\mathbf{b} \in$ range(A), then the linear equation $A\mathbf{x} = \mathbf{b}$ has a unique solution $\hat{\mathbf{x}}$ in null(A)$^\perp$, and any solution can be written as $\hat{\mathbf{x}} + \mathbf{y}$, where $\mathbf{y} \in$ null(A).

We will see in Chapter 4 how we can characterize this unique solution $\hat{\mathbf{x}}$ in a somewhat different way.

The Adjoint Operator

If $\mathbf{A}: R \to R$ is a linear operator on an inner product space R, then the *adjoint* of $\mathbf{A}$ is the linear operator $\mathbf{B}$, which satisfies

$$(\mathbf{Ax}, \mathbf{y}) = (\mathbf{x}, \mathbf{By}) \qquad \text{for all } \mathbf{x}, \mathbf{y} \in R \tag{2.4.16}$$

It is not immediately clear that such a linear operator always exists, but the following development will ensure that it does for finite-dimensional spaces and will give insight into its nature.

Let $\mathbf{x}_1, \ldots, \mathbf{x}_n$ be an orthonormal basis for R, and let A be the matrix representation of $\mathbf{A}$ in this basis. Then $\mathbf{Ax}_j = \sum_{i=1}^n a_{ij}\mathbf{x}_i$ and

$$\begin{aligned}(\mathbf{Ax}, \mathbf{y}) &= \left(\mathbf{A} \sum_{j=1}^n x_j\mathbf{x}_j, \sum_{j=1}^n y_j\mathbf{x}_j\right) = \left(\sum_{j=1}^n x_j \sum_{i=1}^n a_{ij}\mathbf{x}_i, \sum_{j=1}^n y_j\mathbf{x}_j\right) \\ &= \left(\sum_{i=1}^n \sum_{j=1}^n a_{ij}x_j\mathbf{x}_i, \sum_{j=1}^n y_j\mathbf{x}_j\right) = \sum_{i,j=1}^n \bar{a}_{ij}\bar{x}_j y_i\end{aligned} \tag{2.4.17}$$

since $(\mathbf{x}_i, \mathbf{x}_j) = 0$ if $i \neq j$. Similarly, if the adjoint operator $\mathbf{B}$ exists, let B be its matrix representation. If R is a complex space, $(\mathbf{x}, \mathbf{By}) = \overline{(\mathbf{By}, \mathbf{x})}$, and a calculation identical to (2.4.17), with $\mathbf{B}$ replacing $\mathbf{A}$ and the $\mathbf{x}_i$ and $\mathbf{y}_i$ interchanged, gives

$$(\mathbf{x}, \mathbf{By}) = \overline{(\mathbf{By}, \mathbf{x})} = \sum_{i,j=1}^n b_{ij}\bar{x}_i y_j \tag{2.4.18}$$

By taking vectors **x** and **y** with coordinate vectors $\mathbf{e}_i$ and $\mathbf{e}_j$, we conclude from (2.4.17) and (2.4.18) that if (2.4.16) holds, then $\bar{a}_{ij} = b_{ji}, \ldots, j = 1, \ldots, n$. Therefore,

$$B = A^* \quad (R \text{ complex}), \qquad B = A^T \quad (R \text{ real}) \qquad (2.4.19)$$

On the other hand, if we define **B** to be the linear operator such that

$$\mathbf{B}\mathbf{x}_j = \sum_{i=1}^{n} \bar{a}_{ji}\mathbf{x}_i, \qquad j = 1, \ldots, n \qquad (2.4.20)$$

then **B** satisfies (2.4.16). We summarize this as follows:

2.4.10. If R is a finite-dimensional inner product space, and $\mathbf{A}: R \to R$ *is a linear operator, then the adjoint operator* $\mathbf{B}: R \to R$ *that satisfies* (2.4.16) *exists, and the matrix representations of* **A** *and* **B** *in any orthonormal basis satisfy* (2.4.19).

Note that the representation of the adjoint by the conjugate transpose gives additional insight into the role of the transpose matrix. In the sequel we will denote the adjoint operator by $\mathbf{A}^*$.

Self-Adjoint Operators

If the matrix representation A of the linear operator **A** is Hermitian (or symmetric if the space R is real) in some orthonormal basis, then the same is true in any orthonormal basis. This follows from 2.4.5, because if $\hat{A}$ is the matrix representation of **A** in another orthonormal basis, then $\hat{A} = Q^*AQ$ for some unitary matrix Q (orthogonal if real). Hence,

$$\hat{A}^* = (Q^*AQ)^* = Q^*A^*Q = Q^*AQ = \hat{A}$$

It then follows from (2.4.19) that the adjoint operator **B** is **A** itself.

2.4.11. If **A** *is a linear operator for which* (2.4.16) *holds with* $\mathbf{B} = \mathbf{A}$, *then* **A** *is* self-adjoint. *The matrix representation of a self-adjoint operator in an orthonormal basis is a Hermitian matrix* (*symmetric if the space is real*), *and* (*by* 2.3.11 *and* 1.4.6) *the eigenvalues of a self-adjoint operator are real.*

Self-adjoint linear operators play a major role in the study of linear operators in infinite-dimensional spaces. Our concern will be Hermitian and real symmetric matrices, which will be treated extensively in Chapters 3 and 4.

Orthogonal and Unitary Operators

Another type of linear operator $\mathbf{P}: R \to R$ that plays a distinguished role in inner product spaces satisfies the property

$$(\mathbf{Pr}, \mathbf{Pr}) = (\mathbf{r}, \mathbf{r}) \qquad \text{for all } \mathbf{r} \in R \tag{2.4.21}$$

An operator with this property is called *orthogonal* if R is real, and *unitary* if R is complex; such operators generalize orthogonal and unitary matrices. Indeed, if P is a unitary matrix on C^n with the inner product $\mathbf{x}^*\mathbf{y}$, then

$$(P\mathbf{r})^* P\mathbf{r} = \mathbf{r}^* P^* P\mathbf{r} = \mathbf{r}^*\mathbf{r}$$

which is (2.4.21). As might be expected from the previous results of this section, the matrix representation of an orthogonal (unitary) linear operator in any orthogonal basis on a real (complex) inner product space is an orthogonal (unitary) matrix. The verification of this is left to Exercise 2.4-16.

Summary

In this section we have considered the basic implications of an inner product. The main concepts and results are as follows:

- Definition of inner product spaces and orthogonal vectors;
- The Grammian matrix and linear independence;
- The Gram–Schmidt and QR orthogonalization procedures;
- Changes of orthogonal bases correspond to orthogonal or unitary transformations;
- Orthogonal projection and projectors;
- Orthogonal complements of subspaces;
- $C^n = \text{null}(A) \oplus \text{range}(A^*)$, $\text{range}(A^*) = \text{range}(A^*A)$;
- Decomposition of solutions of $A\mathbf{x} = \mathbf{b}$ into components in $\text{null}(A)$ and $\text{null}(A)^\perp$;
- The adjoint operator and its representation as the conjugate transpose;
- Self-adjoint operators and Hermitian matrices;
- Orthogonal operators and orthogonal matrices.

Exercises 2.4

1. Let A be a real symmetric positive definite matrix. Show that $(\mathbf{x}, \mathbf{y}) \equiv \mathbf{x}^T A\mathbf{y}$ satisfies the axioms of 2.4.1 and, hence, is an inner product on R^n. Similarly, if A is Hermitian positive definite, show that $\mathbf{x}^* A\mathbf{y}$ is an inner product on C^n.

2. Show, without using 2.4.3, that if $\mathbf{x}_1, \ldots, \mathbf{x}_m$ are orthogonal, then they are linearly independent. (See Exercise 1.1-7.)
3. Verify that the Gram-Schmidt process (2.4.6) produces a set of orthogonal vectors $\mathbf{y}_1, \ldots, \mathbf{y}_n$ if $\mathbf{x}_1, \ldots, \mathbf{x}_n$ are linearly independent.
4. Let

$$\mathbf{x}_1 = \begin{bmatrix} 1 \\ 2 \end{bmatrix}, \qquad \mathbf{x}_2 = \begin{bmatrix} 2 \\ 3 \end{bmatrix}.$$

Compute the Grammian matrix $(\mathbf{x}_i^T \mathbf{x}_j)$.

5. Let B be a real $m \times n$ matrix and A a real symmetric positive definite matrix.
   (a) For what inner product is $B^T AB$ a Grammian matrix?
   (b) Apply 2.4.3 to conclude that $B^T AB$ is positive semidefinite and positive definite if and only if the columns of B are linearly independent.
   (c) Prove the result of (b) directly without using 2.4.3.
   (d) Formulate and prove the corresponding results for complex matrices.
6. Let $A = QR$, where Q is orthogonal. Show that RQ is similar to QR and, hence, has the same eigenvalues as A.
7. Let

$$\mathbf{x}_1 = \begin{bmatrix} 1 \\ 0 \\ 1 \end{bmatrix}, \qquad \mathbf{x}_2 = \begin{bmatrix} 1 \\ 1 \\ 0 \end{bmatrix}, \qquad \mathbf{x}_3 = \begin{bmatrix} 1 \\ 2 \\ 3 \end{bmatrix}.$$

In the inner product $\mathbf{x}^T \mathbf{y}$, orthogonalize these vectors by the Gram-Schmidt and QR procedures. Do the same for Gram-Schmidt in the inner product defined by $(\mathbf{x}, \mathbf{y}) = x_1 y_1 + 2x_2 y_2 + x_3 y_3$.

8. Let P be the matrix of (2.4.12), where X is an $n \times m$ matrix. Show that $P^* = P$ and $P^2 = P$.
9. Show that any plane rotation matrix is orthogonal.
10. Let S be a subspace of an inner product space R. Show that the orthogonal complement $S^\perp$ is a linear space and that (2.4.14) holds.
11. For the vectors and two inner products of Exercise 7, compute the orthogonal projection of $\mathbf{x}_3$ onto the subspace spanned by $\mathbf{x}_1$ and $\mathbf{x}_2$.
12. If P is an $n \times m$ matrix with orthonormal columns and $m < n$, show that $P^* P = I_m$, but $PP^* \neq I_n$.
13. Let $A = \left[\begin{smallmatrix} 0 & 1 \\ 0 & 0 \end{smallmatrix}\right]$. Show that $\text{null}(A) = \text{range}(A) = \text{span}\left[\begin{smallmatrix} 1 \\ 0 \end{smallmatrix}\right]$.
14. If P is an orthogonal projector, show that $I - P$ is an orthogonal projector onto $\text{range}(P)^\perp$.
15. Let R be the space of real column 2-vectors with an inner product defined by

$$(\mathbf{x}, \mathbf{y}) = x_1 y_1 + 2x_2 y_2$$

Let $\mathbf{A}: R \to R$ be the linear operator whose matrix representation in the basis $\mathbf{e}_1, \mathbf{e}_2$ is

$$A = \begin{bmatrix} 2 & 1 \\ 1 & 2 \end{bmatrix}$$

Find the matrix representation of the adjoint of $\mathbf{A}$ in the basis $\mathbf{e}_1, \mathbf{e}_2$, and ascertain if $\mathbf{A}$ is self-adjoint.

16. Let $\mathbf{P}: R \to R$ be an orthogonal (unitary) linear operator on a real (complex) inner product space R. Show that the matrix representation of $\mathbf{P}$ in any orthonormal basis for R is an orthogonal (unitary) matrix.
17. Let $\mathbf{x}_1, \ldots, \mathbf{x}_n$ be orthonormal vectors in R^n. Show that $A\mathbf{x}_1, \ldots, A\mathbf{x}_n$ are also orthonormal if and only if the $n \times n$ matrix A is orthogonal.
18. Let A be an $m \times n$ matrix. Show that $\text{null}(A^*) = \text{range}(A)^\perp$.
19. Show that the eigenvalues of

$$\begin{bmatrix} \cos\theta & \sin\theta \\ -\sin\theta & \cos\theta \end{bmatrix}$$

are $\cos\theta \pm i \sin\theta$. Use this to show that the eigenvalues of any $n \times n$ plane rotation matrix are $\cos\theta \pm i \sin\theta$, and 1 with a multiplicity of $n - 2$.

20. If P is a (not necessarily orthogonal) projection matrix ($P^2 = P$), show that $\text{null}(P)$ and $\text{range}(P)$ are (not necessarily orthogonal) complements.
21. Let P be an $n \times n$ orthogonal projector onto $S = \text{span}(\mathbf{x}_1, \ldots, \mathbf{x}_m)$, where $\dim S = m$. Find the eigenvalues of P, including their multiplicities, and describe the corresponding eigenvectors.

2.5. Normed Linear Spaces

On R^n the usual Euclidean length of a vector $\mathbf{x}$ is $(\sum_{i=1}^n x_i^2)^{1/2}$. It has the following three basic properties, which we associate with distance.

2.5.1. Definition. *Norms.* A real-valued function on a linear space R is called a *norm*, and is denoted $\|\;\|$, if the following properties hold:

(a) $\|\mathbf{x}\| \geq 0$ and $\|\mathbf{x}\| = 0$ only if $\mathbf{x} = 0$;
(b) $\|\alpha\mathbf{x}\| = |\alpha|\,\|\mathbf{x}\|$ for all $\mathbf{x} \in R$ and scalars α;
(c) $\|\mathbf{x} + \mathbf{y}\| \leq \|\mathbf{x}\| + \|\mathbf{y}\|$ for all $\mathbf{x}, \mathbf{y} \in R$.

A linear space with a norm is a *normed linear space.*

If R is an inner product space, we can define a norm in terms of the inner product by

$$\|\mathbf{x}\| = (\mathbf{x}, \mathbf{x})^{1/2} \tag{2.5.1}$$

It is immediately verified that properties (a) and (b) of 2.5.1 are satisfied. Using the Cauchy-Schwarz inequality (2.4.4), we can prove the "triangle inequality" (c) by

$$\begin{aligned}\|\mathbf{x}+\mathbf{y}\|^2 &= (\mathbf{x}+\mathbf{y}, \mathbf{x}+\mathbf{y}) = (\mathbf{x}, \mathbf{x}) + (\mathbf{x}, \mathbf{y}) + (\mathbf{y}, \mathbf{x}) + (\mathbf{y}, \mathbf{y}) \\ &\leq (\mathbf{x}, \mathbf{x}) + 2(\mathbf{x}, \mathbf{x})^{1/2}(\mathbf{y}, \mathbf{y})^{1/2} + (\mathbf{y}, \mathbf{y}) = (\|\mathbf{x}\| + \|\mathbf{y}\|)^2\end{aligned} \tag{2.5.2}$$

Thus, we have shown the following:

2.5.2. *An inner product space is a normed linear space with the norm defined by* (2.5.1).

We mentioned in Section 2.4 that if B is an $n \times n$ real symmetric positive definite matrix, then $(\mathbf{x}, \mathbf{y}) \equiv \mathbf{x}^T B \mathbf{y}$ defines an inner product on R^n. Hence,

$$\|\mathbf{x}\| = (\mathbf{x}^T B \mathbf{x})^{1/2} \tag{2.5.3}$$

is a norm on R^n, and the special case when B is the identity matrix gives the usual Euclidean norm

$$\|\mathbf{x}\|_2 = \left(\sum_{i=1}^{n} x_i^2\right)^{1/2} \tag{2.5.4}$$

On C^n we would assume in (2.5.3) that B is Hermitian positive definite and replace $\mathbf{x}^T$ by $\mathbf{x}^*$.

In addition to the inner product norms on R^n or C^n, a variety of other norms can be defined. The two most commonly used are the l_1 *norm*

$$\|\mathbf{x}\|_1 = \sum_{i=1}^{n} |x_i| \tag{2.5.5}$$

sometimes called the sum norm or the 1 norm, and the l_∞ *norm*

$$\|\mathbf{x}\|_\infty = \max_{1 \leq i \leq n} |x_i| \tag{2.5.6}$$

also called the max norm or ∞ norm. The verification of the axioms of 2.5.1 for (2.5.5) and (2.5.6) is immediate and is left to Exercise 2.5-3.

The norms (2.5.4) and (2.5.5) are special cases of the general family of l_p norms defined by

$$\|\mathbf{x}\|_p = \left(\sum_{i=1}^{n} |x_i|^p\right)^{1/p} \tag{2.5.7}$$

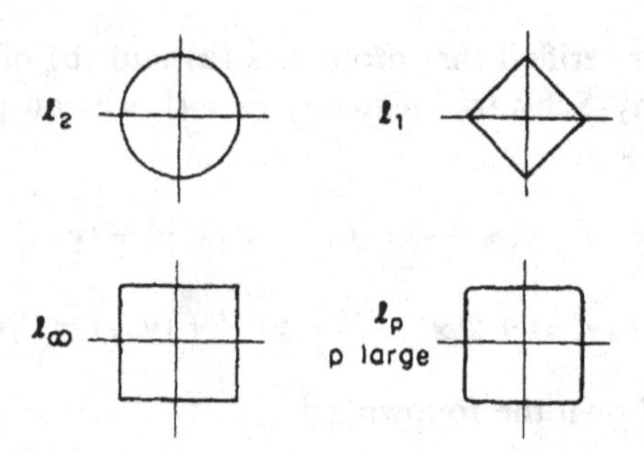

Figure 2.6. Unit balls and unit spheres in several norms.

for any real number $p \geq 1$. The norm (2.5.6) is the limiting case of (2.5.7) as $p \to \infty$ (Exercises 2.5-4 and 2.5-5).

The *unit ball* on a normed linear space R is the set $\{\mathbf{x} \in R: \|\mathbf{x}\| \leq 1\}$, and the *unit sphere* is $\{\mathbf{x} \in R: \|\mathbf{x}\| = 1\}$. These sets are depicted in Figure 2.6 for several norms on R^2 (see Exercise 2.5-6).

Additional norms can be defined on R^n or C^n, or on a general normed linear space, in the following way, the proof of which is left to Exercise 2.5-7.

2.5.3. If $\| \ \|$ is a norm on R^n (or C^n) and P is a nonsingular real (or complex) $n \times n$ matrix, then $\|\mathbf{x}\|_P \equiv \|P\mathbf{x}\|$ is a norm. More generally, if R is a normed linear space with norm $\| \ \|$ and $\mathbf{P}: R \to R$ is a one-to-one linear operator, then $\|\mathbf{x}\|_\mathbf{P} \equiv \|\mathbf{P}\mathbf{x}\|$ is also a norm on R.

Convergence of a sequence of vectors $\{\mathbf{x}_k\}$ in a normed linear space is defined by

$$\mathbf{x}_k \to \mathbf{x} \text{ as } k \to \infty \quad \text{if and only if} \quad \|\mathbf{x}_k - \mathbf{x}\| \to 0 \text{ as } k \to \infty \tag{2.5.8}$$

Although R^n or C^n have a large number of possible norms, convergence of a sequence is independent of the particular norm.

2.5.4. NORM EQUIVALENCE THEOREM. If $\| \ \|$ and $\| \ \|'$ are any two norms on R^n (or C^n) and $\{\mathbf{x}_k\}$ is a sequence, then

$$\|\mathbf{x}_k\| \to 0 \text{ as } k \to \infty \quad \text{if and only if} \quad \|\mathbf{x}_k\|' \to 0 \text{ as } k \to \infty \tag{2.5.9}$$

Proof. We will show that there exist constants $c_1, c_2 > 0$ so that

$$c_1\|\mathbf{x}\| \leq \|\mathbf{x}\|' \leq c_2\|\mathbf{x}\| \tag{2.5.10}$$

Then (2.5.9) is an immediate consequence of (2.5.10). In order to prove (2.5.10) it suffices to assume that $\|\ \|'$ is the l_2 norm, because if the two relations

$$d_1\|\mathbf{x}\| \leq \|\mathbf{x}\|_2 \leq d_2\|\mathbf{x}\|, \qquad d_1'\|\mathbf{x}\|' \leq \|\mathbf{x}\|_2 \leq d_2'\|\mathbf{x}\|'$$

both hold with nonzero constants, then (2.5.10) is true with $c_1 = d_1/d_2'$ and $c_2 = d_2/d_1'$.

Let $\mathbf{e}_i$, $i = 1, \ldots, n$, be the natural basis for C^n. Then, by the Cauchy-Schwarz inequality,

$$\|\mathbf{x}\| = \left\|\sum_{i=1}^n x_i\mathbf{e}_i\right\| \leq \sum_{i=1}^n |x_i|\,\|\mathbf{e}_i\| \leq \beta\|\mathbf{x}\|_2, \qquad \beta = \left(\sum_{i=1}^n \|\mathbf{e}_i\|^2\right)^{1/2} \tag{2.5.11}$$

Hence, the left inequality of (2.5.10) holds with $c_1 = 1/\beta$. On the other hand, using (2.5.11) and Exercise 2.5-8,

$$|\,\|\mathbf{x}\| - \|\mathbf{y}\|\,| \leq \|\mathbf{x} - \mathbf{y}\| \leq \beta\|\mathbf{x} - \mathbf{y}\|_2$$

which shows that $\|\ \|$ is a continuous function (with respect to the l_2 norm). Therefore, since the unit sphere $S = \{\mathbf{x}\colon \|\mathbf{x}\|_2 = 1\}$ is closed and bounded, $\|\ \|$ is bounded away from zero on S; that is, $\|\mathbf{x}\| \geq \alpha$ for some $\alpha > 0$ and all $\mathbf{x} \in S$. Hence, for arbitrary $\mathbf{x}$,

$$\|\mathbf{x}\| = \|\mathbf{x}\|_2\left\|\frac{\mathbf{x}}{\|\mathbf{x}\|_2}\right\| \geq \alpha\|\mathbf{x}\|_2$$

so that the right-hand inequality of (2.5.10) holds with $c_2 = 1/\alpha$. □

Operator Norms

Let R and S be normed linear spaces, with norms $\|\ \|_R$ and $\|\ \|_S$, and let $\mathbf{A}: R \to S$ be a linear operator. Then we can define the norm of $\mathbf{A}$ by

$$\|\mathbf{A}\| = \max_{\mathbf{x}\neq 0} \frac{\|\mathbf{Ax}\|_S}{\|\mathbf{x}\|_R} = \max_{\|\mathbf{x}\|_R=1} \|\mathbf{Ax}\|_S \tag{2.5.12}$$

The geometric interpretation of an operator norm is that it is the maximum norm of all vectors of norm one after transformation by $\mathbf{A}$. This is illustrated in Figure 2.7 for vectors in R^2 with the l_2 norm.

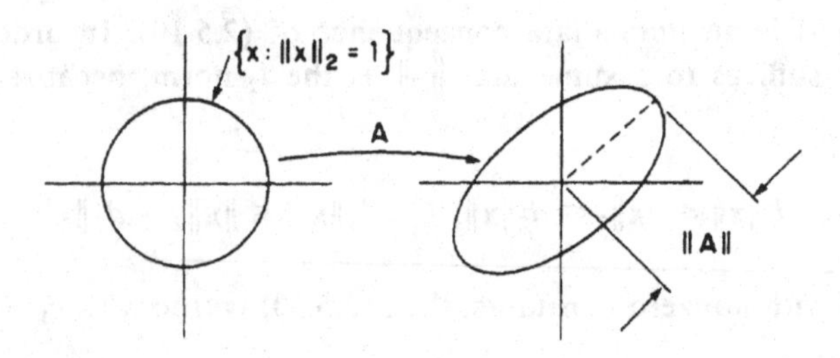

Figure 2.7. Geometry for an operator norm.

The following properties of operator norms follow immediately from the corresponding properties of vector norms and the definition (2.5.12). Their verification is left to Exercise 2.5-9.

$$\begin{aligned} &\text{(a)}\quad \|\mathbf{A}\| \geq 0 \text{ and } \|\mathbf{A}\| = 0 \text{ if and only if } \mathbf{A} = 0. \\ &\text{(b)}\quad \|\alpha\mathbf{A}\| = |\alpha|\,\|\mathbf{A}\| \text{ for all scalars } \alpha. \\ &\text{(c)}\quad \|\mathbf{A}+\mathbf{B}\| \leq \|\mathbf{A}\| + \|\mathbf{B}\|. \end{aligned} \tag{2.5.13}$$

A fourth basic property holds for products of operators $\mathbf{B}: R \to S$ and $\mathbf{A}: S \to T$:

$$\|\mathbf{AB}\|_{R,T} \leq \|\mathbf{A}\|_{S,T}\|\mathbf{B}\|_{R,S} \tag{2.5.14}$$

Here, we have indicated by the subscripts the spaces over which the norms are taken. The property (2.5.14) follows immediately from the definition (2.5.12) (Exercise 2.5-9). In the case of matrices A and B, and if the underlying spaces all have the same norm, we write (2.5.14) simply as

$$\|AB\| \leq \|A\|\,\|B\|$$

As with vectors, convergence of a sequence of linear operators $\mathbf{A}_k: R \to S$, $k = 1, 2, \ldots$, to an operator $\mathbf{A}: R \to S$ is defined in terms of a norm by $\|\mathbf{A}_k - \mathbf{A}\| \to 0$ as $k \to \infty$. If $\{A_k\}$ is a sequence of $m \times n$ matrices, then convergence to a matrix A may be defined by any norms on C^n and C^m. We recall from Section 2.2 (see Exercise 2.2-10) that the set of all linear operators from R to S is also a linear space and, hence, a normed linear space with the norm defined by (2.5.12). In particular, the collection of all $m \times n$ matrices is also a normed linear space with the norm defined in terms of the norms chosen on C^n and C^m. By 2.5.4, then, a sequence of $m \times n$ matrices converges in one norm if and only if it converges in any norm. This is equivalent to the convergence of the individual components of the sequence of matrices.

If A is an $n \times n$ matrix, we will usually consider it as generating a linear transformation from R^n (or C^n) into itself and use only one norm in the definition (2.5.13); that is,

$$\|A\| = \max_{\|\mathbf{x}\|=1} \|A\mathbf{x}\| \tag{2.5.15}$$

We next compute the norms of $n \times n$ matrices in terms of the l_1, l_2, and l_∞ norms on C^n. We begin with the l_1 and l_∞ norms.

2.5.5. Let A be an n × n real or complex matrix. Then

$$\|A\|_\infty \equiv \max_{\|\mathbf{x}\|_\infty=1} \|A\mathbf{x}\|_\infty = \max_{1\le i\le n} \sum_{j=1}^{n} |a_{ij}| \tag{2.5.16}$$

and

$$\|A\|_1 \equiv \max_{\|\mathbf{x}\|_1=1} \|A\mathbf{x}\|_1 = \max_{1\le j\le n} \sum_{i=1}^{n} |a_{ij}| \tag{2.5.17}$$

Proof. Consider first the l_1 norm; then for any n-vector $\mathbf{x}$

$$\|A\mathbf{x}\|_1 = \sum_{i=1}^{n} \left| \sum_{j=1}^{n} a_{ij}x_j \right| \le \sum_{i=1}^{n} \sum_{j=1}^{n} |a_{ij}||x_j| = \sum_{j=1}^{n} |x_j| \sum_{i=1}^{n} |a_{ij}|$$

$$\le \left(\max_{1\le j\le n} \sum_{i=1}^{n} |a_{ij}| \right) \|\mathbf{x}\|_1$$

To show that there is some $\mathbf{x} \neq 0$ for which equality is attained in (2.5.17), let k be such that

$$\max_{1\le j\le n} \sum_{i=1}^{n} |a_{ij}| = \sum_{i=1}^{n} |a_{ik}|$$

Then if $\mathbf{e}_k$ is the kth unit vector and $\mathbf{a}_k$ is the kth column of A, we have

$$\|A\mathbf{e}_k\|_1 = \|\mathbf{a}_k\|_1 = \sum_{i=1}^{n} |a_{ik}| = \|A\|_1$$

The proof for $\|A\|_\infty$ is similar. In this case the maximum is taken on for the vector defined by

$$x_i = \begin{cases} \dfrac{a_{ki}}{|a_{ki}|}, & a_{ki} \neq 0, \\ 1, & a_{ki} = 0, \end{cases}$$

where, again, k is chosen so that the maximum in (2.5.16) is achieved. □

We note that the quantities in (2.5.16) and (2.5.17) are sometimes called the maximum row and column sums of A, respectively.

For the l_2 norm we state the result below and postpone the proof until Section 3.1. The *spectral radius* of an $n \times n$ matrix A with eigenvalues $\lambda_1, \ldots, \lambda_n$ is

$$\rho(A) = \max_{1 \le i \le n} |\lambda_i|.$$

2.5.6. *If A is an $n \times n$ matrix, then*

$$\|A\|_2 = [\rho(A^*A)]^{1/2} \tag{2.5.18}$$

We note that, in contrast to $\|A\|_1$ and $\|A\|_\infty$, the computation of $\|A\|_2$ is a fairly formidable task. For this reason the l_1 and l_∞ norms are used extensively in computational work, whereas the l_2 norm plays a major theoretical role.

As a different kind of example of a norm computation, recall that in 2.5.3 we noted that we can define a vector norm by $\|\mathbf{x}\|_P = \|P\mathbf{x}\|$, where $\|\ \|$ is a given norm and P is a nonsingular matrix. In this case the corresponding matrix norm is given by

$$\|A\|_P = \max_{\|\mathbf{x}\|_P = 1} \|A\mathbf{x}\|_P = \max_{\|P\mathbf{x}\| = 1} \|PA\mathbf{x}\| = \max_{\|\mathbf{y}\| = 1} \|PAP^{-1}\mathbf{y}\| = \|PAP^{-1}\|$$

We state this result as follows:

2.5.7. *Let A be an $n \times n$ matrix, $\|\ \|$ any norm on C^n, and P a nonsingular $n \times n$ matrix. If a vector norm is defined on C^n by $\|\mathbf{x}\|_P = \|P\mathbf{x}\|$, then the corresponding matrix norm is $\|A\|_P = \|PAP^{-1}\|$.*

An important special case of 2.5.6 is when A is real and symmetric or Hermitian. Then

$$\|A\|_2 = [\rho(A^2)]^{1/2} = [\rho(A)^2]^{1/2} = \rho(A) \tag{2.5.19}$$

so that $\|A\|_2$ equals the maximum of the absolute value of the eigenvalues of A. In general, for any norm, $\|A\|$ will be an upper bound for the eigenvalues of A, because if λ is any eigenvalue and $\mathbf{x}$ is a corresponding eigenvector, then $\|A\mathbf{x}\| \le \|A\|\,\|\mathbf{x}\|$ or

$$\|A\| \ge \frac{\|A\mathbf{x}\|}{\|\mathbf{x}\|} = |\lambda|$$

Thus

$$\rho(A) \leq \|A\| \tag{2.5.20}$$

In the special case of symmetric matrices and the l_2 norm, (2.5.19) shows that equality holds in (2.5.20), but in general this will not be true. For example, the matrix

$$A = \begin{bmatrix} 0 & 1 \\ 0 & 0 \end{bmatrix}$$

has $\rho(A) = 0$, but $\|A\| > 0$ in any norm. We will see in Chapter 5 when equality can be achieved in (2.5.20).

Matrices that satisfy

$$\|A\mathbf{x}\| = \|\mathbf{x}\| \qquad \text{for all } \mathbf{x} \tag{2.5.21}$$

are called *norm preserving* (or *isometric*). Whether a given matrix is norm preserving depends, of course, on the norm. Conversely, for a given norm it is, in general, difficult to determine the class of matrices that are norm preserving. However, in the l_2 norm, the answer is quite simple, and the following statement complements the definition (2.4.21) of orthogonal or unitary operators.

2.5.8. An $n \times n$ matrix A is norm preserving in the l_2 norm if and only if it is orthogonal (or unitary if A is complex).

Proof. If A is real and orthogonal, then

$$\|A\mathbf{x}\|_2^2 = \mathbf{x}^T A^T A\mathbf{x} = \mathbf{x}^T\mathbf{x} = \|\mathbf{x}\|_2^2$$

Conversely, if A is norm preserving, then $\mathbf{x}^T A^T A\mathbf{x} = \mathbf{x}^T\mathbf{x}$ for all $\mathbf{x}$. By taking $\mathbf{x} = \mathbf{e}_i$ it follows that $\mathbf{a}_i^T\mathbf{a}_i = 1$, where $\mathbf{a}_i$ is the ith column of A. Thus, $A^TA = I + B$, where B is a symmetric matrix with zero main diagonal, and $\mathbf{x}^T B\mathbf{x} = 0$ for all $\mathbf{x}$. In particular, for $\mathbf{x} = \mathbf{e}_i + \mathbf{e}_j$,

$$(\mathbf{e}_i + \mathbf{e}_j)^T B(\mathbf{e}_i + \mathbf{e}_j) = b_{ij} + b_{ji} = 0$$

which implies, since $b_{ij} = b_{ji}$, that $b_{ij} = 0$. Thus, $B = 0$ and $A^TA = I$. The proof in the complex case is similar (Exercise 2.5-13). □

We note that orthogonal matrices also preserve angles between vectors, because if A is orthogonal, then

$$(A\mathbf{x})^T A\mathbf{y} = \mathbf{x}^T\mathbf{y}$$

Orthogonal transformations of R^n are sometimes called *rigid motions* because lengths of vectors and angles between vectors are preserved. One type of such a transformation is a rotation. Another is a *reflection*, which can be defined in terms of a column vector $\mathbf{w}$ as a matrix of the form

$$W = I - 2\mathbf{w}\mathbf{w}^T, \qquad \mathbf{w}^T\mathbf{w} = 1 \tag{2.5.22}$$

Matrices of this type are also called *Householder transformations.* One can see immediately (Exercise 2.5-14) that W is orthogonal. Moreover,

$$W\mathbf{w} = \mathbf{w} - 2\mathbf{w} = -\mathbf{w} \tag{2.5.23}$$

and

$$W\mathbf{u} = \mathbf{u} \qquad \text{if } \mathbf{u}^T\mathbf{w} = 0 \tag{2.5.24}$$

These two relations show that under multiplication by W all vectors in span($\mathbf{w}$) reverse direction, while all vectors in the orthogonal complement of span($\mathbf{w}$) remain the same. Thus, W causes a reflection across the $(n - 1)$-dimensional subspace span$(\mathbf{w})^\perp$; this is illustrated in Figure 2.8.

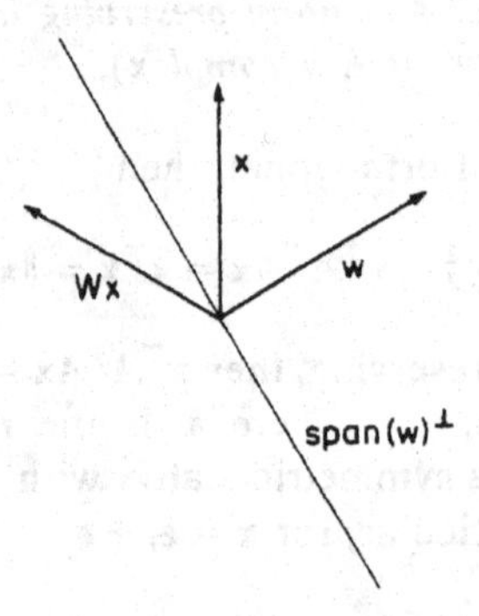

Figure 2.8. A reflection.

Summary

The main points of this section are

- Definition of vector and matrix norms;
- Particular properties of the l_2 norm;

- The equivalence, with respect to sequence convergence, of all norms on C^n;
- The computation of $\|A\|_1$, $\|A\|_\infty$, $\|A\|_2$;
- Orthogonal matrices are norm preserving.

Exercises 2.5

1. Verify that 2.5.1(a) and (b) hold for (2.5.1).
2. Let B be an $n \times n$ real symmetric and indefinite matrix. Show that $(\mathbf{x}^T B\mathbf{x})^{1/2}$ is not a norm on R^n.
3. Verify that (2.5.5) and (2.5.6) satisfy the axioms of 2.5.1. More generally, if $\alpha_1, \ldots, \alpha_n$ are positive numbers show that

$$\sum_{i=1}^{n} \alpha_i |x_i| \qquad \text{and} \qquad \max_{1 \le i \le n} \alpha_i |x_i|$$

define norms on R^n and C^n.
4. For fixed $\mathbf{x}$ show that (2.5.6) is the limit of (2.5.7) as $p \to \infty$.
5. For any real p verify the *Hölder inequality*

$$\left| \sum_{i=1}^{n} x_i y_i \right| \le \left(\sum_{i=1}^{n} |x_i|^p \right)^{1/p} \left(\sum_{i=1}^{n} |y_i|^q \right)^{1/q}$$

where $p^{-1} + q^{-1} = 1$. Use this inequality to show that (2.5.7) is a norm. (*Hint*: For any $a > 0$, $b > 0$, $\alpha > 0$, $\beta > 0$ with $\alpha + \beta = 1$, show that

$$a^\alpha b^\beta \le \alpha a + \beta b$$

and apply this with $\alpha = p^{-1}$, $\beta = q^{-1}$, $a = |x_i|^p / \sum |x_i|^p$, $b = |y_i|^q / \sum |y_i|^q$.)
6. Verify that the unit balls and unit spheres of Figure 2.6 are correct.
7. Verify 2.5.3.
8. Let R be a normed linear space. Show that

$$\big| \|\mathbf{x}\| - \|\mathbf{y}\| \big| \le \|\mathbf{x} - \mathbf{y}\| \qquad \text{for all } \mathbf{x}, \mathbf{y} \in R$$

9. Apply the definition (2.5.12) and the properties of vector norms to verify (2.5.13) and (2.5.14).

10. Let A be an $n \times n$ matrix and define the *Frobenius norm*

$$\|A\| = \left(\sum_{i,j=1}^{n} |a_{ij}|^2 \right)^{1/2}$$

Show that the properties (2.5.13) hold, but that this is not a norm in the sense of (2.5.15) since (2.5.15) implies that $\|I\| = 1$ in any norm.

11. Compute $\|A\|_1$, $\|A\|_2$, $\|A\|_\infty$ and $\rho(A)$ for the matrices

(a) $\begin{bmatrix} 2 & 1 \\ 1 & 2 \end{bmatrix}$

(b) $\begin{bmatrix} 1 & 2 \\ 2 & 0 \end{bmatrix}$

(c) $\begin{bmatrix} 1 & 2 \\ 0 & 1 \end{bmatrix}$

12. Ascertain if $f(\mathbf{x}) \equiv (x_1^2 + 2x_1x_2 + 4x_2^2)^{1/2}$ defines a norm on R^2.
13. Let B be an $n \times n$ complex Hermitian matrix. If $\mathbf{x}^*B\mathbf{x} = 0$ for all $\mathbf{x}$, show that $B = 0$ and use this to prove 2.5.8 when A is complex.
14. For a real vector $\mathbf{w}$ show that a matrix of the form $W = I - 2\mathbf{w}\mathbf{w}^T$ is symmetric and that W is orthogonal if and only if $\mathbf{w}^T\mathbf{w} = 1$. Similarly, show that $W = I - 2\mathbf{w}\mathbf{w}^*$ is Hermitian and that W is unitary if and only if $\mathbf{w}^*\mathbf{w} = 1$.
15. Use (2.5.23) and (2.5.24) to conclude for the matrix W of (2.5.22) that $\mathbf{w}$ is an eigenvector of W with eigenvalue -1, and any vector orthogonal to $\mathbf{w}$ is an eigenvector with eigenvalue 1.
16. Let $\mathbf{u}$ and $\mathbf{v}$ be vectors in C^n. Show that $\|\mathbf{u}\mathbf{v}^*\|_2 = \|\mathbf{u}\|_2\|\mathbf{v}\|_2$.
17. Show that the spectral radius of an $n \times n$ matrix satisfies

$$\rho(A) = \lim_{m \to \infty} \|A^m\|^{1/m}$$

in any norm.

18. Verify the *polarization formula*

$$(\mathbf{x}, \mathbf{y}) = \tfrac{1}{4}(\|\mathbf{x} + \mathbf{y}\|^2 - \|\mathbf{x} - \mathbf{y}\|^2)$$

on any real inner product space with the norm defined in terms of the inner product.

19. A *seminorm* s has the same properties as a norm except that $s(\mathbf{x})$ may be zero even if $\mathbf{x} \neq 0$. Let $\| \ \|$ be any norm on R^n and P any $n \times n$ singular matrix. Show that $s(\mathbf{x}) \equiv \|P\mathbf{x}\|$ is a seminorm.
20. For the l_1 and l_∞ norms on C^n, show that $\|\mathbf{x}\|_\infty \leq \|\mathbf{x}\|_1 \leq n\|\mathbf{x}\|_\infty$.

Review Questions—Chapter 2

Answer whether the following statements are true or false, in general, and justify your assertion.

1. If A is an $m \times n$ matrix for which $A\mathbf{x} = 0$ has only the solution $\mathbf{x} = 0$, then A^{-1} exists.
2. A permutation matrix has only real eigenvalues.
3. If U is the row echelon form of an $m \times n$ matrix A and the first r columns of U are linearly independent, then the first r columns of A are linearly independent.
4. If A is Hermitian positive definite, then $\text{null}(A) = \{0\}$.
5. If A is an $m \times n$ matrix, then A^*A is positive definite if and only if $\text{rank}(A) = n$.
6. If $\mathbf{x}_1, \mathbf{x}_2, \ldots, \mathbf{x}_p$ are linearly independent elements of a vector space V, then they are a basis for V.
7. The number of linearly independent vectors in a linear space V can never exceed $\dim V$.
8. If V_1 and V_2 are subspaces of V such that $V_1 \cap V_2 = \{0\}$, then $V_1 + V_2 = V$.
9. If $\mathbf{x}_1, \ldots, \mathbf{x}_n$ and $\mathbf{y}_1, \ldots, \mathbf{y}_n$ are two bases for V, and if $\boldsymbol{\alpha}$ and $\boldsymbol{\beta}$ are the coordinate vectors in these bases for a given element $\mathbf{x}$, then $\boldsymbol{\alpha} = \boldsymbol{\beta}$.
10. If $\mathbf{A}: R \to S$ is a linear operator with $\dim R = n$ and $\dim S = m$, then any $m \times n$ matrix A is a matrix representation of $\mathbf{A}$ for the right bases of R and S.
11. If the $m \times n$ matrix A is a matrix representation of $\mathbf{A}$, and if P and Q are $m \times m$ and $n \times n$ nonsingular matrices, then PAQ is also a matrix representation of $\mathbf{A}$.
12. If A is an $m \times n$ matrix, then $\dim \text{null}(A) + \dim \text{null}(A^*) = n$.
13. If A and B are $n^2 \times n^2$ matrices each of rank n, then $\text{rank}(AB) \geqslant n$.
14. If B is $m \times p$ and C is $p \times n$, then $\text{rank}(BC) < p$ unless $m = n = p$.
15. If $\mathbf{x}$ and $\mathbf{y}$ are column n-vectors and $A = \mathbf{x}\mathbf{y}^T$, then $\dim \text{null}(A) = n - 1$.
16. An $n \times m$ matrix B is a left inverse of the $m \times n$ matrix A if $AB = I_m$.
17. If A and $\hat{A}$ are two matrix representations of a linear operator $\mathbf{A}: R \to R$, then A and $\hat{A}$ have the same eigenvectors.
18. If A and B are $n \times n$ matrices and A is nonsingular, then AB and BA have the same eigenvalues.
19. Any n linearly independent vectors in an inner product space are orthogonal.
20. If X is an $n \times m$ matrix with orthogonal columns, then XX^* is an orthogonal projector.
21. If A is an $n \times n$ matrix, then $\text{null}(A)$ and $\text{range}(A)$ are orthogonal complements in C^n.
22. If S_1 and S_2 are orthogonal subspaces of C^n, then they are orthogonal complements.

23. If A is a singular $n \times n$ matrix, then $A\mathbf{x} = 0$ has infinitely many solutions in $\text{null}(A)^{\perp}$.
24. If $\| \ \|$ is a norm on C^n and B an $n \times n$ matrix, then $\|\mathbf{x}\|' = \|B\mathbf{x}\|$ defines another norm on C^n.
25. If A is nonsingular, then $\|A\| \geq \|A^{-1}\|^{-1}$ in any operator norm.
26. Let A be an $m \times n$ matrix. If $m < n$, then any m-vector $\mathbf{b}$ is always in range(A), and, thus, $A\mathbf{x} = \mathbf{b}$ always has solutions.
27. If A is a 3×4 matrix with dim null(A) = 2, then dim range(A) = 1.
28. If P_n is the linear space of all polynomials of degree n or less, then dim $P_n = n$.
29. If $\mathbf{A}: R^n \to R^m$ is a one-to-one linear operator, then any matrix representation of $\mathbf{A}$ is nonsingular.
30. If S_1 and S_2 are two orthogonal subspaces of an inner product space R, then $R = S_1 \oplus S_2$.
31. If $\mathbf{A}$ is a self-adjoint linear operator on a real inner product space R, then any matrix representation of $\mathbf{A}$ is symmetric.
32. $f(\mathbf{x}) \equiv (x_1^2 + 2x_1x_2 + 4x_2^2)^{1/2}$ defines a norm on R^2.
33. If the $m \times n$ matrix A has a left inverse, then it also has a right inverse.
34. If $\mathbf{x}_1, \ldots, \mathbf{x}_m$ are m vectors in C^n and $S = \text{span}(\mathbf{x}_1, \ldots, \mathbf{x}_m)$, then dim $S = m$.
35. If S is a subspace of a finite-dimensional inner product space R, then any vector $\mathbf{x} \in R$ can be written as $\mathbf{x} = \mathbf{u} + \mathbf{v}$, where $\mathbf{u} \in S$ and $\mathbf{v} \in S^{\perp}$.
36. If A, B, C are $n \times n$ matrices such that A is similar to B and B is similar to C, then A is similar to C.
37. If A is an $m \times n$ matrix, then A^*A is singular if and only if $m < n$.

References and Extensions: Chapter 2

1. The concepts of linear space and linear operators are fundamental in much of mathematics and applied mathematics, especially in the study of ordinary and partial differential equations and integral equations. A standard reference is Dunford and Schwartz [1958]. The book by Halmos [1958], although addressed specifically to finite-dimensional spaces, gives a thorough dimension-free treatment whenever possible. We note that the real or complex scalars in our definition of a linear space may be replaced by elements of a field.
2. An inner product or normed linear space R is *complete* if whenever $\{\mathbf{x}_i\}$ is a Cauchy sequence of elements in R, its limit is also in R. Finite-dimensional spaces are always complete. Complete inner product spaces are called *Hilbert spaces*; two standard examples are the collection of infinite vectors

$$l_2 = \left\{(x_1, x_2, \ldots): \sum_{i=1}^{\infty} |x_i|^2 < \infty\right\}$$

and the set of functions

$$L_2[a, b] = \left\{\text{integrable functions } x \text{ on } [a, b] \text{ such that } \int_a^b |x(t)|^2\, dt < \infty\right\}$$

In the second example, Riemann integration does not suffice to guarantee completeness; the Lebesgue integral is used. In the first example the inner product is defined by

$$(\mathbf{x}, \mathbf{y}) = \sum_{i=1}^{\infty} x_i y_i$$

and in the second example it is defined by

$$(\mathbf{x}, \mathbf{y}) = \int_a^b x(t)y(t)\, dt$$

3. Complete normed linear spaces are called *Banach spaces*. The spaces l_2 and L_2 are special cases of the Banach spaces l_p and L_p defined by

$$l_p = \left\{(x_1, x_2, \ldots): \sum_{i=1}^{\infty} |x_i|^p < \infty\right\}$$

and

$$L_p[a, b] = \left\{\text{set of integrable functions } x \text{ on } [a, b] \text{ such that } \int_a^b |x(t)|^p\, dt < \infty\right\}$$

where p is any real number greater than or equal to 1. The norms in these two cases are defined by

$$\|\mathbf{x}\|_p = \left(\sum_{i=1}^{\infty} |x_i|^p\right)^{1/p}, \qquad \|\mathbf{x}\|_p = \left(\int_a^b |x(t)|^p\, dt\right)^{1/p}$$

The limiting cases as $p \to \infty$ are also defined, and the norms are

$$\|\mathbf{x}\|_\infty = \sup_{1 \le i < \infty} |x_i|, \qquad \|\mathbf{x}\|_\infty = \sup_{a \le t \le b} |x(t)|$$

where it is assumed that these supremums are finite.

Another standard example of a Banach space is $C[a, b]$, the collection of continuous functions on the interval $[a, b]$ with the norm defined by

$$\|\mathbf{x}\| = \max_{a \le t \le b} |x(t)|$$

For a good introduction to Banach and Hilbert spaces, see Dunford and Schwartz [1958].

4. The Gram–Schmidt orthogonalization procedure, while useful theoretically, is not recommended as a computational technique. Because of rounding error, there can be a severe loss of orthogonality in the computed orthogonal vectors. The *QR* process or other techniques are preferred for large-scale computation. For a thorough discussion and additional references, see Golub and van Loan [1983].

3

Canonical Forms

In this chapter we consider the following question: Given a linear operator $\mathbf{A}: R \to S$, where R and S are finite-dimensional linear spaces, what is the "simplest" form that a matrix representation of $\mathbf{A}$ can take by judicious choice of bases in R and S? By the results of Section 2.2, this question is equivalent to the following one: Given an $m \times n$ matrix A, what is the "simplest" form that the matrix PAQ can take by judicious choice of nonsingular matrices P and Q?

An extremely important, and much more difficult, question arises when $S = R$, so that $\mathbf{A}: R \to R$. Then the question is: What is the "simplest" form that a matrix representation of $\mathbf{A}$ can take by judicious choice of a basis for R? Or, equivalently, what is the "simplest" form that an $n \times n$ matrix A can take under a similarity transformation $P^{-1}AP$?

An important special case of the similarity problem is when P is restricted to be an orthogonal or unitary matrix, and it is here that we start our development. We show first that a Hermitian matrix is unitarily similar to a diagonal matrix, and this establishes our goal of "simplest"; that is, we will attempt to reduce the given matrix by similarity or equivalence transformations to a diagonal matrix, or as close to that as possible. Under orthogonal or unitary transformations, we can reduce any $n \times n$ matrix to a triangular one, but diagonal matrices are achievable only when A is normal.

In Section 3.2 we consider general similarity transformations and first give some simple conditions that ensure that the matrix is similar to a diagonal matrix. This is not achievable in general, however, and the best possible result is the Jordan canonical form, which is almost diagonal.

In Section 3.3 we consider equivalence transformations and show first that any square matrix A is equivalent not only to a diagonal matrix but also to an identity matrix of the same rank as A. We then consider special equivalence transformations. Congruences are important in the study of

Hermitian matrices, and orthogonal or unitary equivalences lead to the singular value decomposition.

3.1. Orthogonal and Unitary Similarity Transformations

We consider in this section similarity transformations of the form

$$B = P^* A P \tag{3.1.1}$$

where A is a given $n \times n$ matrix and P is orthogonal or unitary. The matrices A and B are then called *unitarily* (or *orthogonally*) *similar.* We start with Hermitian matrices and the following theorem, which is one of the most important in matrix theory. Recall, from Section 1.4, that a Hermitian matrix has real eigenvalues and that we use the notation $\mathrm{diag}(\lambda_1, \ldots, \lambda_n)$ for a diagonal matrix.

3.1.1. DIAGONAL FORM FOR HERMITIAN MATRICES. *Let A be an $n \times n$ Hermitian matrix with eigenvalues $\lambda_1, \ldots, \lambda_n$. Then there is a unitary matrix P so that $P^*AP = D$, where $D = \mathrm{diag}(\lambda_1, \ldots, \lambda_n)$.*

Proof. Let $\mathbf{x}_1$, with $\mathbf{x}_1^*\mathbf{x}_1 = 1$, be an eigenvector of A corresponding to λ_1. Then, by 2.4.4, there is a basis $\mathbf{x}_1, \mathbf{u}_1, \ldots, \mathbf{u}_{n-1}$ for C^n of orthonormal vectors. Let $U_1 = (\mathbf{u}_1, \ldots, \mathbf{u}_{n-1})$ be the $n \times (n-1)$ matrix whose columns are the $\mathbf{u}_i$. Then the matrix $P_1 = (\mathbf{x}_1, U_1)$ is unitary, and $U_1^*\mathbf{x}_1 = 0$ since $\mathbf{x}_1$ is orthogonal to the $\mathbf{u}_i$. Thus

$$P_1^*AP_1 = \begin{bmatrix} \mathbf{x}_1^* \\ U_1^* \end{bmatrix} (\lambda_1\mathbf{x}_1, AU_1) = \begin{bmatrix} \lambda_1 & \mathbf{x}_1^*AU_1 \\ 0 & U_1^*AU_1 \end{bmatrix}.$$

But $P_1^*AP_1$ is Hermitian, so that $\mathbf{x}_1^*AU_1 = 0$, and thus

$$P_1^*A_1P_1 = \begin{bmatrix} \lambda_1 & 0 \\ 0 & U_1^*AU_1 \end{bmatrix}$$

Since $P_1^*AP_1$ is similar to A, it has the same eigenvalues. Therefore, by (1.4.6), $A_2 = U_1^*AU_1$ has the eigenvalues $\lambda_2, \ldots, \lambda_n$ of A.

Since A_2 is Hermitian, we can do exactly the same as above and conclude that there is an $(n-1) \times (n-1)$ unitary matrix Q_2 such that

$$Q_2^*A_2Q_2 = \begin{bmatrix} \lambda_2 & 0 \\ 0 & A_3 \end{bmatrix}$$

where A_3 is $(n-2)\times(n-2)$, Hermitian, and has eigenvalues $\lambda_3, \ldots, \lambda_n$. Continuing in this way, we obtain unitary matrices $Q_3, \ldots, Q_{n-1}$, each of dimension one less than the previous one, such that

$$Q_i^* A_i Q_i = \begin{bmatrix} \lambda_i & 0 \\ 0 & A_{i+1} \end{bmatrix}.$$

Here, $A_i, \ldots, A_n$ are Hermitian matrices each of dimension one less than the previous one, and A_i has the eigenvalues $\lambda_i, \ldots, \lambda_n$ of A. We then define $n \times n$ unitary matrices

$$P_i = \begin{bmatrix} I_{i-1} & 0 \\ 0 & Q_i \end{bmatrix}, \qquad i = 2, \ldots, n-1$$

where I_i is the $i \times i$ identity matrix. By the construction of the Q_i, we have that

$$P^* A P = D \tag{3.1.2}$$

where $P = P_1 \cdots P_{n-1}$ and $D = \text{diag}(\lambda_1, \ldots, \lambda_n)$. Since the product of unitary matrices is unitary, P is the unitary matrix that was sought. □

We next examine several ramifications of this basic theorem. Writing P in terms of its columns and multiplying (3.1.2) by P, we have

$$A(\mathbf{p}_1, \ldots, \mathbf{p}_n) = (\mathbf{p}_1, \ldots, \mathbf{p}_n)D = (\lambda_1 \mathbf{p}_1, \ldots, \lambda_n \mathbf{p}_n) \tag{3.1.3}$$

Equating the columns of (3.1.3) gives $A\mathbf{p}_i = \lambda_i \mathbf{p}_i$, $i = 1, \ldots, n$. Thus, the columns of P are eigenvectors of A, and an equivalent way of stating 3.1.1 is as follows:

3.1.2. An $n \times n$ Hermitian matrix A has n orthogonal eigenvectors.

Theorem 3.1.1, of course, applies also to real symmetric matrices. It is important to note that in this case the eigenvectors may be chosen to be real so that the unitary matrix of 3.1.1 is orthogonal. Because of its importance, we state this as a separate result.

3.1.3. If A is an $n \times n$ real symmetric matrix, then there is an orthogonal matrix P, whose columns are eigenvectors of A, such that $P^T A P = \text{diag}(\lambda_1, \ldots, \lambda_n)$, where $\lambda_1, \ldots, \lambda_n$ are the eigenvalues of A.

As a simple example, consider the matrix

$$A = \begin{bmatrix} 2 & 1 \\ 1 & 2 \end{bmatrix}$$

which has the eigenvalues and eigenvectors

$$\lambda_1 = 1, \qquad \mathbf{p}_1 = \frac{1}{\sqrt{2}}\begin{bmatrix} 1 \\ -1 \end{bmatrix}, \qquad \lambda_2 = 3, \qquad \mathbf{p}_2 = \frac{1}{\sqrt{2}}\begin{bmatrix} 1 \\ 1 \end{bmatrix}$$

so that

$$P = \frac{1}{\sqrt{2}}\begin{bmatrix} 1 & 1 \\ -1 & 1 \end{bmatrix}, \qquad D = \begin{bmatrix} 1 & 0 \\ 0 & 3 \end{bmatrix}$$

Note that in this example, as well as in general, the order of the eigenvalues on the diagonal of D is unimportant, and we could have taken $D = \text{diag}(3, 1)$. However, this would then require interchanging the columns of P since the ith column of P must be an eigenvector corresponding to the eigenvalue in the ith diagonal position of D.

We next ask, What other matrices besides Hermitian ones have real eigenvalues and n orthogonal eigenvectors? The answer is none.

3.1.4. Let A be an $n \times n$ matrix with real eigenvalues and n orthogonal eigenvectors. Then A is Hermitian. If the eigenvectors are all real, then A is real and symmetric.

The proof of this follows from writing the relations $A\mathbf{p}_i = \lambda_i \mathbf{p}_i$ in the form (3.1.3). We may assume that the eigenvectors $\mathbf{p}_i$ are normalized ($\mathbf{p}_i^* \mathbf{p}_i = 1$), so that the matrix $P = (\mathbf{p}_1, \ldots, \mathbf{p}_n)$ is unitary. Thus, multiplying (3.1.3) by P^* on the right gives $A = PDP^*$. Then $A^* = PD^*P^* = PDP^*$, so that A is Hermitian. If all the eigenvectors are real, then P is real and orthogonal, and thus $A = PDP^T$ is real and symmetric. Note that one consequence of 3.1.4 is that a complex Hermitian matrix *must* have at least one complex eigenvector.

Spectral Theorems

We next give a linear space interpretation of 3.1.1. Let $\mathbf{A}: R \to R$ be a self-adjoint linear operator on an inner product space R, and let A be the matrix representation of $\mathbf{A}$ in an orthonormal basis $\mathbf{r}_1, \ldots, \mathbf{r}_n$. By Theorem 2.4.11, A is Hermitian, and the eigenvalues of $\mathbf{A}$ are equal to the eigenvalues $\lambda_1, \ldots, \lambda_n$ of A and are real. Let $P = (p_{ij})$ be the unitary matrix of Theorem

3.1.1. Then by the definition (2.2.3) of a matrix representation and using $A\mathbf{p}_j = \lambda_j \mathbf{p}_j$ in component form, we have

$$\mathbf{A}\left(\sum_{i=1}^{n} p_{ij}\mathbf{r}_i\right) = \sum_{i=1}^{n} p_{ij}\mathbf{A}\mathbf{r}_i = \sum_{i=1}^{n} p_{ij} \sum_{k=1}^{n} a_{ki}\mathbf{r}_k = \sum_{k=1}^{n}\left(\sum_{i=1}^{n} a_{ki}p_{ij}\right)\mathbf{r}_k = \lambda_j \sum_{k=1}^{n} p_{kj}\mathbf{r}_k$$

Hence, $\sum_{i=1}^{n} p_{ij}\mathbf{r}_i$ is an eigenvector of $\mathbf{A}$ corresponding to the eigenvalue λ_j. Note that the columns of P, the eigenvectors of A, are just the coordinate vectors of the eigenvectors of $\mathbf{A}$ in the basis $\mathbf{r}_1, \ldots, \mathbf{r}_n$. Since the columns of P are orthogonal, we also have

$$\left(\sum_{i=1}^{n} p_{ij}\mathbf{r}_i, \sum_{i=1}^{n} p_{ik}\mathbf{r}_i\right) = \sum_{i=1}^{n} p_{ij}p_{ik} = \delta_{jk}$$

so that the eigenvectors of $\mathbf{A}$ are orthonormal and, hence, form an orthonormal basis. We summarize this as follows:

3.1.5. Spectral Theorem for Self-Adjoint Operators. *If* $\mathbf{A}$ *is a self-adjoint linear operator on a finite-dimensional inner product space R, then* $\mathbf{A}$ *has real eigenvalues and corresponding eigenvectors that form an orthonormal basis for R. In this basis of eigenvectors, the matrix representation of* $\mathbf{A}$ *is a diagonal matrix.*

The previous theorem says that a self-adjoint linear operator really corresponds to a very simple matrix, a diagonal one, if we are only sensible enough to choose the right basis for the space. The catch, of course, is that the right basis requires knowing the eigenvectors of $\mathbf{A}$.

The action of a real symmetric matrix, considered as a linear operator, is rather simple to describe in terms of the eigensystem $\lambda_1, \ldots, \lambda_n$ and $\mathbf{p}_1, \ldots, \mathbf{p}_n$ of A. If $0 < \lambda_i < 1$, A contracts by the factor λ_i in the direction $\mathbf{p}_i$, whereas if $\lambda_i > 1$, there is expansion by the factor λ_i. On the other hand, if $\lambda_i < 0$, then there is a reflection in the direction $\mathbf{p}_i$ together with expansion or contraction as before. Thus, if $\mathbf{x} = \sum_{i=1}^{n} c_i\mathbf{p}_i$, then $A\mathbf{x} = \sum_{i=1}^{n} c_i\lambda_i\mathbf{p}_i$, and the vector $A\mathbf{x}$ can be obtained, in principle, by the expansions, contractions, and reflections in the individual orthogonal directions defined by the eigenvectors. Note that this kind of geometric action is markedly different from that of, say, a rotation matrix.

Another interpretation of Theorem 3.1.1 is the following. Multiply Equation (3.1.3) on the right by P^* to obtain

$$A = (\lambda_1\mathbf{p}_1, \ldots, \lambda_n\mathbf{p}_n)\begin{bmatrix}\mathbf{p}_1^* \\ \vdots \\ \mathbf{p}_n^*\end{bmatrix} = \sum_{i=1}^{n} \lambda_i\mathbf{p}_i\mathbf{p}_i^* \tag{3.1.4}$$

This represents the matrix A as a sum of rank one matrices $\lambda_i \mathbf{p}_i \mathbf{p}_i^*$ formed from the eigenvalues and eigenvectors of A, and it is called the *spectral representation* of A. A simple example is given by

$$\begin{bmatrix} 2 & 1 \\ 1 & 2 \end{bmatrix} = \frac{1}{2}\begin{bmatrix} 1 \\ -1 \end{bmatrix}(1, -1) + \frac{3}{2}\begin{bmatrix} 1 \\ 1 \end{bmatrix}(1, 1)$$

We now discuss more fully the eigenvectors of A, which will lead us to a somewhat different form of (3.1.4). Let $\lambda_i \neq \lambda_j$ be eigenvalues of A with corresponding eigenvectors $\mathbf{q}_i$ and $\mathbf{q}_j$:

$$A\mathbf{q}_i = \lambda_i \mathbf{q}_i, \qquad A\mathbf{q}_j = \lambda_j \mathbf{q}_j$$

Multiplying the first equation by $\mathbf{q}_j^*$, the second by $\mathbf{q}_i^*$, conjugating the second equality, and then subtracting gives, since A is Hermitian,

$$(\lambda_i - \lambda_j)\mathbf{q}_j^* \mathbf{q}_i = \mathbf{q}_j^* A \mathbf{q}_i - \mathbf{q}_j^* A \mathbf{q}_i = 0 \tag{3.1.5}$$

Thus, eigenvectors corresponding to distinct eigenvalues are necessarily orthogonal. Now suppose that λ_1 is an eigenvalue of multiplicity m. Then m columns of the matrix P are eigenvectors corresponding to λ_1, and we may assume that these are $\mathbf{p}_1, \ldots, \mathbf{p}_m$. For any constants $c_1, \ldots, c_m$, we have

$$A\left(\sum_{i=1}^{m} c_i \mathbf{p}_i\right) = \sum_{i=1}^{m} c_i A\mathbf{p}_i = \lambda_1 \sum_{i=1}^{m} c_i \mathbf{p}_i$$

so that any linear combination of $\mathbf{p}_1, \ldots, \mathbf{p}_m$ is also an eigenvector. Thus, corresponding to λ_1 is an m-dimensional subspace $S_1 = \text{null}(A - \lambda_1 I)$ of eigenvectors. The construction of the matrix P of 3.1.1 has produced a particular orthonormal basis for S_1, but any other orthonormal basis would have been just as good. Moreover, any m linearly independent vectors in S_1 constitute a basis of eigenvectors for S_1 whether they are orthogonal or not.

Next, let $\lambda_1, \ldots, \lambda_p$ be the distinct eigenvalues of A with multiplicities $m_1, \ldots, m_p$ such that $m_1 + \cdots + m_p = n$, and let $S_1, \ldots, S_p$ be the corresponding subspaces[1] of eigenvectors of A. By (3.1.5) these subspaces are mutually orthogonal. Hence, we can write C^n as the direct sum (see Section 2.4) of the orthogonal subspaces $S_1, \ldots, S_p$:

$$C^n = S_1 \oplus \cdots \oplus S_p \tag{3.1.6}$$

[1]A subspace of eigenvectors is, strictly speaking, incorrect since 0 is not an eigenvector. This language is commonly used, however.

Moreover, these subspaces are invariant in the following sense.

3.1.6. DEFINITION: *Invariant Subspaces.* If $\mathbf{A}: R \to R$ is a linear operator, then a subspace S of R is *invariant* under $\mathbf{A}$ (or is an *invariant subspace* of $\mathbf{A}$) *if* $\mathbf{Ax} \in S$ whenever $\mathbf{x} \in S$.

Now recall (Section 2.4) that a matrix of the form $\mathbf{pp}^*$, where $\mathbf{p}^*\mathbf{p} = 1$, is an orthogonal projector onto span($\mathbf{p}$). Hence, the spectral representation (3.1.4) expresses A as a sum of multiples of orthogonal projectors. If A has multiple eigenvalues, we may put the corresponding rank 1 projectors together to form an orthogonal projector onto the subspace S_i. For example, if again λ_1 has multiplicity m and $\mathbf{p}_1, \ldots, \mathbf{p}_m$ are orthonormal eigenvectors corresponding to λ_1, let P_1 be the matrix with columns $\mathbf{p}_1, \ldots, \mathbf{p}_m$. Then $P_1^* P_1 = I_m$, and the matrix

$$\mathbf{p}_1\mathbf{p}_1^* + \cdots + \mathbf{p}_m\mathbf{p}_m^* = P_1 P_1^*$$

is an orthogonal projector onto the subspace S_1 spanned by $\mathbf{p}_1, \ldots, \mathbf{p}_m$.

We summarize the above in the following theorem:

3.1.7. SPECTRAL REPRESENTATION OF SYMMETRIC AND HERMITIAN MATRICES. *Let A be an $n \times n$ Hermitian matrix with distinct eigenvalues $\lambda_1, \ldots, \lambda_p$ of multiplicities $m_1, \ldots, m_p$. Let $S_1, \ldots, S_p$ be the subspaces of corresponding eigenvectors with $\dim S_i = m_i$. Then $C^n = S_1 \oplus \cdots \oplus S_p$ and*

$$A = \sum_{i=1}^{p} \lambda_i Q_i \tag{3.1.7}$$

where Q_i is an orthogonal projector onto S_i. If A is real and symmetric, the subspaces S_i may be taken in R^n, so that $R^n = S_1 \oplus \cdots \oplus S_p$.

If we wish, we may use (2.4.12) and represent the orthogonal projectors of (3.1.7) as the matrices $P_i(P_i^* P_i)^{-1} P_i^*$, where the columns of P_i are any basis for S_i. Then (3.1.7) becomes

$$A = \sum_{i=1}^{n} \lambda_i P_i (P_i^* P_i)^{-1} P_i^* \tag{3.1.8}$$

If the columns of the P_i are orthonormal, then (3.1.8) reduces to

$$A = \sum_{i=1}^{p} \lambda_i P_i P_i^* \tag{3.1.9}$$

Because of its importance, we also state explicitly the linear operator formulation of 3.1.7.

3.1.8. SPECTRAL REPRESENTATION OF SELF-ADJOINT OPERATORS. *Let $A: R \to R$ be a self-adjoint linear operator on an inner product space R. If $\lambda_1, \ldots, \lambda_p$ are the distinct eigenvalues of A and $S_1, \ldots, S_p$ are the corresponding subspaces of eigenvectors, then $R = S_1 \oplus \cdots \oplus S_p$ and*

$$A = \sum_{i=1}^{p} \lambda_i Q_i$$

where Q_i is an orthogonal projection operator onto S_i.

Criteria for Definiteness

We next give a useful corollary of 3.1.1, which will allow us to characterize definiteness properties of Hermitian matrices in terms of eigenvalues.

3.1.9. Let A be an $n \times n$ Hermitian matrix with eigenvalues $\lambda_1 \leq \cdots \leq \lambda_n$. Then for all $\mathbf{x} \in C^n$,

$$\lambda_1 \mathbf{x}^*\mathbf{x} \leq \mathbf{x}^* A \mathbf{x} \leq \lambda_n \mathbf{x}^*\mathbf{x} \tag{3.1.10}$$

Proof. By 3.1.1 there is a unitary matrix P such that

$$P^*AP = \operatorname{diag}(\lambda_1, \ldots, \lambda_n)$$

Hence, with $\mathbf{y} = P^*\mathbf{x}$, we have $\mathbf{y}^*\mathbf{y} = \mathbf{x}^*\mathbf{x}$ and

$$\mathbf{x}^*A\mathbf{x} = \mathbf{y}^*P^*AP\mathbf{y} = \sum_{i=1}^{n} \lambda_i |y_i^2| \leq \lambda_n \mathbf{y}^*\mathbf{y} = \lambda_n \mathbf{x}^*\mathbf{x}$$

The other inequality is proved analogously. □

In Section 1.4 we noted that a positive definite Hermitian matrix must have positive eigenvalues. By 3.1.9 the converse also holds, because if $\lambda_1 > 0$, then (3.1.10) shows that $\mathbf{x}^*A\mathbf{x} > 0$ for all $\mathbf{x} \neq 0$. Similarly, if $\lambda_1 \geq 0$, then $\mathbf{x}^*A\mathbf{x} \geq 0$ for all $\mathbf{x}$, so that A is positive semidefinite. On the other hand, if $\lambda_n < 0$, then $\mathbf{x}^*A\mathbf{x} < 0$ for all $\mathbf{x} \neq 0$, so that A is negative definite; and if $\lambda_n \leq 0$, A is negative semidefinite. We summarize this as follows.

3.1.10. Let A be an $n \times n$ Hermitian matrix. Then

(a) *A is positive (negative) definite if and only if all its eigenvalues are positive (negative);*
(b) *A is positive (negative) semidefinite if and only if all its eigenvalues are nonnegative (nonpositive);*
(c) *A is indefinite if and only if it has both positive and negative eigenvalues.*

We next digress briefly to use 3.1.9 to give the proof, which was omitted in Chapter 2, of Theorem 2.5.6:

$$\|A\|_2 = \rho(A^*A)^{1/2} \tag{3.1.11}$$

that is, the l_2 norm of an $n \times n$ matrix A is the square root of the spectral radius of the hermitian matrix A^*A.

Proof of 2.5.6. Set $\mu^2 = \rho(A^*A)$. Then, for any $\mathbf{x}$, 3.1.9 shows that

$$\|A\mathbf{x}\|_2^2 = \mathbf{x}^*A^*A\mathbf{x} \leq \mu^2\mathbf{x}^*\mathbf{x}$$

so that $\|A\|_2 \leq \mu$. On the other hand, if $\mathbf{u}$ is an eigenvector of A^*A corresponding to μ^2, then

$$\mathbf{u}^*A^*A\mathbf{u} = \mu^2\mathbf{u}^*\mathbf{u}$$

which shows that equality holds in (3.1.11). □

Triangular Form

We now consider matrices A that are not Hermitian, and our goal is still to find a unitary matrix P so that P^*AP is as simple as possible. As a first step, we show that any $n \times n$ matrix is unitarily similar to a triangular matrix.

3.1.11. Schur's Theorem. *If A is an $n \times n$ matrix, then there is a unitary matrix P such that $P^*AP = T$, where T is upper triangular.*

Proof. The proof is a minor modification of that of 3.1.1. At the first stage we have

$$P_1^*AP_1 = \begin{bmatrix} \mathbf{x}_1^* \\ U_1^* \end{bmatrix} (\lambda_1\mathbf{x}_1, AU_1) = \begin{bmatrix} \lambda_1 & \mathbf{x}_1^*AU_1 \\ 0 & U_1^*AU_1 \end{bmatrix}$$

where the vector $\mathbf{x}_1^* A U_1$ is now, in general, not zero. We continue the process, just as in 3.1.1, to obtain matrices $Q_2, \ldots, Q_{n-1}$ and unitary matrices $P_2, \ldots, P_{n-1}$; the only difference is that the upper triangular portions of the reduced matrices are not necessarily zero. Thus, we obtain

$$P^*AP = T$$

where $P = P_1 \cdots P_{n-1}$ and T is upper triangular. □

Note that the diagonal elements of T are the eigenvalues of A. However, the columns of P, other than the first one, are not, in general, eigenvectors of A. These columns are sometimes called *Schur vectors.* It is easy to see (Exercise 3.1-12) that the subspaces $\text{span}(\mathbf{p}_1, \ldots, \mathbf{p}_k)$, $k = 1, \ldots, n$, are all invariant subspaces of A. We also note that if A is real with real eigenvalues, then the matrix P of 3.1.11 may be taken to be orthogonal.

Normal Matrices

We now ask when the triangular matrix of Schur's theorem can be diagonal; that is, what matrices are unitarily similar to diagonal matrices?

3.1.12. DIAGONALIZATION OF NORMAL MATRICES. *An $n \times n$ matrix A is unitarily similar to a diagonal matrix if and only if it commutes with its conjugate transpose:*

$$AA^* = A^*A \tag{3.1.12}$$

Matrices satisfying (3.1.12) (or $AA^T = A^TA$ if A is real) are called *normal.* Examples of normal matrices are (Exercise 3.1-8)

Hermitian (or real and symmetric) (3.1.13a)

skew-Hermitian (or real and skew-symmetric) (3.1.13b)

unitary (or real and orthogonal) (3.1.13c)

Proof of 3.1.12. If $A = PDP^*$, where D is diagonal and P is unitary, then since diagonal matrices commute, we have

$$AA^* = PDP^*PD^*P^* = PDD^*P^* = PD^*DP^* = PD^*P^*PDP^* = A^*A$$

so that (3.1.12) holds. For the converse, by Schur's theorem, let $A = PTP^*$, where P is unitary and T is triangular. If (3.1.12) holds, then

$$TT^* = P^*APP^*A^*P = P^*AA^*P = P^*A^*AP = P^*A^*PP^*AP = T^*T$$

Hence T is normal. But it is easy to see (Exercise 3.1-5) that a normal triangular matrix must be diagonal. □

Note that if $A = PDP^*$, then $A^* = PD^*P^*$, so that the columns of P are eigenvectors of both A and A^*. Therefore, if A is normal, then A and A^* have a common set of eigenvectors. The converse also holds, and this is another characterization of normal matrices. Thus, if A is not normal, then A^* and A necessarily have at least some different eigenvectors. However, these eigenvectors have the following orthogonality property. If

$$A\mathbf{x} = \lambda\mathbf{x}, \qquad A^*\mathbf{y} = \mu\mathbf{y}$$

and $\bar{\mu} \neq \lambda$, then, as in (3.1.5),

$$(\lambda - \bar{\mu})\mathbf{y}^*\mathbf{x} = \mathbf{y}^*A\mathbf{x} - \mathbf{y}^*A\mathbf{x} = 0$$

so that $\mathbf{x}$ and $\mathbf{y}$ are orthogonal. Recall that in Section 1.4 we defined a left eigenvector of A as a column vector $\mathbf{y}$ such that $\mathbf{y}^*A = \mu\mathbf{y}^*$, which is equivalent to $A^*\mathbf{y} = \bar{\mu}\mathbf{y}$. Then the above result states that the left and right eigenvectors of A corresponding to distinct eigenvalues are orthogonal.

Theorem 3.1.12 shows that if we restrict ourselves to unitary similarity transformations, then only very special (but important) matrices can be reduced to a diagonal matrix. In the next section we will see that a much larger class of matrices can be reduced to diagonal matrices if we allow general similarity transformations.

Summary

The main results and concepts of this section are as follows:

- A Hermitian matrix is unitarily similar to a diagonal matrix and a real symmetric matrix is orthogonally similar to a diagonal matrix.
- A Hermitian matrix has n orthogonal eigenvectors. Conversely, a matrix with real eigenvalues and n orthogonal eigenvectors must be Hermitian.
- A self-adjoint linear operator on a finite-dimensional inner product space has an orthonormal basis of eigenvectors in which the matrix representation is diagonal.
- A Hermitian matrix may be written as a linear combination of orthogonal projectors onto invariant subspaces.
- The eigenvalues of a Hermitian matrix determine its definiteness properties.
- Any matrix is unitarily similar to a triangular matrix.
- A matrix is normal if and only if it commutes with its conjugate transpose.
- A matrix is unitarily similar to a diagonal matrix if and only if it is normal.

Exercises 3.1

1. Compute the eigenvalues λ_1, λ_2 and normalized eigenvectors $\mathbf{p}_1, \mathbf{p}_2$ of

$$A = \begin{bmatrix} 2 & 1 \\ 1 & 2 \end{bmatrix}$$

and find orthogonal and diagonal matrices P and D such that $A = PDP^T$.

2. Let A be a Hermitian matrix with the spectral representation (3.1.4). Show that $A^k = \sum \lambda_i^k \mathbf{p}_i \mathbf{p}_i^*$ for any positive integer k. More generally, show that, for any polynomial $f(\lambda), f(A) = \sum f(\lambda_i)\mathbf{p}_i\mathbf{p}_i^*$. Show also, if A^{-1} exists, that $A^{-1} = \sum \lambda_i^{-1}\mathbf{p}_i\mathbf{p}_i^*$.

3. Let $\mathbf{w}$ be a given n-vector and $A = \mathbf{w}\mathbf{w}^*$. Show that
   (a) A is Hermitian.
   (b) A has one eigenvalue equal to $\mathbf{w}^*\mathbf{w}$ with eigenvector $\mathbf{w}$.
   (c) The remaining $n - 1$ eigenvalues are zero and span$(\mathbf{w})^\perp$, the orthogonal complement of span$(\mathbf{w})$, is an $(n - 1)$-dimensional invariant subspace of eigenvectors.

4. Let A be an $n \times n$ Hermitian matrix. Show that
   (a) If A is nonsingular, then A^{-1} is Hermitian.
   (b) If A is positive definite, then A^{-1} is positive definite.
   (c) If A is nonsingular and indefinite, then A^{-1} is indefinite.

5. Let T be a normal triangular matrix. Equate the elements of T^*T and TT^* to show that T must be diagonal.

6. Show that any $n \times n$ matrix can be written in the form

$$A = H_1 + iH_2, \qquad H_1 = \tfrac{1}{2}(A + A^*), \qquad H_2 = \frac{1}{2i}(A - A^*)$$

where H_1 and H_2 are Hermitian. Show also that A is normal if and only if H_1 and H_2 commute: $H_1H_2 = H_2H_1$.

7. Let A be a Hermitian matrix. Show that iA is normal.

8. Verify that Hermitian, skew-Hermitian, and unitary matrices are all normal. Give an example of a 2×2 normal matrix that is not Hermitian, skew-Hermitian, or unitary.

9. Let $A = (a_{ij})$ be an $n \times n$ matrix, and let $\|A\| = (\sum_{i,j=1}^n |a_{ij}|^2)^{1/2}$ be the Frobenius norm of Exercise 2.5-10. If P and Q are $n \times n$ unitary matrices, show that $\|PA\| = \|AQ\| = \|A\|$. Use this together with 3.1.11 and 3.1.12 to show that if $\lambda_1, \ldots, \lambda_n$ are the eigenvalues of A, then

$$\sum_{i=1}^n |\lambda_i|^2 = \sum_{i,j=1}^n |a_{ij}|^2$$

if and only if A is normal.

10. Show that an $n \times n$ matrix A is normal if and only if $\|A\mathbf{x}\|_2 = \|A^*\mathbf{x}\|_2$ for all $\mathbf{x}$.
11. Let A be an $n \times n$ normal matrix. Show that $\rho(A) = \|A\|_2$. Is the converse true?
12. Let $P^*AP = T$, where P is unitary and T is triangular. Let $\mathbf{p}_1, \ldots, \mathbf{p}_n$ be the columns of P. Show that the subspaces span$(\mathbf{p}_1, \ldots, \mathbf{p}_k)$, $k = 1, \ldots, n$, are all invariant subspaces of A.
13. Let $A = PDP^*$ be a Hermitian positive definite matrix, where P is unitary and $D = \text{diag}(\lambda_1, \ldots, \lambda_n)$. Define $D^{1/2} = \text{diag}(\lambda_1^{1/2}, \lambda_2^{1/2}, \ldots, \lambda_n^{1/2})$. Show that $A^{1/2} = PD^{1/2}P^*$ is the unique Hermitian positive definite square root of A.
14. Let A be Hermitian and positive definite. Use the Cauchy-Schwarz inequality and the previous exercise to show that

$$|\mathbf{x}^*\mathbf{y}|^2 \leq \mathbf{x}^*A\mathbf{x}\mathbf{y}^*A^{-1}\mathbf{y}$$

15. Let A be nonsingular, and let P be a unitary matrix such that $P^*(A^*A)P = D$, where D is diagonal. Define $S = PD^{1/2}P^*$ and $Q = AS^{-1}$. Show that Q is unitary and thus conclude that any nonsingular matrix can be written as a product $A = QS$, where Q is unitary and S is Hermitian and positive definite.
16. Let

$$A = \begin{bmatrix} 3 & 0 & -1 \\ 0 & 4 & 0 \\ -1 & 0 & 3 \end{bmatrix}$$

Find the spectral representations (3.1.4) and (3.1.7) of A.

17. Show that if P is unitary, then $\|A\|_2 = \|PAP^*\|_2$.

3.2. The Jordan Canonical Form

We now consider similarity transformations

$$B = P^{-1}AP \tag{3.2.1}$$

of an $n \times n$ matrix A, where P is not necessarily unitary. Motivated by Theorem 3.1.1, in which we saw that a symmetric or a Hermitian matrix was always similar to a diagonal matrix, we ask if the same can be achieved for an arbitrary $n \times n$ matrix A; that is, given A, can we always find a nonsingular matrix P so that the matrix B of (3.2.1) is diagonal? Unfortunately, the answer in general is no, and the purpose of this section is to ascertain what can be accomplished.

We begin with a basic result which, in a sense, gives a complete answer as to when a matrix is similar to a diagonal matrix. Such a matrix is called *diagonalizable* or *diagonable.*

3.2.1. *Let A be an $n \times n$ matrix. Then A is similar to a diagonal matrix if and only if A has n linearly independent eigenvectors.*

Proof. Let (3.2.1) hold with $B = D = \mathrm{diag}(\lambda_1, \ldots, \lambda_n)$. Then

$$AP = PD = (\lambda_1 \mathbf{p}_1, \ldots, \lambda_n \mathbf{p}_n) \tag{3.2.2}$$

where $\mathbf{p}_1, \ldots, \mathbf{p}_n$ are the columns of P. Equating the columns of both sides of (3.2.2) shows that $A\mathbf{p}_i = \lambda_i \mathbf{p}_i$, $i = 1, \ldots, n$. Thus, $\mathbf{p}_1, \ldots, \mathbf{p}_n$ are eigenvectors and are linearly independent since P is nonsingular. Conversely, if A has n linearly independent eigenvectors $\mathbf{p}_1, \ldots, \mathbf{p}_n$ corresponding to the eigenvalues $\lambda_1, \ldots, \lambda_n$, the matrix P with columns $\mathbf{p}_1, \ldots, \mathbf{p}_n$ is nonsingular, and (3.2.2) shows that A is similar to D. □

Using the results of Chapter 2, we can restate 3.2.1 in terms of matrix representations of linear operators. The proof is left to Exercise 3.2-1.

3.2.2. *Let $\mathbf{A}: R \to R$ be a linear operator on a finite-dimensional linear space R. Then $\mathbf{A}$ has a diagonal matrix representation if and only if R has a basis consisting of eigenvectors of $\mathbf{A}$.*

A sufficient condition that an $n \times n$ matrix A have n linearly independent eigenvectors is that its eigenvalues are all distinct. We first prove the following basic result:

3.2.3. *If $\lambda_1, \ldots, \lambda_m$ are distinct eigenvalues of an $n \times n$ matrix A, and $\mathbf{x}_1, \ldots, \mathbf{x}_m$ are corresponding eigenvectors, then $\mathbf{x}_1, \ldots, \mathbf{x}_m$ are linearly independent.*

Proof. Suppose that $\mathbf{x}_1, \ldots, \mathbf{x}_m$ are linearly dependent so that

$$\sum_{i=1}^{m} c_i \mathbf{x}_i = 0$$

with not all constants $c_1, \ldots, c_m$ equal to zero. We may assume that $c_m \neq 0$ by renumbering the eigenvalues if necessary. Then

$$0 = A \sum_{i=1}^{m} c_i \mathbf{x}_i = \sum_{i=1}^{m} c_i \lambda_i \mathbf{x}_i$$

If we subtract $\lambda_1 \sum c_i \mathbf{x}_i$ from this and multiply by $A - \lambda_2 I$, we obtain

$$0 = (A - \lambda_2 I) \sum_{i=2}^{m} c_i(\lambda_i - \lambda_1)\mathbf{x}_i = \sum_{i=3}^{m} c_i(\lambda_i - \lambda_1)(\lambda_i - \lambda_2)\mathbf{x}_i$$

Multiplying by $A - \lambda_3 I$, then $A - \lambda_4 I$, and so on, we conclude that

$$0 = c_m(\lambda_m - \lambda_{m-1})(\lambda_m - \lambda_{m-2}) \cdots (\lambda_m - \lambda_1)\mathbf{x}_m$$

But this is a contradiction if the λ_i are distinct, since $c_m \neq 0$ and $\mathbf{x}_m \neq 0$. □

Note that in 3.2.3 it is not assumed that the eigenvalues are simple. (Recall that an eigenvalue is simple if its multiplicity is 1.) However, in the case that $m = n$, then the λ_i are simple, and the corresponding n eigenvectors are linearly independent. Hence, we conclude from 3.2.1 that

3.2.4. If the $n \times n$ matrix A has n distinct eigenvalues, then A is similar to a diagonal matrix.

Having n distinct eigenvalues is only a sufficient condition for A to be similar to a diagonal matrix. For example, Theorem 3.1.1 shows that a symmetric matrix is always similar to a diagonal matrix, regardless of the multiplicities of the eigenvalues; the identity matrix is the extreme example of this.

Theorems 3.2.2 and 3.2.4 together show that if there are matrices that are not similar to a diagonal matrix, they necessarily have multiple eigenvalues and less than n linearly independent eigenvectors. The simplest example of such a matrix is of the form

$$A = \begin{bmatrix} \lambda & 1 \\ 0 & \lambda \end{bmatrix} \tag{3.2.3}$$

Here, λ is an eigenvalue of multiplicity 2, and the eigenvectors of A must satisfy

$$(A - \lambda I)\mathbf{x} = \begin{bmatrix} 0 & 1 \\ 0 & 0 \end{bmatrix} \begin{bmatrix} x_1 \\ x_2 \end{bmatrix} = 0$$

or x_1 = arbitrary, $x_2 = 0$. Thus, all eigenvectors are multiples of $\mathbf{e}_1$, and 3.2.1 shows that A cannot be similar to a diagonal matrix. More generally, one can show in an analogous way (Exercise 3.2-2) that all eigenvectors of the matrix

$$J_i = \begin{bmatrix} \lambda_i & 1 & & & \\ & \lambda_i & \cdot & & \\ & & \ddots & \cdot & \\ & & & \ddots & 1 \\ & & & & \lambda_i \end{bmatrix} \tag{3.2.4}$$

are again multiples of $\mathbf{e}_1$.

If not all matrices are similar to a diagonal matrix, what is the simplest form that a matrix can, in general, take under a similarity transformation? Matrices of the form (3.2.4) provide the building blocks for this simplest form, which is a diagonal matrix except for 1's in certain positions in the diagonal above the main diagonal. This is the matrix as given in the next theorem.

3.2.5. JORDAN CANONICAL FORM THEOREM. *An $n \times n$ matrix A is similar to a matrix of the form*

$$J = \operatorname{diag}(J_1, \ldots, J_p) \tag{3.2.5}$$

where each J_i is an $r_i \times r_i$ matrix of the form (3.2.4), *and $\sum_{i=1}^{p} r_i = n$. Equivalently, if $\mathbf{A}: R \to R$ is a linear operator on a finite-dimensional complex linear space R, there is a basis for R in which the matrix representation of $\mathbf{A}$ is* (3.2.5).

The matrix J of (3.2.5) is called the *Jordan canonical form of A*. Before proving Theorem 3.2.5, we will discuss a few basic implications of the theorem. The $r_i \times r_i$ matrices J_i are called *Jordan blocks*, and the r_i may be any integers from 1 to n such that $\sum_{i=1}^{p} r_i = n$. Thus, a special case of (3.2.5) is when $p = n$ and $r_i = 1, i = 1, \ldots, n$, so that J is a diagonal matrix. At the other extreme is the case $p = 1$ and $r_1 = n$, so that J is itself a Jordan block of dimension n. All possibilities between these are allowable. To illustrate this, we list below the possible Jordan forms of a 4×4 matrix with an eigenvalue 2 of multiplicity 4.

In Figure 3.1 we have indicated by dashed lines the different Jordan blocks. In (a) and (e) we have the diagonal and single Jordan block cases

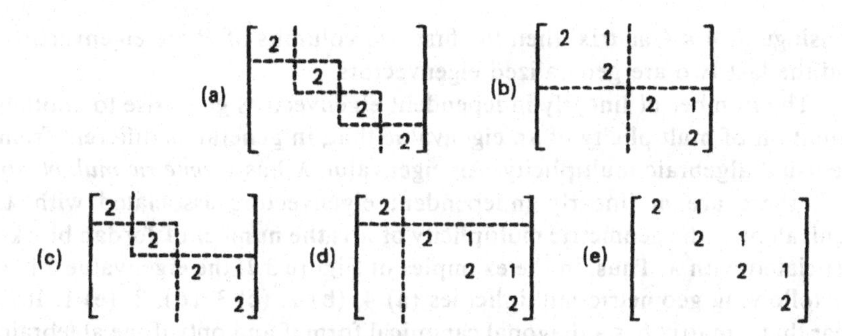

Figure 3.1. Possible Jordan forms.

discussed above. For (b), $p = 2$ and $r_1 = r_2 = 2$, whereas for (d) $p = 2$ also, but $r_1 = 1$ and $r_2 = 3$. Finally, in (c), $p = 3$, $r_1 = r_2 = 1$, and $r_3 = 2$. These are all the possible Jordan forms for this matrix up to permutations of the Jordan blocks in (c) and (d).

We noted previously that a matrix of the form (3.2.4) has only one linearly independent eigenvector, a vector with a 1 in the first position and 0's elsewhere. Hence, the unit vectors with 1's in positions $1, r_1 + 1, r_1 + r_2 + 1, \ldots, r_1 + \cdots + r_{p-1} + 1$ are linearly independent eigenvectors of J. Moreover, J can have no more linearly independent eigenvectors (Exercise 3.2-3). For example, linearly independent eigenvectors of the matrices of Figure 3.1 are given by

$$\text{(a)} \quad \begin{bmatrix} 1 \\ 0 \\ 0 \\ 0 \end{bmatrix} \begin{bmatrix} 0 \\ 1 \\ 0 \\ 0 \end{bmatrix} \begin{bmatrix} 0 \\ 0 \\ 1 \\ 0 \end{bmatrix} \begin{bmatrix} 0 \\ 0 \\ 0 \\ 1 \end{bmatrix}, \qquad \text{(b)} \quad \begin{bmatrix} 1 \\ 0 \\ 0 \\ 0 \end{bmatrix} \begin{bmatrix} 0 \\ 0 \\ 1 \\ 0 \end{bmatrix}$$

$$\text{(c)} \quad \begin{bmatrix} 1 \\ 0 \\ 0 \\ 0 \end{bmatrix} \begin{bmatrix} 0 \\ 1 \\ 0 \\ 0 \end{bmatrix} \begin{bmatrix} 0 \\ 0 \\ 1 \\ 0 \end{bmatrix}, \qquad \text{(d)} \quad \begin{bmatrix} 1 \\ 0 \\ 0 \\ 0 \end{bmatrix} \begin{bmatrix} 0 \\ 1 \\ 0 \\ 0 \end{bmatrix}, \qquad \text{(e)} \quad \begin{bmatrix} 1 \\ 0 \\ 0 \\ 0 \end{bmatrix}$$

Now let P be the nonsingular matrix such that $P^{-1}AP = J$. Then

$$AP = PJ \tag{3.2.6}$$

and the columns of P in the $1, r_1 + 1, r_1 + r_2 + 1, \ldots, r_1 + \cdots + r_{p-1} + 1$ positions of P are eigenvectors of A, and these are linearly independent eigenvectors since P is nonsingular. Again, there can be no more linearly independent eigenvectors of A (Exercise 3.2-3). The other columns of P are called ***principal vectors*** or ***generalized eigenvectors.*** For example, if $A = PJP^{-1}$, where J is the 4×4 matrix (d) of Figure 3.1 and P is a nonsingular 4×4 matrix, then the first two columns of P are eigenvectors, and the last two are generalized eigenvectors.

The number of linearly independent eigenvectors gives rise to another definition of multiplicity of an eigenvalue that, in general, is different from the usual algebraic multiplicity. An eigenvalue λ has *geometric multiplicity* m if there are m linearly independent eigenvectors associated with λ. Equivalently, the geometric multiplicity of λ is the number of Jordan blocks associated with λ. Thus, in the examples of Figure 3.1, the eigenvalue 2 has the following geometric multiplicities (a) 4, (b) 2, (c) 3, (d), 2, (e) 1. It is clear that a matrix has a diagonal canonical form if and only if the algebraic and geometric multiplicities are the same for each eigenvalue. Otherwise, the matrix is sometimes called *defective.*

We next discuss the generalized eigenvectors in more detail. Equate the first r_1 columns of (3.2.6) corresponding to the first Jordan block J_1:

$$A\mathbf{p}_1 = \lambda_1\mathbf{p}_1, \qquad A\mathbf{p}_i = \lambda_1\mathbf{p}_i + \mathbf{p}_{i-1}, \quad i = 2, \ldots, r_1 \tag{3.2.7}$$

The vectors $\mathbf{p}_1, \ldots, \mathbf{p}_{r_1}$ are called a *chain* (or *Jordan chain*) of length r_1. The vector $\mathbf{p}_{r_1}$ is the *lead* vector of the chain, while $\mathbf{p}_i$ is a generalized eigenvector of *degree* i, if $i \geq 2$. Note that the generalized eigenvectors $\mathbf{p}_2, \ldots, \mathbf{p}_{r_1}$ do not satisfy an eigenvalue equation $A\mathbf{p} = \lambda\mathbf{p}$ but something very close to it.

We can rewrite (3.2.7) in the form

$$(A - \lambda_1 I)\mathbf{p}_1 = 0, \qquad (A - \lambda_1 I)\mathbf{p}_i = \mathbf{p}_{i-1}, \quad i = 2, \ldots, r_1 \tag{3.2.8}$$

which shows that $A - \lambda_1 I$ maps $\mathbf{p}_i$ into the next element $\mathbf{p}_{i-1}$ of the chain. In particular, the lead element $\mathbf{p}_{r_1}$ generates the chain by repeated multiplication by $A - \lambda_1 I$. The same holds for each of the eigenvalues of A, and we have

3.2.6. *If A is an $n \times n$ matrix such that $A = PJP^{-1}$, where J is the Jordan form* (3.2.5), *then the columns of P form p Jordan chains*

$$\{\mathbf{p}_1, \ldots, \mathbf{p}_{r_1}\}, \{\mathbf{p}_{r_1+1}, \ldots, \mathbf{p}_{r_1+r_2}\}, \ldots, \{\mathbf{p}_{n-r_p+1}, \ldots, \mathbf{p}_n\} \tag{3.2.9}$$

We illustrate 3.2.6 by the examples of Figure 3.1. For (a) there are four chains, each of length 1 and consisting of only an eigenvector. For (b) there are two chains, each of length 2. For (c) there are three chains, two of length 1, consisting of eigenvectors, and one of length 2. For (d) there are two chains, one of length 1 consisting of an eigenvector, and one of length 3. Finally, for (e) there is a single chain of length 4. It is left to Exercise 3.2-11 to exhibit explicitly those chains. Note that these examples illustrate that, in general, there will be as many chains associated with a given eigenvalue as there are Jordan blocks with that eigenvalue.

The relations (3.2.8) imply that

$$(A - \lambda_1 I)^i \mathbf{p}_j = 0, \qquad i = 1, \ldots, r_1, j = 1, \ldots, i \tag{3.2.10}$$

If λ_1 is distinct from the other eigenvalues, this shows that the linearly independent vectors $\mathbf{p}_1, \ldots, \mathbf{p}_i$, $i \le r_1$, form a basis for null$((A - \lambda_1 I)^i)$; that is, $\mathbf{p}_1$ is a basis for the eigenspace null$(A - \lambda_1 I)$, $\mathbf{p}_1$, $\mathbf{p}_2$ is a basis for null$((A - \lambda_1 I)^2)$, and so on. If several Jordan blocks have the same eigenvalue, then the corresponding chains must be combined to obtain the bases for the null spaces. For example, suppose that $\lambda_1 = \lambda_2 = \lambda_3$. Then $\mathbf{p}_1$, $\mathbf{p}_{r_1+1}$ and $\mathbf{p}_{r_1+r_2+1}$ are a basis for null$(A - \lambda_1 I)$; these vectors plus the next three vectors in the three chains are a basis for null$((A - \lambda_1 I)^2)$; and so on.

Either (3.2.7) or (3.2.10) may be used, in principle (but, we stress, not in practice) to compute the generalized eigenvectors and, hence, the Jordan form. For example, the matrix

$$A = \frac{1}{2}\begin{bmatrix} 5 & -1 \\ 1 & 3 \end{bmatrix}$$

has the eigenvalues 2 and 2. An eigenvector is $(1, 1)^T$. Using (3.2.7), we solve the system

$$\frac{1}{2}\begin{bmatrix} 5-4 & -1 \\ 1 & 3-4 \end{bmatrix}\begin{bmatrix} x_1 \\ x_2 \end{bmatrix} = \begin{bmatrix} 1 \\ 1 \end{bmatrix}$$

which has the solutions $(2 + c, c)^T$, for arbitrary c. Choosing, say, $c = 1$ gives

$$P = \begin{bmatrix} 1 & 3 \\ 1 & 1 \end{bmatrix}, \qquad P^{-1}AP = \begin{bmatrix} 2 & 1 \\ 0 & 2 \end{bmatrix}$$

which is the Jordan form. The first column of P is an eigenvector, and the second column a generalized eigenvector.

We now return to the proof of the Jordan canonical form theorem.

Proof of 3.2.5. The proof is by induction on the size of the matrix. It is clearer to work with linear operators, and the induction hypothesis will be that for any linear operator **A** on a complex linear space of dimension less than n, there is a basis in which the matrix representation has the Jordan form (3.2.5). Clearly, this is trivially true for $n = 1$.

Now let $\mathbf{A}: R \to R$ be a linear operator on an n-dimensional space R, and let λ be any eigenvalue. Then $r = \dim \text{range}(\mathbf{A} - \lambda I) < n$. Let $\mathbf{A}_\lambda$ be the restriction of $\mathbf{A} - \lambda I$ to range$(\mathbf{A} - \lambda I)$; that is, $\mathbf{A}_\lambda x = (\mathbf{A} - \lambda I)x$ for all $x \in \text{range}(\mathbf{A} - \lambda I)$. Then, by the induction hypothesis, there are r linearly independent vectors $\mathbf{p}_i$ such that in this basis $\mathbf{A}_\lambda$ has a Jordan form matrix representation.

Next, let $Q = \text{null}(\mathbf{A} - \lambda I) \cap \text{range}(\mathbf{A} - \lambda I)$ and $q = \dim Q$. Every nonzero vector in Q is an eigenvector of $\mathbf{A}_\lambda$ corresponding to the eigenvalue zero. Therefore, $\mathbf{A}_\lambda$ has q Jordan chains ending in linearly independent eigenvectors, and these vectors form a basis for Q. Let $\mathbf{p}_1, \ldots, \mathbf{p}_q$ be the lead vectors for these chains. These lead vectors must be in range$(\mathbf{A} - \lambda I)$; hence, there are vectors $\mathbf{y}_1, \ldots, \mathbf{y}_q$ such that $(\mathbf{A} - \lambda I)\mathbf{y}_i = \mathbf{p}_i$, $i = 1, \ldots, q$. Finally, since, by 2.3.3, $\dim \text{null}(\mathbf{A} - \lambda I) = n - r$, $s = n - r - q$ is the number of linearly independent vectors in null$(\mathbf{A} - \lambda I)$ that are not in Q. Let $\mathbf{z}_1, \ldots, \mathbf{z}_s$ be such vectors. Then the n vectors $\mathbf{p}_1, \ldots, \mathbf{p}_r, \mathbf{y}_1, \ldots, \mathbf{y}_q, \mathbf{z}_1, \ldots, \mathbf{z}_s$ are all either eigenvectors or generalized eigenvectors of $\mathbf{A} - \lambda I$ and, hence, of $\mathbf{A}$ itself. It remains only to show that these n vectors are linearly independent. Suppose that

$$\sum_{i=1}^{r} a_i \mathbf{p}_i + \sum_{i=1}^{q} b_i \mathbf{y}_i + \sum_{i=1}^{s} c_i \mathbf{z}_i = 0 \qquad (3.2.11)$$

Then we wish to show that all the constants a_i, b_i, and c_i must be zero. Multiply (3.2.11) by $\mathbf{A} - \lambda I$; then, since the $\mathbf{z}_i$ are in null$(\mathbf{A} - \lambda I)$, we have

$$\sum_{i=1}^{r} a_i (\mathbf{A} - \lambda I)\mathbf{p}_i + \sum_{i=1}^{q} b_i (\mathbf{A} - \lambda I)\mathbf{y}_i = 0 \qquad (3.2.12)$$

By construction, $(\mathbf{A} - \lambda I)\mathbf{y}_i = \mathbf{p}_i$, $i = 1, \ldots, q$, and the second sum of (3.2.12) is $\sum_{i=1}^{q} b_i \mathbf{p}_i$. The $\mathbf{p}_i$ in this second sum correspond to the eigenvalue λ and are lead vectors of a Jordan chain. Therefore, for all of these $\mathbf{p}_i$, either $(\mathbf{A} - \lambda I)\mathbf{p}_i = 0$ or $(\mathbf{A} - \lambda I)\mathbf{p}_i$ is another $\mathbf{p}_j$, which is not a lead vector of a chain. Thus, none of the $\mathbf{p}_i$ in the second sum of (3.2.12) appear in the first sum. Since the $\mathbf{p}$'s are all linearly independent, it follows that all the b_i must be zero, so that (3.2.11) becomes

$$\sum_{i=1}^{r} a_i \mathbf{p}_i + \sum_{i=1}^{s} c_i \mathbf{z}_i = 0$$

But the $\mathbf{p}_i$ are linearly independent vectors in range$(\mathbf{A} - \lambda I)$, and the $\mathbf{z}_i$ are linearly independent vectors in null$(\mathbf{A} - \lambda I)$ but not in range$(\mathbf{A} - \lambda I)$. Hence, all the a_i and c_i must be zero. □

We now consider other ramifications of the Jordan form. If the Jordan form is diagonal, then there are n linear independent eigenvectors $\mathbf{p}_1, \ldots, \mathbf{p}_n$. Thus, span($\mathbf{p}_i$), $i = 1, \ldots, n$, are all invariant subspaces of A, and

$$C^n = \text{span}(\mathbf{p}_1) \dot{+} \cdots \dot{+} \text{span}(\mathbf{p}_n)$$

is a direct sum decomposition of C^n into invariant subspaces of A. (Note, however, that these subspaces are not necessarily orthogonal, as was the case in Section 3.1 for symmetric matrices.) If the Jordan form is not diagonal, then the generalized eigenvectors do not generate one-dimensional invariant subspaces because of the second relation of (3.2.8). It is the collection of vectors in each Jordan chain that generates the correct invariant subspaces. Thus, if $S_1, \ldots, S_p$ are the subspaces spanned by the chains (3.2.9), then, clearly, each S_i is an invariant subspace and

$$C^n = S_1 \dot{+} \cdots \dot{+} S_p \tag{3.2.13}$$

is a direct sum decomposition of C^n into the largest possible number of invariant subspaces of A. We summarize this as follows:

3.2.7. Invariant Subspace Decomposition. *Let $A = PJP^{-1}$ be an $n \times n$ matrix, where J is the Jordan form* (3.2.5). *If $S_1, \ldots, S_p$ are the subspaces spanned by the vectors* (3.2.9) *of the p Jordan chains of A, then the S_i are invariant subspaces of A, and* (3.2.13) *holds. If A is real with only real eigenvalues, then all eigenvectors and generalized eigenvectors of A may be taken to be real, and* (3.2.13) *holds with C^n replaced by R^n.*

We next consider some algebraic consequences of the Jordan form. In Chapter 1 we noted that if a matrix A has eigenvalues λ_i, $i = 1, \ldots, n$, and $p(A)$ is a polynomial in A, then $p(\lambda_i)$, $i = 1, \ldots, n$, are eigenvalues of $p(A)$. We can now show that these are, in fact, all of the eigenvalues of $p(A)$.

3.2.8. *If A is an $n \times n$ matrix with eigenvalues $\lambda_1, \ldots, \lambda_n$ and p is any polynomial, then the n eigenvalues of $p(A)$ are $p(\lambda_i)$, $i = 1, \ldots, n$.*

Proof. If $A = PJP^{-1}$, where $J = \text{diag}(J_1, \ldots, J_p)$ is the Jordan form, then $A^m = (PJP^{-1})^m = PJ^mP^{-1}$, so that

$$p(A) = Pp(J)P^{-1} = P\,\text{diag}(p(J_1), \ldots, p(J_p))P^{-1} \tag{3.2.14}$$

Since $p(J)$ is triangular, the eigenvalues of $p(A)$ are the diagonal elements of $p(J)$, which are $p(\lambda_i)$, $i = 1, \ldots, n$. □

Note that (3.2.14) shows that $p(A)$ has at least as many linearly independent eigenvectors as A itself, since there will be at least one eigenvector associated with each block $p(J_i)$. However, $p(A)$ may have more eigenvectors and a different Jordan form. For example, the Jordan form of

$$A = \begin{bmatrix} 0 & 1 \\ 0 & 0 \end{bmatrix}$$

is A itself, but A^2 has a diagonal Jordan form.

We next let $p(\lambda) = (\lambda - \lambda_1) \cdots (\lambda - \lambda_n)$ be the characteristic polynomial of A and consider the corresponding matrix polynomial

$$p(A) = (A - \lambda_1 I)(A - \lambda_2 I) \cdots (A - \lambda_n I) \tag{3.2.15}$$

Suppose first that the Jordan form is diagonal. Then, since $A = PDP^{-1}$,

$$\begin{aligned} p(A) = P\, &\mathrm{diag}(0, \lambda_2 - \lambda_1, \ldots, \lambda_n - \lambda_1) \\ &\times \mathrm{diag}(\lambda_1 - \lambda_2, 0, \lambda_3 - \lambda_2, \ldots, \lambda_n - \lambda_2) \cdots P^{-1} \end{aligned} \tag{3.2.16}$$

Clearly, the product of these diagonal matrices is zero, so that $p(A) = 0$. If the Jordan form is not diagonal, then, corresponding to (3.2.16), we have

$$\begin{aligned} p(A) = P\, &\mathrm{diag}((J_1 - \lambda_1 I)^{r_1}, (J_2 - \lambda_1 I)^{r_1}, \ldots) \\ &\times \mathrm{diag}((J_1 - \lambda_2 I)^{r_2}, (J_2 - \lambda_2 I)^{r_2}, \ldots) \cdots P^{-1} \end{aligned} \tag{3.2.17}$$

Now

$$(J_i - \lambda_i I)^{r_i} = \begin{bmatrix} 0 & 1 & & \\ & \ddots & \ddots & \\ & & \ddots & 1 \\ & & & 0 \end{bmatrix}^{r_i} = 0$$

since J_i is $r_i \times r_i$. Therefore, each of the block diagonal matrices of (3.2.17) has at least one zero block in such a fashion that again the product is zero. Thus, we have proved the following result.

3.2.9. CAYLEY–HAMILTON THEOREM. *If p is the characteristic polynomial of an $n \times n$ matrix A, then $p(A) = 0$.*

An $n \times n$ matrix may well satisfy a polynomial of degree less than n. For example, the identity matrix satisfies the polynomial $\lambda - 1$ of degree 1, since $I - I = 0$. The *minimal polynomial* of an $n \times n$ matrix A is the monic (leading coefficient equal to one) polynomial p of least degree such that $p(A) = 0$. From the Jordan form we can easily ascertain the minimal polynomial. Before doing this in general, consider again the matrices of Figure 3.1. The minimal polynomials are (a) $p(\lambda) = \lambda - 2$, (b) $p(\lambda) = (\lambda - 2)^2$, (c) $p(\lambda) = (\lambda - 2)^2$, (d) $p(\lambda) = (\lambda - 2)^3$, (e) $p(\lambda) = (\lambda - 2)^4$. For example, in case (d)

$$(A - 2I)^3 = \begin{bmatrix} 0 & & & \\ & 0 & 1 & 0 \\ & & 0 & 1 \\ & & & 0 \end{bmatrix}^3 = 0$$

and it is easy to see that A cannot satisfy any polynomial of degree 2. It is left to Exercise 3.2-12 to verify the other cases. More generally, let A have the Jordan form (3.2.5), and let $\lambda_1, \ldots, \lambda_q$ be the distinct eigenvalues of A. Let s_i be the dimension of the largest Jordan block associated with λ_i. Then $(J_k - \lambda_i I)^{s_i} = 0$ for all Jordan blocks J_k associated with λ_i, and by an argument exactly analogous to that which led to the Cayley-Hamilton theorem we can conclude that $p(A) = 0$, where

$$p(\lambda) = (\lambda - \lambda_1)^{s_1} \cdots (\lambda - \lambda_q)^{s_q} \tag{3.2.18}$$

We summarize this as follows:

3.2.10. MINIMAL POLYNOMIAL. *Let A be an $n \times n$ matrix with Jordan form* (3.2.5). *Let $\lambda_1, \ldots, \lambda_q$ be the distinct eigenvalues of A, and let s_i be the dimension of the largest Jordan block associated with λ_i. Then the minimal polynomial of A has degree $s_1 + \cdots + s_q$ and is given by* (3.2.18).

Although the Jordan form of a matrix A determines the minimal polynomial of A, the converse is not, in general, true. Again, the examples of Figure 3.1 are instructive. In case (a) the minimal polynomial is $p(\lambda) = \lambda - 2$, which tells us that the Jordan form must be diagonal. Similarly, in cases (d) and (e) the minimal polynomial tells us what the Jordan form must be (Exercise 3.2-13). However, in both cases (b) and (c) the minimal polynomial is $(\lambda - 2)^2$ and does not differentiate between the two different Jordan forms. More information is needed, and this is the following.

3.2.11. DEFINITION: *Elementary Divisors.* Let A be an $n \times n$ matrix with Jordan form (3.2.5). Then the *elementary divisors* of A are the characteristic polynomials of the Jordan blocks of A.

As an example we list the elementary divisors and minimal polynomial of the matrices of Figure 3.1:

(a) elementary divisors: $\lambda - 2, \lambda - 2, \lambda - 2, \lambda - 2$
  minimal polynomial: $\lambda - 2$
(b) elementary divisors: $(\lambda - 2)^2, (\lambda - 2)^2$
  minimal polynomial: $(\lambda - 2)^2$
(c) elementary divisors: $(\lambda - 2)^2, \lambda - 2, \lambda - 2$
  minimal polynomial: $(\lambda - 2)^2$
(d) elementary divisors: $(\lambda - 2)^3, \lambda - 2$
  minimal polynomial: $(\lambda - 2)^3$
(e) elementary divisors: $(\lambda - 2)^4$
  minimal polynomial: $(\lambda - 2)^4$

In cases (b) and (c), where the minimal polynomials are the same, the elementary divisors differentiate between the two different canonical forms.

We note that there is a natural relationship between the elementary divisors and the smallest invariant subspaces of a matrix; namely, the degree of each elementary divisor is the dimension of the corresponding invariant subspace. In particular, if all elementary divisors are linear, as in case (a) above, then the Jordan form is diagonal.

We end this section by noting that the 1's in the superdiagonal positions in the Jordan form are important only insofar as they denote nonzero elements. Many times it is useful to have other nonzero quantities in those positions, and the basis for doing this is the following lemma.

3.2.12. Let A be an $n \times n$ matrix with Jordan canonical form J. Then, for an arbitrary scalar $\alpha \neq 0$, A is similar to a matrix that is identical to its Jordan form, except that each 1 *in a superdiagonal position is replaced by α.*

Proof. Let $A = PJP^{-1}$, and let $D = \text{diag}(1, \alpha, \ldots, \alpha^{n-1})$. Then it is easy to see that $\hat{J} = D^{-1}JD$ is exactly the same as J, except that each 1 in a superdiagonal position has been replaced by α. Thus, if $Q = PD$, then $A = Q\hat{J}Q^{-1}$. □

Summary

The main concepts and results of this section are

- An $n \times n$ matrix with n distinct eigenvalues is similar to a diagonal matrix.
- An $n \times n$ matrix is similar to a diagonal matrix if and only if it has n linearly independent eigenvectors.
- Every $n \times n$ matrix is similar to a matrix that is diagonal except for possible 1's in certain superdiagonal positions (the Jordan canonical form).

- Each Jordan block corresponds to a chain of eigenvectors and generalized eigenvectors that span an invariant subspace. C^n may be written as a direct sum of these invariant subspaces.
- Every $n \times n$ matrix satisfies its characteristic equation (Cayley-Hamilton theorem). The monic polynomial of least degree that a matrix satisfies is the minimal polynomial, and its degree is the sum of the dimensions of the largest Jordan blocks associated with each distinct eigenvalue.
- The elementary divisors of an $n \times n$ matrix are the characteristic polynomials of the Jordan blocks.

Exercises 3.2

1. Use 3.2.1 and the results of Section 2.2 to prove 3.2.2.
2. Show that all eigenvectors of the matrix of (3.2.4) are multiples of e_1.
3. Show that the Jordan canonical form J of (3.2.5) can have no more than p linearly independent eigenvectors. Conclude from this that $A = PJP^{-1}$ has only p linearly independent eigenvectors.
4. Find the number of linearly independent eigenvectors for each of the following matrices:

(a) $$\begin{bmatrix} 1 & 1 & & \\ & 1 & & \\ & & 2 & 1 \\ & & & 2 \end{bmatrix}$$

(b) $$\begin{bmatrix} 1 & 1 & & \\ & 1 & & \\ & & 2 & \\ & & & 2 \end{bmatrix}$$

(c) $$\begin{bmatrix} 1 & & & \\ & 1 & & \\ & & 2 & \\ & & & 2 \end{bmatrix}$$

5. Ascertain if the following matrices are similar to a diagonal matrix:

(a) $\begin{bmatrix} 2 & 1 \\ 1 & 2 \end{bmatrix}$, (b) $\begin{bmatrix} 2 & 1 \\ 0 & 2 \end{bmatrix}$,

(c) $\begin{bmatrix} 1 & 2 \\ 3 & 4 \end{bmatrix}$, (d) $\begin{bmatrix} 0 & 1 \\ -1 & 2 \end{bmatrix}$

6. Determine the Jordan canonical form and the eigenvectors and generalized eigenvectors of the matrices of Exercise 5.
7. Let A and B be two similar $n \times n$ matrices and assume that A is similar to a diagonal matrix. Show that B is also similar to a diagonal matrix.
8. Let A be a 5×5 matrix with an eigenvalue $\lambda = 3$ of multiplicity 5. Give all possible Jordan canonical forms of A up to permutations of the Jordan blocks. Do the same if $\lambda = 3$ is an eigenvalue of multiplicity 2 and $\lambda = 1$ is of multiplicity 3.
9. Find the Jordan canonical form of the matrix

$$\begin{bmatrix} 1 & 1 & 1 & & & \\ & 2 & 2 & & & \\ & & 3 & & & \\ & & & 1 & 1 & 1 \\ & & & & 2 & 2 \\ & & & & & 3 \end{bmatrix}$$

10. Let $A = \mathbf{uv}^*$ be a rank 1 matrix, where $\mathbf{u}$ and $\mathbf{v}$ are column vectors. Show that $\mathbf{v}^*\mathbf{u}$ is an eigenvalue and 0 is an eigenvalue of multiplicity $n - 1$ (or of multiplicity n if $\mathbf{v}^*\mathbf{u} = 0$). Find the eigenvectors and Jordan canonical form of A.
11. Compute eigenvectors and generalized eigenvectors for each matrix of Figure 3.1. Use these to exhibit the Jordan chains for each matrix.
12. Verify that the minimal polynomials of the matrices of Figure 3.1 are (a) $\lambda - 2$, (b) $(\lambda - 2)^2$, (c) $(\lambda - 2)^2$, (d) $(\lambda - 2)^3$, (e) $(\lambda - 2)^4$.
13. Suppose that a 4×4 matrix with eigenvalue 2 of multiplicity 4 has minimal polynomial $\lambda - 2$. Conclude that the Jordan form must be diagonal. Similarly, if the minimal polynomial is $(\lambda - 2)^3$ or $(\lambda - 2)^4$, conclude that the Jordan form must be (d) or (e) of Figure 3.1, respectively; but if the minimal polynomial is $(\lambda - 2)^2$, the Jordan form can be either (b) or (c) of Figure 3.1.
14. An $n \times n$ matrix A is *nilpotent* if $A^p = 0$ for some integer p. Show that a strictly upper or lower triangular matrix (the main diagonal elements are zero) is nilpotent. Show also that a matrix is nilpotent if and only if all its eigenvalues are zero.
15. Let A be an $n \times n$ nonsingular matrix. Use the Cayley-Hamilton theorem to express A^{-1} as polynomial in A.
16. Let A be an $n \times n$ upper triangular matrix of the form

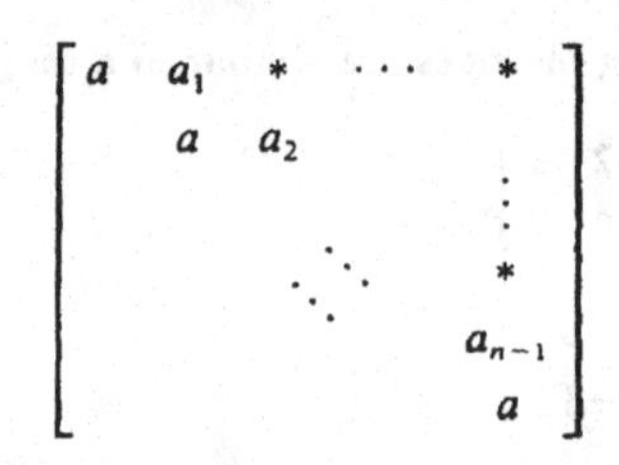

where $a_i \neq 0$, $i = 1, \ldots, n-1$. Show that A has only one linearly independent eigenvector, and, hence, its Jordan form has only a single Jordan block.

17. Suppose that A is a nonsingular matrix with a single Jordan block in its Jordan canonical form. Show that A^{-1} has one linearly independent eigenvector, and hence, its Jordan form also has only a single block.
18. Use the result of Exercise 17 to show that if A is nonsingular, then A^{-1} has the same block structure in its Jordan canonical form as does A.
19. Let A be a triangular matrix. Show that there is a diagonal matrix D so that the off-diagonal elements of $D^{-1}AD$ are bounded by $\varepsilon > 0$ in absolute value.
20. Let $\mathbf{u}$ and $\mathbf{v}$ be real column vectors such that $\beta = \mathbf{u}^T\mathbf{v} + 1 \neq 0$. Then by the Sherman-Morrison formula (Exercise 1.3-14), $(I + \mathbf{u}\mathbf{v}^T)^{-1} = I - \beta^{-1}\mathbf{u}\mathbf{v}^T$. If $A = (I + \mathbf{u}\mathbf{v}^T)^{-1}D(I + \mathbf{u}\mathbf{v}^T)$, where D is diagonal, show how integer matrices A with prescribed integer eigenvalues and known integer eigenvectors may be generated.

3.3. Equivalence Transformations

We turn now to seeking the simplest form that a given $m \times n$ matrix A can take under an equivalence transformation

$$B = PAQ \tag{3.3.1}$$

where P and Q are nonsingular $m \times m$ and $n \times n$ matrices, respectively. The following theorem gives a complete answer to this question.

3.3.1. If A is an $m \times n$ real or complex matrix of rank r, then there is a nonsingular $m \times m$ matrix P and a nonsingular $n \times n$ matrix Q such that

$$PAQ = \begin{bmatrix} I_r & 0 \\ 0 & 0 \end{bmatrix} \tag{3.3.2}$$

where I_r is the $r \times r$ identity matrix.

Proof. Since $\operatorname{rank}(A) = r$, A has r linearly independent rows, and there is an $m \times m$ permutation matrix P_1 so that the first r rows of P_1A are linearly independent. If $\hat{A}$ is the $r \times n$ matrix with these rows, then $\operatorname{rank}(\hat{A}) = r$, and $\hat{A}$ has r linearly independent columns. Therefore, there is an $n \times n$ permutation matrix Q_1 such that the first r columns of $\hat{A}Q_1$ are linearly independent. Thus, if

$$A_1 = P_1AQ_1 = \begin{bmatrix} A_{11} & A_{12} \\ A_{21} & A_{22} \end{bmatrix}$$

then the $r \times r$ submatrix A_{11} is nonsingular. We can then do the calculations

$$A_2 = LA_1 = \begin{bmatrix} A_{11}^{-1} & 0 \\ -A_{21}A_{11}^{-1} & I_{n-r} \end{bmatrix} A_1 = \begin{bmatrix} I_r & A_{11}^{-1}A_{12} \\ 0 & A_{22} - A_{21}A_{11}^{-1}A_{12} \end{bmatrix}$$

and

$$A_3 = A_2 U = A_2 \begin{bmatrix} I_r & -A_{11}^{-1}A_{12} \\ 0 & I_{n-r} \end{bmatrix} = \begin{bmatrix} I_r & 0 \\ 0 & C \end{bmatrix}$$

Since the matrices P_1, Q_1, L, and U are all nonsingular, Theorem 2.3.2 shows that $\text{rank}(A_3) = r$. Hence $C = 0$, for otherwise we would have $\text{rank}(A_3) > r$. Therefore, (3.3.2) holds with $P = LP_1$ and $Q = Q_1U$. □

We next delineate some important special cases of 3.3.1. Recall that matrices A and B are said to be equivalent if (3.3.1) holds

3.3.2. Let A be an $m \times n$ matrix.

(a) *If $\text{rank}(A) = m < n$, then A is equivalent to a matrix of the form $[I_m \ \ 0]$, where I_m is the $m \times m$ identity matrix.*

(b) *If $\text{rank}(A) = n < m$, then A is equivalent to a matrix of the form*

$$\begin{bmatrix} I_n \\ 0 \end{bmatrix}$$

(c) *If $m = n$ and A is nonsingular, then A is equivalent to the identity matrix.*

As another corollary, we restate 3.3.1 in terms of matrix representations of a linear operator.

3.3.3. Let $\mathbf{A}: R \to S$ be a linear operator, where $\dim R = n$ and $\dim S = m$, and assume that $\text{rank}(\mathbf{A}) = r$. Then there are bases for R and S for which the matrix representation of $\mathbf{A}$ in these bases has the form of the right-hand side of (3.3.2).

Proof. Let $\mathbf{r}_1, \ldots, \mathbf{r}_n$ and $\mathbf{s}_1, \ldots, \mathbf{s}_m$ be any bases for R and S, and let A be the matrix representation of $\mathbf{A}$ in these bases. Let P and Q be the matrices of (3.3.2), and let $T = P^{-1}$. If we define new bases by

$$\hat{\mathbf{r}}_j = \sum_{i=1}^{n} q_{ij}\mathbf{r}_i, \quad j = 1, \ldots, n, \qquad \hat{\mathbf{s}}_j = \sum_{i=1}^{m} t_{ij}\mathbf{s}_i, \quad j = 1, \ldots, m$$

then Theorem 2.2.2 shows that the matrix representation of **A** in these new bases is given by

$$\hat{A} = T^{-1}AQ = PAQ = \begin{bmatrix} I_r & 0 \\ 0 & 0 \end{bmatrix} \qquad \square$$

The basic results 3.3.1 and 3.3.3 give complete answers to the questions, What is the simplest form a matrix can take under an equivalence transformation? or, equivalently, What is the simplest matrix representation of a linear operator? The answer in both cases is the identity matrix, or as close to the identity matrix as the rank and dimensions of the matrix will allow. Note that we are able to achieve a simpler canonical form under equivalence than under similarity because we are allowed to change bases in two spaces rather than just one.

Congruences and the Inertia Theorem

We next consider special types of equivalence transformations. The first is when $P = Q^*$, so that

$$\hat{A} = Q^*AQ \tag{3.3.3}$$

where, again, Q is nonsingular. The relation (3.3.3) is called a *congruence* or *congruence transformation* of A, and A and $\hat{A}$ are said to be *congruent*. As we will see in the next chapter, congruence transformations arise naturally in dealing with quadratic forms, in which A is assumed to be real and symmetric, and we will assume now that A is Hermitian. Then the congruent matrix (3.3.3) is also Hermitian.

If Q were a unitary matrix, then (3.3.3) would be a similarity transformation, and $\hat{A}$ would have the same eigenvalues as A. If Q is not unitary, then this is no longer the case, but the following remarkable theorem shows that the *signs* of the eigenvalues of a Hermitian matrix are preserved under any congruence. We first give the following definition.

3.3.4. DEFINITION: *Inertia.* Let A be an $n \times n$ Hermitian matrix. Then the *inertia* of A, denoted by $\text{In}(A)$, is (i, j, k), where i, j, and k are the numbers of positive, negative, and zero eigenvalues of A, respectively.

For example, if $A = \text{diag}(3, -1, 0)$, then $\text{In}(A) = (1, 1, 1)$, and if $A = \text{diag}(1, 2, 3)$, then $\text{In}(A) = (3, 0, 0)$.

3.3.5. INERTIA THEOREM. *Let A be an $n \times n$ Hermitian matrix and Q a nonsingular $n \times n$ matrix. Then*

$$In(A) = In(Q^*AQ) \tag{3.3.4}$$

Proof. Let $\hat{A} = Q^*AQ$. Then $\hat{A}$ is also Hermitian, and by Theorem 3.1.1 there are unitary matrices U and $\hat{U}$ so that, with $\mathbf{y} = U^*\mathbf{x}$, $\mathbf{z} = Q^{-1}\mathbf{x}$, and $\mathbf{w} = \hat{U}^*\mathbf{z}$ we have

$$\mathbf{x}^*A\mathbf{x} = \sum_{s=1}^{n} \lambda_s y_s^2, \qquad \mathbf{x}^*A\mathbf{x} = \mathbf{z}^*\hat{A}\mathbf{z} = \sum_{s=1}^{n} \hat{\lambda}_s w_s^2 \tag{3.3.5}$$

where the λ_s and $\hat{\lambda}_s$ are the eigenvalues of A and $\hat{A}$, respectively. We may assume that the eigenvalues of A are ordered so that

$$\begin{gathered}\lambda_1 \geqslant \cdots \geqslant \lambda_i > 0, \quad \lambda_{i+1} \leqslant \cdots \leqslant \lambda_{i+j} < 0, \\ \lambda_{i+j+1} = \cdots = \lambda_{i+j+k} = 0\end{gathered} \tag{3.3.6}$$

Thus, $\text{In}(A) = (i, j, k)$, and similarly for $\hat{A}$ with $\text{In}(\hat{A}) = (\hat{i}, \hat{j}, \hat{k})$. Now suppose that $i \neq \hat{i}$, say $\hat{i} > i$. Let $\mathbf{u}_1, \ldots, \mathbf{u}_i$ be the first i columns of U, let $\mathbf{p}_{\hat{i}+1}, \ldots, \mathbf{p}_n$ be the indicated rows of $P = \hat{U}^*Q^{-1}$, and let B be the matrix whose first i rows are $\mathbf{u}_1^*, \ldots, \mathbf{u}_i^*$, and whose remaining rows are $\mathbf{p}_{\hat{i}+1}, \ldots, \mathbf{p}_n$. Then B has $i + n - \hat{i} < n$ rows and n columns, and therefore the system of equations $B\mathbf{x} = 0$ has a nonzero solution $\hat{\mathbf{x}}$. For this solution, by (3.3.5),

$$\hat{\mathbf{x}}^*A\hat{\mathbf{x}} = \sum_{s=i+1}^{n} \lambda_s y_s^2 < 0$$

since, by construction of $\hat{\mathbf{x}}$, $y_s = \mathbf{u}_s^*\hat{\mathbf{x}} = 0$, $s = 1, \ldots, i$. Similarly, $w_s = 0$, $s = \hat{i} + 1, \ldots, n$, and therefore

$$\hat{\mathbf{x}}^*A\hat{\mathbf{x}} = \sum_{s=1}^{i} \hat{\lambda}_s w_s^2 > 0$$

This is a contradiction, and thus $i = \hat{i}$. It can be shown similarly that $j = \hat{j}$ and, therefore, $\text{In}(A) = \text{In}(\hat{A})$. □

The *signature*, sig(A), of an $n \times n$ Hermitian matrix is defined as $i - j$, the difference of the numbers of positive and negative eigenvalues. Since the rank of a Hermitian matrix is the number of nonzero eigenvalues, and rank is preserved by an equivalence transformation, an equivalent way of stating 3.3.5 is that sig(A) and rank(A) are preserved under a congruence transformation.

We note that a trivial special case of 3.3.5 is that if A is positive definite, then Q^*AQ is also, which is easily proved directly (Exercise 3.3-4).

We turn now to the canonical form under congruence. Again, let U be the unitary matrix of 3.1.1 such that

$$U^*AU = \text{diag}(\lambda_1, \ldots, \lambda_n)$$

where we assume that the eigenvalues of A are again ordered according to (3.3.6). Let

$$D = \operatorname{diag}(\lambda_1^{-1/2}, \ldots, \lambda_i^{-1/2}, (-\lambda_{i+1})^{-1/2}, \ldots, (-\lambda_{i+j})^{-1/2}, 1, \ldots, 1)$$

Then, with $Q = UD$,

$$Q^*AQ = \operatorname{diag}(1, \ldots, 1, -1, \ldots, -1, 0, \ldots, 0) \tag{3.3.7}$$

and we take this as the canonical form of a Hermitian matrix A under congruence.

3.3.6. CANONICAL FORM UNDER CONGRUENCE. *Let A be an $n \times n$ Hermitian matrix. Then there is a nonsingular matrix Q such that* (3.3.7) *holds, where i and j are the numbers of positive and negative eigenvalues of A, respectively. If A is real, then Q may be taken to be real.*

We will see the geometric interpretation of this canonical form in the next chapter.

The Singular Value Decomposition

We consider one more special type of equivalence transformation and restrict the matrices P and Q to be orthogonal or unitary. Recall that under a unitary similarity transformation, the simplest form we can achieve, in general, is a triangular matrix (Schur's theorem, 3.1.11). Under a unitary equivalence transformation, we will show that we can achieve a diagonal matrix but not the identity matrix as with a general equivalence.

Recall that for any matrix A, A^*A is Hermitian and positive semidefinite and thus has nonnegative eigenvalues. The nonnegative square roots of the eigenvalues of A^*A are called the *singular values* of A. By 2.3.4, $\operatorname{rank}(A^*A) = \operatorname{rank}(A)$, and thus the number of positive singular values is equal to $\operatorname{rank}(A)$. If A itself is Hermitian, then $A^*A = A^2$, and the singular values are the absolute values of the eigenvalues of A.

3.3.7. SINGULAR VALUE DECOMPOSITION. *Let A be an $m \times n$ matrix of rank r. Then there is an $m \times m$ unitary matrix P and an $n \times n$ unitary matrix Q such that*

$$A = PDQ \tag{3.3.8}$$

where D is $m \times n$ and zero except for the diagonal entries $\sigma_1, \ldots, \sigma_r$ which are the positive singular values of A. In the case that $m = n$, $D = \operatorname{diag}(\sigma_1, \ldots, \sigma_r, 0, \ldots, 0)$.

Proof. Let $\lambda_1, \ldots, \lambda_n$ and $\mathbf{x}_1, \ldots, \mathbf{x}_n$ be the eigenvalues and corresponding orthonormal eigenvectors of A^*A, and set $\sigma_i = \lambda_i^{1/2}$, $i = 1, \ldots, n$. We may assume that the λ_i are ordered so that $\sigma_i > 0$, $i = 1, \ldots, r$ and $\sigma_i = 0$, $i = r+1, \ldots, n$. Let $\mathbf{y}_i = \sigma_i^{-1} A\mathbf{x}_i$, $i = 1, \ldots, r$. Then

$$\mathbf{y}_i^*\mathbf{y}_j = \frac{1}{\sigma_i\sigma_j}\mathbf{x}_i^* A^* A \mathbf{x}_j = \frac{\lambda_j}{\sigma_i\sigma_j}\mathbf{x}_i^*\mathbf{x}_j = \delta_{ij}, \qquad i, j = 1, \ldots, r$$

where δ_{ij} is the Kronecker δ. By 2.4.4 we can extend $\mathbf{y}_1, \ldots, \mathbf{y}_r$ to an orthonormal basis $\mathbf{y}_1, \ldots, \mathbf{y}_m$ for C^m, and we then set

$$P = (\mathbf{y}_1, \ldots, \mathbf{y}_m), \qquad Q^* = (\mathbf{x}_1, \ldots, \mathbf{x}_n)$$

The matrices P and Q are unitary, and

$$\mathbf{y}_i^* A\mathbf{x}_j = \sigma_i^{-1}\mathbf{x}_i^* A^* A\mathbf{x}_j = \sigma_i^{-1}\lambda_j \mathbf{x}_i^*\mathbf{x}_j = \sigma_i^{-1}\lambda_j\delta_{ij}, \qquad i = 1, \ldots, r, j = 1, \ldots, n$$

Since $\mathbf{x}_j^* A^* A\mathbf{x}_j = 0$ for $j > r$, we have that $A\mathbf{x}_j = 0$ for $j > r$, and, hence,

$$\mathbf{y}_i^* A\mathbf{x}_j = \begin{cases} 0, & i > r, j > r, \\ \sigma_j \mathbf{y}_i^*\mathbf{y}_j = 0, & i > r, j \leq r \end{cases}$$

Therefore, we have shown that $P^*AQ^* = (\mathbf{y}_i^* A\mathbf{x}_j) \equiv D$, where D is zero except in the first r diagonal positions, which contain the positive singular values of A. Thus, (3.3.8) follows. □

A simple example of a singular value decomposition is

$$A = \begin{bmatrix} 1 & 0 & -1 \\ -1 & 0 & 1 \end{bmatrix}, \qquad P = \frac{1}{\sqrt{2}}\begin{bmatrix} 1 & 1 \\ -1 & 1 \end{bmatrix}, \qquad D = \begin{bmatrix} 2 & 0 & 0 \\ 0 & 0 & 0 \end{bmatrix},$$

$$Q = \frac{1}{\sqrt{2}}\begin{bmatrix} 1 & 0 & -1 \\ 0 & \sqrt{2} & 0 \\ -1 & 0 & -1 \end{bmatrix}$$

Here the singular values are 2 and 0.

The proof of 3.3.7 showed that the matrix Q^* is a unitary matrix that puts A^*A into diagonal form. Similarly, from (3.3.8) we have that

$$AA^* = PDQQ^*D^*P^* = PDD^*P^* \tag{3.3.9}$$

and DD^* is the $m \times m$ diagonal matrix whose nonzero diagonal entries are $\sigma_i^2 = \lambda_i$, $i = 1, \ldots, r$. Thus, P is a unitary matrix that puts AA^* in diagonal form. This also shows the following, which can be proved directly (see Exercise 3.3-5 and Theorem 6.4.4).

3.3.8. If A is an $m \times n$ matrix, then A^*A and AA^* have the same nonzero eigenvalues.

Equation (3.3.9) also suggests how unique the singular value decomposition is. Suppose that (3.3.8) holds for unitary matrices P and Q and any $m \times n$ matrix D that is zero except for its main diagonal entries. Then (3.3.9) shows that the diagonal matrix DD^* must have the eigenvalues λ_i of AA^* as its diagonal entries; hence, D can only be a matrix whose diagonal entries are $\pm\lambda_i^{1/2}$, and the singular value decomposition makes the choice of the positive square roots. The matrices P and Q are also unique up to different possible bases for the eigenspaces of AA^* and A^*A. That is, suppose first that $m = n$ and A^*A has n distinct eigenvalues. Then P and Q are unique up to the signs of their rows (or columns). Otherwise, P and Q are unique up to the choice of an orthonormal basis for each eigenspace corresponding to a multiple eigenvalue.

We note that the columns of P are sometimes called the *left singular vectors* of A, and the columns of Q^* are called the *right singular vectors*. We also note that any unitary equivalence transformation of a matrix A preserves the singular values of A, because if $\hat{A} = PAQ$, where P and Q are unitary, then $\hat{A}^*\hat{A}$ and A^*A are similar.

Summary

We have considered in this chapter a variety of matrix transformations. We summarize these, together with properties of the matrix that are preserved under the transformations and the different canonical forms, in the following table:

Transformation	Matrix A	Preserves	Canonical form
Equivalence: PAQ	$m \times n$	Rank	$\begin{bmatrix} I_r & 0 \\ 0 & 0 \end{bmatrix}$
Congruence: Q^*AQ	$n \times n$, Hermitian	Inertia	$\text{diag}(1, \ldots, 1, -1, \ldots, -1, 0, \ldots, 0)$
Similarity: PAP^{-1}	$n \times n$	Eigenvalues	Jordan form
Unitary similarity	$n \times n$	$\|A\|_2$	Triangular
Unitary similarity	$n \times n$, normal	$\|A\|_2$	$\text{diag}(\lambda_1, \ldots, \lambda_n)$
Unitary equivalence	$m \times n$	Singular values	$\text{diag}(\sigma_1, \ldots, \sigma_r, 0, \ldots, 0)$

Exercises 3.3

1. Ascertain if the matrices

$$A = \begin{bmatrix} 1 & 1 & 1 \\ 2 & 2 & 2 \end{bmatrix}, \qquad B = \begin{bmatrix} 1 & 2 & 3 \\ 1 & 2 & 3 \end{bmatrix}$$

are equivalent, and find the canonical form (3.3.2) for each of the matrices.

2. Find the canonical form (3.3.7) for each of the following matrices:

(a) $\begin{bmatrix} 2 & 1 \\ 1 & 2 \end{bmatrix}$, (b) $\begin{bmatrix} 1 & 1 \\ 1 & 1 \end{bmatrix}$, (c) $\begin{bmatrix} 0 & 1 \\ 1 & 0 \end{bmatrix}$

(d) $\begin{bmatrix} -1 & 1-i \\ 1+i & -1 \end{bmatrix}$, (e) $\begin{bmatrix} -2 & 1-i \\ 1+i & -2 \end{bmatrix}$

3. Prove the following extension of 3.3.5. Let Q be an $n \times m$ matrix of rank n, and let $\text{In}(A) = (i, j, k)$ and $\text{In}(Q^*AQ) = (i', j', k')$. Show that $i' = i$, $j' = j$.
4. Let A be an $n \times n$ Hermitian positive semidefinite matrix and Q any $n \times n$ matrix. Show that Q^*AQ is also positive semidefinite, and that it is positive definite if and only if A and Q are nonsingular.
5. Prove 3.3.8 directly without using 3.3.7.
6. Let A be $n \times n$, Hermitian, and positive semidefinite. Show that the singular value decomposition of A is PDP^*, where P is unitary and D is diagonal, and thus is the same as the unitary similarity transformation of 3.1.1.
7. Let A be $n \times n$ Hermitian and indefinite. Find the singular value decomposition of A in terms of the eigenvalues and eigenvectors of A.
8. Find the singular value decompositions of

(a) $\begin{bmatrix} 1 & 1 \\ 0 & 1 \end{bmatrix}$ (b) $\begin{bmatrix} 1 & 1 \\ 0 & 0 \end{bmatrix}$

9. Let A be an $m \times n$ matrix, and let P and Q be $m \times m$ and $n \times n$ unitary matrices. Show that A and PAQ have the same singular values.

Review Questions—Chapter 3

Answer whether the following statements are true or false and justify your assertions. The first eight statements pertain to an $n \times n$ Hermitian matrix A.

1. A has n orthogonal eigenvectors if and only if all n eigenvalues of A are distinct.
2. A is always unitarily similar to a diagonal matrix.
3. If $A = PDP^*$, where D is diagonal and P is unitary, then the rows of P are eigenvectors of A.
4. The spectral representation of A may be viewed as a linear combination of rank 1 orthogonal projectors.
5. Eigenvectors of A corresponding to distinct eigenvalues are necessarily orthogonal.
6. If λ is an eigenvalue of A of multiplicity m, then there is an $(m+1)$-dimensional subspace of eigenvectors associated with λ.
7. C^n may be written as a direct sum of orthogonal invariant subpaces of A.
8. A is positive semidefinite if and only if all its eigenvalues except one are positive.
9. An $n \times n$ matrix A is unitarily similar to a triangular matrix if and only if A is nonsingular.
10. An $n \times n$ matrix A is unitarily similar to a diagonal matrix if and only if it is normal.
11. An orthogonal matrix is always unitarily similar to a real diagonal matrix.
12. An $n \times n$ matrix A is similar to a diagonal matrix if and only if A has n distinct eigenvalues.
13. An $n \times n$ matrix A with an eigenvalue λ of multiplicity m always has an m-dimensional invariant subspace associated with λ.
14. If the Jordan form of A has p blocks, then A has exactly p linearly independent eigenvectors.
15. If A is a 4×4 matrix with eigenvalues 1 and 2, each of multiplicity 2, then the Jordan form of A has two blocks.
16. A Jordan chain always consists of eigenvectors of A.
17. If the Jordan form of A has p blocks, then A has $p + 1$ Jordan chains.
18. For any $n \times n$ matrix A, C^n can be written as a direct sum of n invariant subspaces of A.
19. Any $n \times n$ matrix always satisfies a polynomial of degree n.
20. The degree of the minimal polynomial of A determines the Jordan form of A.
21. Any $n \times n$ matrix is similar to a matrix of the form $D + E$, where D is diagonal and $\|E\|_\infty < 10^{-6}$.

22. If A is an $m \times n$ matrix with $m \leq n$, then A is equivalent to $(I, 0)$, where I is the $m \times m$ identity matrix.
23. If A is Hermitian and P is nonsingular, then P^*AP has the same eigenvalues as A.
24. If one knows the inertia of a Hermitian matrix A, then one also knows whether A is singular.
25. The canonical form under congruence of a Hermitian matrix is the identity matrix.
26. If A is a Hermitian matrix, then its singular value decomposition is the same as its unitary similarity reduction to diagonal form.
27. If A is $n \times n$ with eigenvalues $\lambda_1, \ldots, \lambda_n$, then its singular values are $|\lambda_i|$, $i = 1, \ldots, n$.
28. If $A = PDQ$ is the singular value decomposition of an $n \times n$ nonsingular matrix A, then D is nonsingular.
29. If A is $m \times n$, then AA^* and A^*A have the same eigenvalues.
30. If A and B are $n \times n$ matrices that are unitarily equivalent, then they have the same eigenvalues.

References and Extensions: Chapter 3

1. The spectral theorem for self-adjoint linear operators on a Hilbert space is one of the fundamental results of functional analysis; see, for example, Dunford and Schwartz [1963].
2. If A is a normal matrix, then A and A^* can each be represented as a polynomial in terms of the other; that is, there are polynomials p and q such that $A = p(A^*)$, $A^* = q(A)$. For a proof, see Gantmacher [1959].
3. There are a number of different proofs of the Jordan canonical form theorem. Two of the classical approaches are based on invariant subspaces and elementary divisors; see, for example, Gantmacher [1959] for proofs along these lines. Recently, there have been a number of simpler proofs. We have given one due to the Russian mathematician Filippov in the 1960s, following Strang [1980]. Other proofs are given by Galperin and Waksman [1980] and Fletcher and Sorensen [1983]. Although these recent proofs are technically simpler, they are perhaps not as insightful as the older proofs based on invariant subspaces or elementary divisors.
4. Another approach to the proof of the Inertia Theorem, 3.3.5, is based on the continuity of the eigenvalues of a matrix as functions of its elements. Let A be Hermitian and P nonsingular. Let $P = QR$, where Q is unitary and R is upper triangular and nonsingular, and define

$$Q(t) = (1 - t)Q + tQR = Q[(1 - t)I + tR]$$

Then $Q(t)$ is nonsingular for all $0 \leq t \leq 1$ and $Q(0) = Q$, $Q(1) = P$. Let $A(t) = Q(t)^* A Q(t)$. Then $A(0)$ is similar to A and has the same eigenvalues. If A is nonsingular, then $A(t)$ is also nonsingular for $t \in [0, 1]$, and since the eigenvalues are continuous functions of t, none can be zero. Hence, all eigenvalues of $A(t)$ retain the same sign as $t \to 1$. If A is singular, we can set $\hat{A} = A + \varepsilon I$, and then let $\varepsilon \to 0$.

5. The problem of computing eigenvalues, eigenvectors, and associated canonical forms has been the subject of intensive research for the last 40 years. See Golub and van Loan [1983] for an excellent overall review, and Parlett [1980] for a detailed treatment of symmetric matrices.

4

Quadratic Forms and Optimization

We now begin several applications of the theory developed in the last two chapters. In the present chapter we consider first the geometry of the solution sets of quadratic equations in n variables. These solution sets generalize ellipses, parabolas, and hyperbolas in two variables, and a classification of their geometry in n dimensions is given by the inertia of the coefficient matrix of the quadratic form. In the next section we treat the unconstrained quadratic optimization problem and show that a necessary and sufficient condition for a unique solution is that the coefficient matrix be definite. We then consider the special constrained optimization problem of a quadratic function on the unit sphere and show that the maximum and minimum are just the largest and smallest eigenvalues of the coefficient matrix. This leads to the famous min-max representation of the eigenvalues of a Hermitian matrix. In Section 4.3 we specialize the minimization problem to the very important least squares problem and give the basic result on the existence and uniqueness of a solution. Particular examples are the linear regression and polynomial approximation problems. Then we treat the least squares problem in a more general way and obtain a minimum norm solution in the case that the original problem has infinitely many solutions. We show that this minimum norm solution can be represented in terms of a generalized inverse based on the singular value decomposition.

4.1. The Geometry of Quadratic Forms

A quadratic function of n real variables can be written as

$$f(\mathbf{x}) = \sum_{i,j=1}^{n} a_{ij}x_i x_j - 2\sum_{i=1}^{n} b_i x_i - c$$

for given real constants a_{ij}, b_j, and c, or in matrix form as

$$f(\mathbf{x}) = \mathbf{x}^T A\mathbf{x} - 2\mathbf{b}^T\mathbf{x} - c \tag{4.1.1}$$

The right-hand side of (4.1.1) is known as a *quadratic form*, and the study of its basic properties is one of the main topics of this chapter.

When the vectors $\mathbf{x}$ of (4.1.1) are real, as will usually be the case in dealing with quadratic forms, we may assume that the matrix A is symmetric, because of the following. Any real $n \times n$ matrix may be written as

$$A = A_s + A_{ss}, \qquad A_s = \tfrac{1}{2}(A + A^T), A_{ss} = \tfrac{1}{2}(A - A^T) \tag{4.1.2}$$

It is clear that A_s is symmetric and A_{ss} is skew symmetric ($A_{ss} = -A_{ss}^T$). Thus, any real $n \times n$ matrix may be written as a sum of symmetric and skew-symmetric matrices, which are called the symmetric and skew-symmetric parts of A. If A were complex, we would replace the transpose by conjugate transpose in (4.1.2) and conclude that A can be written as a sum of Hermitian and skew-Hermitian matrices.

Now, if S is any real skew-symmetric matrix, then for real vectors $\mathbf{x}$,

$$(\mathbf{x}^T S\mathbf{x})^T = \mathbf{x}^T S^T\mathbf{x} = -\mathbf{x}^T S\mathbf{x} \tag{4.1.3}$$

But $(\mathbf{x}^T S\mathbf{x})^T$ is just a real number and equal to its transpose; thus, (4.1.3) shows it must be zero. Hence, if A is as in (4.1.2), then for any real vector, $\mathbf{x}^T A\mathbf{x} = \mathbf{x}^T A_s\mathbf{x}$, and only the symmetric part of A contributes to the values of $\mathbf{x}^T A\mathbf{x}$. Thus, it is customary to assume at the outset that A is symmetric.

The first question we will address is this: What are the real solutions of the equation $f(\mathbf{x}) = 0$? For simplicity we will assume at first that $\mathbf{b} = 0$, so that the equation is

$$\mathbf{x}^T A\mathbf{x} = c \tag{4.1.4}$$

where A is a real symmetric $n \times n$ matrix and c is a given scalar. Let us treat first the case $n = 2$, so that (4.1.4) is

$$a_{11}x_1^2 + 2a_{12}x_1x_2 + a_{22}x_2^2 = c \tag{4.1.5}$$

which describes a so-called *conic section*. Recall from elementary analytic geometry that if we make the change of variables

$$x_1 = y_1 \cos\theta - y_2 \sin\theta, \qquad x_2 = y_1 \sin\theta + y_2 \cos\theta \tag{4.1.6}$$

then (4.1.5) becomes

$$\lambda_1 y_1^2 + \lambda_2 y_2^2 = c \tag{4.1.7}$$

provided that θ is chosen to satisfy

$$\tan 2\theta = \frac{2a_{12}}{a_{11} - a_{22}} \tag{4.1.8}$$

The solution of (4.1.7) is then easily ascertained to be an ellipse if λ_1, λ_2, and c all have the same sign, a hyperbola if $\lambda_1\lambda_2 < 0$, and a (degenerate) parabola if $\lambda_1\lambda_2 = 0$.

We now wish to rephrase this analytic geometry in matrix terms, again concentrating for the moment on Equation (4.1.5). The change of variables (4.1.6) is given by

$$\mathbf{x} = P\mathbf{y}, \qquad P = \begin{pmatrix} \cos\theta & -\sin\theta \\ \sin\theta & \cos\theta \end{pmatrix} \tag{4.1.9}$$

and then (4.1.5) becomes

$$\mathbf{y}^T P^T A P \mathbf{y} = \mathbf{y}^T D \mathbf{y}, \qquad D = \text{diag}(\lambda_1, \lambda_2) \tag{4.1.10}$$

assuming that (4.1.8) is satisfied. Recall (Exercise 1.1-8) that the matrix P is orthogonal. The transformation $\mathbf{x} = P\mathbf{y}$, or $\mathbf{y} = P^T\mathbf{x}$, represents a rotation of the plane with the angle θ chosen so that the new coordinate axes align with the principal axes of the conic section. The columns of P are simply the directions of the principal axes, and the quantities $(c/\lambda_i)^{1/2}$ are the lengths of the axes. This is illustrated in Figure 4.1 for the case of an ellipse. A direct trigonometric calculation shows that the quantities λ_1 and λ_2 are simply the eigenvalues of A, but we do not need to make this calculation since we know this by the similarity transformation P^TAP.

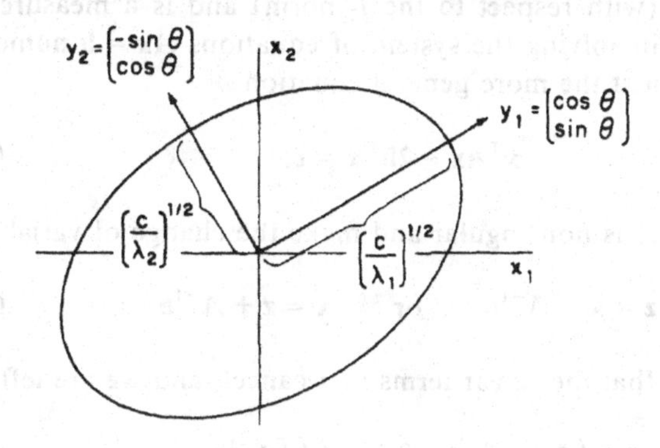

Figure 4.1. An ellipse.

We now wish to extend the above discussion to the $n \times n$ matrix A, assumed real and symmetric. We apply the basic theorem 3.1.1 and let P be the orthogonal matrix such that

$$P^TAP = D = \text{diag}(\lambda_1, \ldots, \lambda_n) \tag{4.1.11}$$

where $\lambda_1, \ldots, \lambda_n$ are the eigenvalues of A. With the change of variable

$$\mathbf{x} = P\mathbf{y}, \qquad \mathbf{y} = P^T\mathbf{x} \tag{4.1.12}$$

Equation (4.1.4) becomes

$$\mathbf{y}^T P^T A P \mathbf{y} = \mathbf{y}^T D \mathbf{y} = \sum_{i=1}^{n} \lambda_i y_i^2 = c \tag{4.1.13}$$

If $\lambda_i > 0$, $i = 1, \ldots, n$, and $c > 0$, (4.1.13) defines an *ellipsoid* in n variables, centered at the origin. In the variables $y_1, \ldots, y_n$ the axes of this ellipsoid are in the coordinate directions, whereas in the original variables $x_1, \ldots, x_n$ the axes are in the directions of the eigenvectors of A. The effect of the change of variable (4.1.12) is, just as in the case of two variables, to align the axes of the new coordinate system with the axes of the ellipsoid, and for this reason (4.1.13) is sometimes called the *reduction to principal axes* of the quadratic form (4.1.4). The lengths of the axes of the ellipsoid are, analogous to Figure 4.1, $(c/\lambda_i)^{1/2}$, $i = 1, \ldots, n$.

Recall that in Section 2.5 we defined an inner product norm on R^n by $\|\mathbf{x}\| = (\mathbf{x}^T A \mathbf{x})^{1/2}$, where A is symmetric and positive definite. The unit sphere in this norm is then the set of all $\mathbf{x}$ such that $\mathbf{x}^T A \mathbf{x} = 1$, which is an ellipsoid. We also note that if λ_1 and λ_n are the maximum and minimum eigenvalues of A, then λ_1/λ_n is a measure of the eccentricity of the ellipsoid; the larger this ratio, the more the ellipsoid deviates from a sphere. In numerical analysis this ratio is called the *condition number* of the symmetric positive definite matrix A (with respect to the l_2 norm) and is a measure of the intrinsic difficulty in solving the system of equations $A\mathbf{x} = \mathbf{b}$ numerically.

We consider next the more general equation

$$\mathbf{x}^T A \mathbf{x} - 2\mathbf{b}^T \mathbf{x} = c \tag{4.1.14}$$

If we assume that A is nonsingular and make the change of variable

$$\mathbf{z} = \mathbf{x} - A^{-1}\mathbf{b} \qquad \text{or} \qquad \mathbf{x} = \mathbf{z} + A^{-1}\mathbf{b} \tag{4.1.15}$$

in (4.1.14), we see that the linear terms in $\mathbf{z}$ cancel, and we are left with

$$\mathbf{z}^T A \mathbf{z} = \hat{c}, \qquad \hat{c} = c + \mathbf{b}^T A^{-1} \mathbf{b} \tag{4.1.16}$$

Thus (4.1.14) is reduced to the form (4.1.4) considered previously. If, again, the eigenvalues of A are all positive, then (4.1.16) describes an ellipsoid centered at the origin in the $\mathbf{z}$ coordinate system. Hence, in the original $\mathbf{x}$ variables, (4.1.14) describes an ellipsoid centered at $A^{-1}\mathbf{b}$.

What if the λ_i are not all positive? Assume that eigenvalues are ordered so that

$$\lambda_1 \geq \cdots \geq \lambda_i > 0, \quad \lambda_{i+1} \leq \cdots \leq \lambda_{i+j} < 0, \quad \lambda_{i+j+1} = \cdots = \lambda_{i+j+k} = 0 \tag{4.1.17}$$

We also assume that $\lambda_1 > 0$ and $c \geq 0$, which rules out redundant or impossible cases. For example, if $\lambda_i < 0$, $i = 1, \ldots, n$, and $c > 0$, then (4.1.13), and hence (4.1.14), has no real solutions. On the other hand, if $c < 0$, and all eigenvalues are negative, multiplying (4.1.14) by -1 returns us to the previous case of an ellipsoid. Now recall from Section 3.3 that the inertia of A is $\text{In}(A) = (i, j, k)$, where, as in (4.1.17), i is the number of positive eigenvalues, j is the number of negative eigenvalues, and k is the number of zero eigenvalues. In(A) completely determines the geometric type of the quadratic form in the following sense. Under the assumption $\lambda_1 > 0$, for $n = 2$, In(A) has only the three instances

(2, 0, 0)	ellipse
(1, 0, 1)	parabola
(1, 1, 0)	hyperbola

whereas for $n = 3$, there are six instances:

(3, 0, 0)	ellipsoid
(2, 0, 1)	elliptic paraboloid
(2, 1, 0)	one-sheeted hyperboloid
(1, 0, 2)	parabolic cylinder
(1, 1, 1)	hyperbolic paraboloid
(1, 2, 0)	two-sheeted hyperboloid

In general, In(A) will have $n(n+1)/2$ possible instances (Exercise 4.1-5), each characterizing a different geometric type in the above sense. As in the case $n = 2$, the solution sets of (4.1.14) are sometimes called *conic sections.*

As mentioned previously, the assumption that $\lambda_1 > 0$ simply rules out redundant cases. For example, if $\text{In}(A) = (0, 2, 0)$, then, provided $c < 0$, (4.1.14) still defines an ellipse. With the assumption that $\lambda_1 > 0$, each possible instance of In(A) corresponds to a different geometric entity. Note also that this classification using In(A) does not distinguish special cases within a general class. For example, if all eigenvalues are equal, then (4.1.14) represents an important special case of an ellipsoid, namely, an n-sphere.

In those cases in which $\text{In}(A) = (i, j, 0)$, the matrix A has no zero eigenvalues and is nonsingular. Hence, the change of variable (4.1.15) is valid, and the conic section has a center at $A^{-1}\mathbf{b}$. For example, if $n = 2$ and

$\mathrm{In}(A) = (1, 1, 0)$, then (4.1.14) describes a hyperbola centered at $A^{-1}\mathbf{b}$. On the other hand, if A is singular, there is no center point. Consider, for example,

$$A = \begin{bmatrix} 1 & \\ & 0 \end{bmatrix}, \qquad \mathbf{b} = \frac{1}{2}\begin{bmatrix} 0 \\ 1 \end{bmatrix}, \qquad c = 1 \tag{4.1.18}$$

so that the quadratic form is

$$x_1^2 - x_2 = 1$$

This is a parabola with a minimum value of -1 whose center line is the x_2 axis. Note that this center line is just $\mathrm{null}(A)$. Note also that linear terms are required to give a true parabola. For example, if $\mathbf{b} = 0$ in (4.1.18), then the quadratic form becomes $x_1^2 = 1$, which is a degenerate parabola consisting of the two lines $x_1 = \pm 1$.

Another example is

$$A = \begin{bmatrix} 4 & & \\ & 1 & \\ & & 0 \end{bmatrix}, \qquad \mathbf{b} = \begin{bmatrix} 2 \\ 1 \\ 1 \end{bmatrix}, \qquad c = 1$$

so that the quadratic form is

$$4x_1^2 + x_2^2 - 4x_1 - 2x_2 - 2x_3 = 1$$

or

$$(2x_1 - 1)^2 + (x_2 - 1)^2 - 2x_3 = 3$$

This is an elliptic paraboloid opening in the x_3 direction with a minimum value of $-\frac{3}{2}$ when $x_1 = \frac{1}{2}$, $x_2 = 1$. The line in the x_3 direction through $x_1 = \frac{1}{2}$, $x_2 = 1$ is a center line and is the affine subspace $S = (\frac{1}{2}, 1, 0)^T + \mathrm{null}(A)$.

More generally, if the inertia of A is $\mathrm{In}(A) = (i, j, k)$ and $A = PDP^T$, then in the variables $\mathbf{y} = P^T\mathbf{x}$, the quadratic form is, with $r = i + j$,

$$\sum_{s=1}^{r} \lambda_s y_s^2 - 2\hat{\mathbf{b}}^T\mathbf{y} = c, \qquad \hat{\mathbf{b}} = P^T\mathbf{b}$$

This may be written as

$$\sum_{s=1}^{r} \lambda_s (y_s - q_s)^2 - 2\sum_{s=r+1}^{n} \hat{b}_s y_s = \hat{c} \tag{4.1.19}$$

where $q_s = \hat{b}_s/\lambda_s$ and $\hat{c} = c + \lambda_1 q_1^2 + \cdots + \lambda_r q_r^2$. For any fixed values of $y_{r+1}, \ldots, y_n$, (4.1.19) represents a conic section in the $y_1, \ldots, y_r$ space with center at $y_i = q_i$, $i = 1, \ldots, r$. All of these conic sections have the same shape and the same center space, which is the $(n - r)$-dimensional affine subspace

$$S = \{y_1 = q_1, \ldots, y_r = q_r, y_s \text{ arbitrary}, s = r, \ldots, n\}$$

Preservation of Type Under Change of Variable

The change of variable $\mathbf{x} = P\mathbf{y}$ used above gives the matrix

$$\hat{A} = P^T A P \tag{4.1.20}$$

as the coefficient matrix of the quadratic form in the $\mathbf{y}$ variables. The changes of variable so far in this section have used only orthogonal matrices, in which case (4.1.20) is a similarity transformation and the eigenvalues of $\hat{A}$ are the same as those of A. If we use an arbitrary real nonsingular matrix P in the change of variable, then (4.1.20) is no longer a similarity transformation. Recall, from Section 3.3, that this is a congruence transformation of A, so that changes of variables in quadratic forms correspond to congruences of the matrix A. For a general congruence the eigenvalues of $\hat{A}$ will, in general, be different from those of A. However, by the Inertia Theorem, 3.3.5, we have that $\text{In}(\hat{A}) = \text{In}(A)$, and we have just seen that $\text{In}(A)$ determines the basic characteristics of the solution set of Equation (4.1.14). That is, $\text{In}(A)$ determines whether (4.1.14) defines an ellipsoid or some other type of conic section. Thus, under an arbitrary change of variable $\mathbf{x} = P\mathbf{y}$, the solution set of

$$\mathbf{y}^T\hat{A}\mathbf{y} - 2\hat{\mathbf{b}}^T\mathbf{y} = c, \qquad \hat{\mathbf{b}} = P^T\mathbf{b} \tag{4.1.21}$$

has the same basic geometric character as that of (4.1.14). For example, if (4.1.14) defines an ellipsoid, then so does (4.1.21) in the new variables $\mathbf{y}$, although the two ellipsoids may have different shapes, reflecting the fact that the eigenvalues of A and $\hat{A}$ will, in general, be different.

Standard Form

If the matrix A is positive definite, so that (4.1.14) defines an ellipsoid, we can always make a change of variable $\mathbf{x} = P\mathbf{y}$, so that (4.1.21) will define a sphere in n dimensions. In particular, as we saw in Section 3.3, there is a nonsingular matrix P such that $P^TAP = I$, so that (4.1.21) becomes

$$\mathbf{y}^T\mathbf{y} - 2\hat{\mathbf{b}}^T\mathbf{y} = c$$

which is the equation of a sphere centered at $\hat{\mathbf{b}}$.

More generally, if $\text{In}(A) = (i, j, k)$, we saw in Section 3.3 that there is a nonsingular matrix P such that

$$P^T AP = \text{diag}(1, \ldots, 1, -1, \ldots, -1, 0, \ldots, 0)$$

so that, in this case, (4.1.21) becomes

$$\sum_{s=1}^{i} y_s^2 - \sum_{s=i+1}^{i+j} y_s^2 - 2\hat{\mathbf{b}}^T \mathbf{y} = c \tag{4.1.22}$$

This is the *standard* (or *canonical*) *form* of Equation (4.1.14). For example, the standard form of the equation of a one-sheeted hyperboloid in three dimensions is

$$y_1^2 + y_2^2 - y_3^2 - \sum_{i=1}^{3} \alpha_i y_i = c$$

Summary

The main results of this section have been

- The solution sets of quadratic equations in n dimensions correspond to conic sections.
- Positive definite quadratic equations correspond to n-dimensional ellipsoids whose principal axes are given by the eigenvectors of the coefficient matrix, and the lengths of the axes are inversely proportional to the square roots of the eigenvalues.
- The inertia of the coefficient matrix determines the type of conic section. The inertia, and hence the type, is preserved under change of variables.

Exercises 4.1

1. Let

$$A = \begin{bmatrix} 1 & 2 & 3 \\ 4 & 5 & 6 \\ 7 & 8 & 9 \end{bmatrix}$$

Find the symmetric and skew symmetric parts of A from (4.1.2).

2. Suppose that the 4×4 symmetric matrix A has eigenvalues 1, 2, 3, 4. What are the lengths of the principal axes of the ellipsoid $\mathbf{x}^T A \mathbf{x} = 1$?
3. Suppose that the 3×3 symmetric matrix A has eigenvalues $-1, 0, 1$. Discuss the geometry of the solution set of $\mathbf{x}^T A \mathbf{x} = 1$. What is the inertia of A?
4. Let

$$A = \begin{bmatrix} 2 & 1 \\ 1 & 2 \end{bmatrix}.$$

Plot the unit sphere of the norm $\|\mathbf{x}\| = (\mathbf{x}^T A \mathbf{x})^{1/2}$, and then plot the ellipse $\mathbf{x}^T A \mathbf{x} - 2\mathbf{b}^T \mathbf{x} - 1 = 0$, where

$$\mathbf{b} = \begin{bmatrix} 1 \\ 1 \end{bmatrix}.$$

5. If A is an $n \times n$ symmetric matrix with $\lambda_1 > 0$, show that $\text{In}(A)$ has $1 + 2 + \cdots + n = n(n+1)/2$ possible instances.
6. Put the ellipse of Exercise 4 in standard form.
7. Let A be a real $n \times n$ symmetric positive definite matrix with eigenvalues $\lambda_1 \leq \cdots \leq \lambda_n$. Show that the two sets

$$\{\mathbf{y} : \mathbf{y} = A\mathbf{x}, \mathbf{x} \in R^n, \mathbf{x}^T \mathbf{x} = 1\}, \qquad \{\mathbf{y} \in R^n : \mathbf{y}^T A^{-2} \mathbf{y} = 1\}$$

define the same ellipsoid with axes whose lengths are $\lambda_1, \ldots, \lambda_n$.
8. Give the center of the ellipsoid $\mathbf{x}^T A \mathbf{x} - 2\mathbf{b}^T \mathbf{x} = 2$, where

$$A = \begin{bmatrix} 1 & 0 & 0 \\ 0 & 2 & 1 \\ 0 & 1 & 2 \end{bmatrix}, \qquad b = \begin{bmatrix} 1 \\ 0 \\ 1 \end{bmatrix}$$

and find the directions and lengths of the axes.
9. For an $n \times n$ nonsingular matrix A, the *condition number* of A is $\|A\| \|A^{-1}\|$ and depends on the norm used. Show that $\|A\|_2 \|A^{-1}\|_2 = \sigma_1 / \sigma_n$, where σ_1 and σ_n are the largest and smallest singular values of A. If A is symmetric and positive definite, show that this becomes λ_1/λ_n, where λ_1 and λ_n are the largest and smallest eigenvalues of A.

4.2. Optimization Problems

One of the most prevalent and important problems in applied mathematics is the optimization (minimization or maximization) of a function f of n variables, usually subject to constraints on the variables. Since we can always convert a maximization problem to one of minimization by minimizing $-f$ instead of maximizing f, we shall usually restrict our discussion to minimization.

Mathematically, the problem is formulated as

$$\text{minimize } f(\mathbf{x}) \text{ subject to constraints on } \mathbf{x} \tag{4.2.1}$$

The constraints may be very simple—for example, some or all of the variables x_j must be nonnegative—or they may be complicated relationships between the variables. Two simple examples are

EXAMPLE 1. minimize $x_1^2 + x_2 + x_2x_3$
subject to $x_1 \geqslant 0,\ x_2^2 + x_3^2 = 1$

EXAMPLE 2. minimize $c_1x_1 + c_2x_2 + c_3x_3$
subject to $a_{11}x_1 + a_{12}x_2 + a_{13}x_3 \leqslant 0$
$a_{21}x_1 + a_{22}x_2 + a_{13}x_3 = 1$

In Example 2 the c_i and a_{ij} are known constants. It is an example of the *linear programming problem* in which we are to minimize a linear function of the n variables $x_1, \ldots x_n$ subject to linear inequality or equality constraints on the variables. The problem (4.2.1) is known as the general *mathematical programming problem.*

If there are no constraints on the variables, one speaks of an *unconstrained minimization problem.* The linear unconstrained problem is trivial, because, without constraints, only a constant function can take on a minimum (Exercise 4.2-1). The simplest nontrivial unconstrained minimization problem is the *quadratic unconstrained minimization problem*, which will be the main subject of this section.

As in the previous section, the general quadratic function of n variables can be written as

$$f(\mathbf{x}) \equiv \mathbf{x}^T A\mathbf{x} - 2\mathbf{b}^T\mathbf{x} + c \tag{4.2.2}$$

where A is a real $n \times n$ symmetric matrix. We know from the calculus that a necessary condition for a continuously differentiable function of n variables to take on a minimum value at a point $\hat{\mathbf{x}}$, called a *minimizer* of f, is that the partial derivatives of the function are all zero at $\hat{\mathbf{x}}$. If we differentiate (4.2.2) and set the partial derivatives to zero, we have

$$\frac{\partial f}{\partial x_k} = 2 \sum_{j=1}^{n} a_{kj}x_j - 2b_k = 0, \qquad k = 1, \ldots, n \tag{4.2.3}$$

or

$$Ax = \mathbf{b} \tag{4.2.4}$$

That is, any $\mathbf{x}$ that minimizes the function (4.2.2) must be a solution of the linear system (4.2.4). [It is now clear why we wrote (4.2.2) with $-2\mathbf{b}$.] On the other hand, a solution of (4.2.4) does not necessarily minimize (4.2.2). Such a solution is called a *stationary point* of f.

We will show that the necessary and sufficient condition on the matrix A of (4.2.2) for f to have a unique minimizer is that A be positive definite. Suppose, first, that the real symmetric matrix A is indefinite, and assume that λ_k is a negative eigenvalue. Let, again, P be the orthogonal matrix of 3.1.1, so that, with $\mathbf{x} = P\mathbf{y}$ and $\hat{\mathbf{b}} = P^T\mathbf{b}$,

$$f(\mathbf{x}) = \mathbf{x}^T A\mathbf{x} - 2\mathbf{b}^T\mathbf{x} + c = \sum_{i=1}^{n} \lambda_i y_i^2 - 2\hat{\mathbf{b}}^T\mathbf{y} + c \equiv \hat{f}(\mathbf{y}) \tag{4.2.5}$$

Now let $y_k \to \pm\infty$ while the other variables y_j are held fixed at zero. Then, since $\lambda_k < 0$, it is clear that $\hat{f}(\mathbf{y})$ in (4.2.5) tends to $-\infty$ because the quadratic term dominates the linear terms. Thus, in the original $\mathbf{x}$ coordinates, the quadratic function tends to $-\infty$ as $\mathbf{x}$ tends to infinity in the direction of an eigenvector corresponding to λ_k. Hence, f can have no minimum, and the minimization problem has no solution.

We next define a *level set* of f to be

$$L_\alpha = \{\mathbf{x}: f(\mathbf{x}) \leq \alpha\}$$

for a given constant α. The boundary of the level set, denoted by $\hat{L}_\alpha$, is the set of all $\mathbf{x}$ such that

$$f(\mathbf{x}) = \mathbf{x}^T A\mathbf{x} - 2\mathbf{b}^T\mathbf{x} + c = \alpha$$

If A is positive definite, we saw in the previous section that $\hat{L}_\alpha$ is an ellipsoid, provided that

$$\alpha - c + \mathbf{b}^T A^{-1}\mathbf{b} > 0$$

and L_α is the volume bounded by this ellipsoid. Then L_α is a closed bounded set, and f must attain a minimum on L_α. Unless $\alpha = c - \mathbf{b}^T A^{-1}\mathbf{b}$, in which case L_α is the single point $A^{-1}\mathbf{b}$, the minimum must be taken on at an interior point of L_α, and, as we saw previously, any minimizing point $\mathbf{x}$ must satisfy the equation $A\mathbf{x} = \mathbf{b}$. Since A is positive definite, it is nonsingular; and this equation has a unique solution that must, therefore, be the unique minimizer of f.

Suppose, finally, that A is positive semidefinite, with $\lambda_1, \ldots, \lambda_r$ its nonzero eigenvalues and $\lambda_{r+1} = \cdots = \lambda_n = 0$. Then (4.2.5) becomes

$$\hat{f}(\mathbf{y}) = \sum_{i=1}^{r} \lambda_i y_i^2 - 2\hat{\mathbf{b}}^T \mathbf{y} + c$$

If $\hat{b}_k \neq 0$ for some $k > r$, then $\hat{f}$ can have no minimum, since $-\hat{b}_k y_k$ tends to $-\infty$ as $\hat{b}_k y_k \to +\infty$, and there are no quadratic terms in y_k to counterbalance this. On the other hand, if $\hat{b}_k = 0$, $k = r+1, \ldots, n$, then $\hat{f}$ becomes a positive definite quadratic form in $y_1, \ldots, y_k$ and, as above, has a minimum. In terms of the original function f, we can then conclude that there is a minimum if and only if $f(\mathbf{x})$ is bounded below (does not tend to $-\infty$) as $\|\mathbf{x}\| \to \infty$. Next, let $\hat{\mathbf{y}}$ be a minimizing point of $\hat{f}$. Then so is $\hat{\mathbf{y}} + \mathbf{w}$, where $\mathbf{w}$ is any vector with zeros in its first r components, since $\hat{f}(\hat{\mathbf{y}}) = \hat{f}(\hat{\mathbf{y}} + \mathbf{w})$. Note that if $\hat{b}_k = 0$, $k = r+1, \ldots, n$, then $\mathbf{b} = P\hat{\mathbf{b}} = \sum_{j=1}^{r} \hat{b}_i \mathbf{p}_i$. Thus, $\mathbf{b}$ is a linear combination of eigenvectors corresponding to the nonzero eigenvalues of A, and this is equivalent to $\mathbf{b} \in \text{range}(A)$. (See Exercise 4.2-5.)

We summarize the above discussions in the following basic theorem.

4.2.1. Minimization of Quadratic Forms. *Let A be a real $n \times n$ symmetric matrix. Then for any $\mathbf{b}$ and c the minimization problem*

$$\text{minimize } \mathbf{x}^T A\mathbf{x} - 2\mathbf{b}^T\mathbf{x} + c \tag{4.2.6}$$

has a unique solution, $A^{-1}\mathbf{b}$, if and only if A is positive definite. If A is indefinite, (4.2.6) *has no solution. If A is positive semidefinite, then* (4.2.6) *has either no solution or infinitely many. There are infinitely many solutions if and only if any one of the following equivalent conditions holds*:

(a) $\mathbf{b}$ *is a linear combination of eigenvectors of A corresponding to nonzero eigenvalues*;
(b) $\mathbf{b} \in$ *range*(A);
(c) $f(\mathbf{x})$ *is bounded below as* $\|\mathbf{x}\| \to \infty$.

We note that the corresponding result can be obtained for the maximization problem if "positive definite" and "positive semidefinite" are changed to "negative" and "negative semidefinite" in 4.2.1. We also note that several examples of 4.2.1 are given in Exercises 4.2-3 and 4.2-4.

The Rayleigh Quotient and Min-Max Theorems

We consider next a very special constrained minimization problem:

$$\text{minimize } \mathbf{x}^T A\mathbf{x} \text{ subject to } \mathbf{x}^T\mathbf{x} = 1 \tag{4.2.7}$$

That is, we wish to minimize $\mathbf{x}^T A\mathbf{x}$ on the l_2 unit sphere in R^n. Assume, as before, that A is symmetric with eigenvalues $\lambda_1 \leq \cdots \leq \lambda_n$. Since the unit sphere is closed and bounded, we know from the calculus that a minimum is achieved. Moreover, from 3.1.9 the minimum value must be at least as large as λ_1. But if $\mathbf{x}_1$ is an eigenvector corresponding to λ_1 and $\mathbf{x}_1^T\mathbf{x}_1 = 1$, then $\mathbf{x}_1^T A\mathbf{x}_1 = \lambda_1$, so that the minimum is achieved at this eigenvector. Note that this minimizer is not unique, since $-\mathbf{x}_1$ is also a minimizer. More generally, if λ_1 is an eigenvalue of multiplicity p, then any normalized eigenvector in the corresponding p-dimensional eigenspace is also a minimizer. The same reasoning holds for the maximum, so that

$$\max_{\mathbf{x}^T\mathbf{x}=1} \mathbf{x}^T A\mathbf{x} = \lambda_n$$

and the maximum is taken on for any normalized eigenvector corresponding to λ_n.

What role do the other eigenvalues and eigenvectors play? The problem (4.2.7) is equivalent (Exercise 4.2-2) to minimizing the function

$$R(\mathbf{x}) = \frac{\mathbf{x}^T A\mathbf{x}}{\mathbf{x}^T\mathbf{x}}, \qquad \mathbf{x} \neq 0 \tag{4.2.8}$$

which is called the *Rayleigh quotient* of A. Thus, the above discussion can be rephrased to say that $\min R(\mathbf{x}) = \lambda_1$, $\max R(\mathbf{x}) = \lambda_n$ and the minimum and maximum are taken on at any eigenvector (not necessarily normalized) corresponding to λ_1 and λ_n, respectively. In order to see the role of the other eigenvalues and eigenvectors, compute the partial derivatives of R:

$$\begin{aligned}\frac{\partial R}{\partial x_i} &= \left(\mathbf{x}^T\mathbf{x}\frac{\partial}{\partial x_i}(\mathbf{x}^T A\mathbf{x}) - \mathbf{x}^T A\mathbf{x}\frac{\partial}{\partial x_i}(\mathbf{x}^T\mathbf{x})\right)\Big/(\mathbf{x}^T\mathbf{x})^2 \\ &= 2\left(\mathbf{x}^T\mathbf{x}\sum_{j=1}^{n} a_{ij}x_j - \mathbf{x}^T A\mathbf{x}x_i\right)\Big/(\mathbf{x}^T\mathbf{x})^2, \qquad i = 1, \ldots, n\end{aligned} \tag{4.2.9}$$

The stationary points of R are those vectors $\mathbf{x}$ for which all of the partial derivatives of (4.2.9) are zero. Setting these partial derivatives to zero gives

$$\mathbf{x}^T\mathbf{x}\sum_{j=1}^{n} a_{ij}x_j = \mathbf{x}^T A\mathbf{x}x_i, \qquad i = 1, \ldots, n$$

or

$$A\mathbf{x} = R(\mathbf{x})\mathbf{x} \tag{4.2.10}$$

This equation is satisfied whenever $\mathbf{x}$ is an eigenvector of A; for example, if $\mathbf{x}_i$ is an eigenvector corresponding to λ_i, then $R(\mathbf{x}_i) = \lambda_i$, and (4.2.10) is $A\mathbf{x}_i = \lambda_i \mathbf{x}_i$.

We summarize the above discussion in the following theorem.

4.2.2. RAYLEIGH QUOTIENT THEOREM. *Let A be a real $n \times n$ symmetric matrix with eigenvalues $\lambda_1 \leq \cdots \leq \lambda_n$. Then the minimum and maximum values of the Rayleigh quotient* (4.2.8) *are λ_1 and λ_n, respectively, and these values are taken on for any eigenvector of A corresponding to λ_1 or λ_n, respectively. Every eigenvector corresponding to an eigenvalue not equal to λ_1 or λ_n is a stationary point of $R(\mathbf{x})$.*

The previous result showed that

$$\lambda_1 = \min_{\mathbf{x}^T\mathbf{x}=1} \mathbf{x}^T A \mathbf{x}, \qquad \lambda_n = \max_{\mathbf{x}^T\mathbf{x}=1} \mathbf{x}^T A \mathbf{x} \tag{4.2.11}$$

Is it possible to characterize the intermediate eigenvalues in terms of minima or maxima? The answer is yes; the idea is to choose proper subspaces over which the maxima and minima are taken.

4.2.3. COURANT–FISCHER MIN-MAX REPRESENTATION. *Let A be an $n \times n$ real symmetric matrix with eigenvalues $\lambda_1 \leq \cdots \leq \lambda_n$. Then*

$$\lambda_k = \min_{V_k} \max\{\mathbf{x}^T A\mathbf{x}: \mathbf{x} \in V_k, \mathbf{x}^T\mathbf{x} = 1\}, \qquad k = 1, \ldots, n \tag{4.2.12}$$

where the minimization is taken over all subspaces V_k of R^n of dimension k.

Before giving the proof of 4.2.3, we note some special cases. For $k = n$ the only subspace V_n of dimension n is R^n itself. Hence, there is no minimization, and (4.2.12) reduces to the second equation of (4.2.11). For $k = 1$ every subspace V_1 is of dimension 1, and with the condition $\mathbf{x}^T\mathbf{x} = 1$ there is no maximization to perform. Hence, (4.2.12) reduces to the first equation of (4.2.11).

PROOF OF 4.2.3. Since A is symmetric, it has an orthonormal basis of eigenvectors $\mathbf{x}_1, \ldots, \mathbf{x}_n$ corresponding to $\lambda_1, \ldots, \lambda_n$. Let $\hat{V}_k = \text{span}(\mathbf{x}_1, \ldots, \mathbf{x}_k)$. Then, for any $\mathbf{x} \in \hat{V}_k$ with $\mathbf{x} = \sum_{i=1}^k \alpha_i \mathbf{x}_i$ and $\mathbf{x}^T\mathbf{x} = \sum_{i=1}^r |\alpha_i|^2 = 1$,

$$\mathbf{x}^T A\mathbf{x} = \sum_{i=1}^k \lambda_i |\alpha_i|^2 \leq \lambda_k \sum_{i=1}^k |\alpha_i|^2 = \lambda_k$$

so that the quantity on the right-hand side of (4.2.12) is no greater than λ_k. On the other hand, if $\hat{V} = \text{span}(\mathbf{x}_k, \ldots, \mathbf{x}_n)$, then for any $\mathbf{x} = \sum_{i=k}^{n} \alpha_i \mathbf{x}_i$ in $\hat{V}$, with $\mathbf{x}^T\mathbf{x} = 1$, we have

$$\mathbf{x}^T A \mathbf{x} = \sum_{i=k}^{n} \lambda_i |\alpha_i|^2 \geq \lambda_k$$

But dim $\hat{V} = n - k + 1$, so that the intersection of $\hat{V}$ with any k-dimensional subspace V_k must be at least one dimensional. Hence, the right-hand side of (4.2.12) is at least as large as λ_k. □

We note that 4.2.3 also holds for Hermitian matrices (Exercise 4.2-7).

Summary

The main results and concepts of this section are as follows:

- A quadratic function has a unique minimizer if and only if the coefficient matrix is positive definite;
- A quadratic function with an indefinite coefficient matrix can have no minimum or maximum;
- A quadratic function with a positive semidefinite coefficient matrix has either no minimizer or infinitely many;
- The Rayleigh quotient of a symmetric matrix is bounded by the largest and smallest eigenvalues; these values are taken on at corresponding eigenvectors, and any other eigenvector is a stationary point;
- All eigenvalues of a symmetric matrix can be characterized by a min-max principle.

Exercises 4.2

1. Let $f(\mathbf{x}) = c_1x_1 + \cdots + c_nx_n + c$. Show that f does not have a minimum on R^n unless the constants $c_1, \ldots, c_n$ are all zero.
2. Show that the problem (4.2.7) is equivalent to minimizing the function R of (4.2.8).
3. Show that $x_1^2 + x_3^2 + x_2$ has no minimum, but $x_1^2 + x_3^2$, considered as a function of x_1, x_2, and x_3, has infinitely many.
4. Find the set of minimizers of the quadratic function $\mathbf{x}^T A\mathbf{x} - 2\mathbf{b}^T\mathbf{x}$ in each of the following situations:

(a) $A = \begin{bmatrix} 2 & 1 \\ 1 & 2 \end{bmatrix}$, $\mathbf{b} = \begin{bmatrix} 1 \\ 1 \end{bmatrix}$

(b) $A = \begin{bmatrix} 1 & 1 \\ 1 & 1 \end{bmatrix}$, $\mathbf{b} = \begin{bmatrix} 1 \\ 1 \end{bmatrix}$

(c) $A = \begin{bmatrix} 1 & 1 \\ 1 & 1 \end{bmatrix}$, $\mathbf{b} = \begin{bmatrix} 1 \\ -1 \end{bmatrix}$

5. Let A be a Hermitian matrix with nonzero eigenvalues $\lambda_1, \ldots, \lambda_r$ and corresponding linearly independent eigenvectors $\mathbf{x}_1, \ldots, \mathbf{x}_r$. Show that a vector $\mathbf{b}$ is a linear combination of $\mathbf{x}_1, \ldots, \mathbf{x}_r$ if and only if $\mathbf{b} \in \text{range}(A)$.
6. Find the minimizers, maximizers, and stationary vectors for $\mathbf{x}^T A\mathbf{x}/\mathbf{x}^T\mathbf{x}$, where

$$A = \begin{bmatrix} 2 & 1 & & \\ 1 & 2 & & \\ & & 4 & \\ & & & 1 \end{bmatrix}$$

7. Formulate and prove 4.2.3 for Hermitian matrices.

4.3. Least Squares Problems and Generalized Inverses

An extremely important special case of the unconstrained minimization problem for quadratic functions is the *least squares problem*, which arises in statistics and in approximation theory. As a simple example of this problem, suppose that we have m observations or measurements of a variable z that we postulate is a linear function

$$z = x_1 y_1 + \cdots + x_n y_n \tag{4.3.1}$$

of variables $y_1, \ldots, y_n$ for constant (but unknown) coefficients $x_1, \ldots, x_n$. The measurements of z correspond to particular values of the variables y_j, so that for a given set $y_1^{(i)}, \ldots, y_n^{(i)}$ of the y_j we have a measurement $z^{(i)}$. We then wish to ascertain the unknown coefficients $x_1, \ldots, x_n$ of (4.3.1) by minimizing the sum of the squares of the deviations of $x_1 y_1^{(i)} + \cdots + x_n y_n^{(i)}$ from $z^{(i)}$; that is, we wish to minimize

$$f(\mathbf{x}) \equiv \sum_{i=1}^{m} \left(z^{(i)} - \sum_{j=1}^{n} x_j y_j^{(i)} \right)^2 \tag{4.3.2}$$

The function f is a measure of how far the relationship (4.3.1) fails to fit the data for a set of coefficients $x_1, \ldots, x_n$. The choice of this criterion of minimizing the sum of squares of the deviations is sometimes justified by a statistical assumption that the measurements $z^{(i)}$ have errors that obey a Gaussian or normal distribution, and sometimes by the mathematical tractability of this problem, which we shall examine shortly.

The problem of minimizing the function f of (4.3.2) is sometimes called the *linear regression* problem and is a special case of the following more general least squares problem. Let $\phi_1 \cdots \phi_n$ be n given functions of $\mathbf{y} = (y_1, \ldots, y_n)^T$ and assume now that z is a linear combination of these given functions for unknown coefficients $x_1, \ldots, x_n$:

$$z = x_1\phi_1(\mathbf{y}) + \cdots + x_n\phi_n(\mathbf{y}) \tag{4.3.3}$$

By repeating the steps leading to the function f of (4.3.2), we are then led to

$$f(\mathbf{x}) = \sum_{i=1}^{m} \left(z^{(i)} - \sum_{j=1}^{n} x_j\phi_j^{(i)} \right)^2 \tag{4.3.4}$$

where $\phi_j^{(i)} = \phi_j(\mathbf{y}^{(i)})$. Thus, this least squares problem is to minimize the function f of (4.3.4) over $x_1, \ldots, x_n$ for given values of $z^{(i)}$ and $\phi_j^{(i)}$. Note that the function f of (4.3.2) is the special case of (4.3.4) when $\phi_i(\mathbf{y}) = y_i$. Note also that (4.3.4) is a quadratic function of the x_i; we shall see its matrix formulation shortly.

Another particularly important special case of (4.3.4) is when $\phi_i(\mathbf{y}) = y_1^{i-1}$, so that the ϕ_i are just powers of a single variable y_1, which we shall call y. Thus, the relation (4.3.3) becomes

$$z = x_1 + x_2y + x_3y^2 + \cdots + x_ny^{n-1}$$

which is just an $(n-1)$st degree polynomial in y. Minimizing f is then the *least squares polynomial approximation* problem.

If we introduce the matrix

$$B = \begin{bmatrix} \phi_1^{(1)} & \cdots & \phi_n^{(1)} \\ \vdots & & \vdots \\ \phi_1^{(m)} & \cdots & \phi_n^{(m)} \end{bmatrix} \tag{4.3.5}$$

then it is easy to see (Exercise 4.3-2) that (4.3.4) may be written as

$$f(\mathbf{x}) = (B\mathbf{x} - \mathbf{z})^T(B\mathbf{x} - \mathbf{z}) = \mathbf{x}^TB^TB\mathbf{x} - 2\mathbf{x}^TB^T\mathbf{z} + \mathbf{z}^T\mathbf{z} \tag{4.3.6}$$

where $\mathbf{z}^T = (z^{(1)}, \ldots, z^{(m)})$. This is a quadratic form with coefficient matrix $A = B^TB$ and constant vector $\mathbf{b} = B^T\mathbf{z}$. Hence, by the results of the previous section, minimizers of f must satisfy the system of equations

$$B^TB\mathbf{x} = B^T\mathbf{z} \tag{4.3.7}$$

which are known as the *normal equations* of the least squares problem.

By Theorem 4.2.1 the quadratic function (4.3.6) has a unique minimizer if and only if the $n \times n$ coefficient matrix B^TB is positive definite. Since, by Theorem 2.3.4, $\text{rank}(B^TB) = \text{rank}(B)$, it follows that B^TB is positive definite if and only if $\text{rank}(B) = n$. Thus we have the following basic theorem.

4.3.1. LEAST SQUARES THEOREM. *If B is a real $m \times n$ matrix and $\mathbf{z}$ a given real m-vector, then the quadratic function* (4.3.6) *has a unique minimizer if and only if rank$(B) = n$.*

We note that 4.3.1 extends to complex B and $\mathbf{z}$ (Theorem 4.3.6). We note also that by (4.3.7) the minimizer may be represented as $(B^TB)^{-1}B^T\mathbf{z}$. (We stress, however, that this is usually *not* the way to perform the calculation in practice, and we will return to this point later.) Finally, we note that in the special case $m = n$, the condition $\text{rank}(B) = n$ is equivalent to B being nonsingular. Thus, the solution is

$$(B^TB)^{-1}B^T\mathbf{z} = B^{-1}\mathbf{z}$$

and the function of (4.3.6) is zero at the minimizer; in this case we obtain an exact solution rather than just a least squares solution.

Let us consider 4.3.1 applied to the least squares polynomial approximation problem

$$\underset{x_1,\ldots,x_n}{\text{minimize}} \sum_{i=1}^{m} \left(z_i - \sum_{j=1}^{n} x_j y_i^{j-1} \right)^2 \tag{4.3.8}$$

where $z_1, \ldots, z_m$ and $y_1, \ldots, y_m$ are given. In this case the ϕ_i are powers of a single variable, y, so that $\phi_j^{(i)} = y_i^{j-1}$, $j = 1, \ldots, n$, and the matrix B of (4.3.5) is

$$B = \begin{bmatrix} 1 & y_1 & y_1^2 & \cdots & y_1^{n-1} \\ 1 & y_2 & \vdots & & \vdots \\ \vdots & \vdots & \vdots & & \vdots \\ 1 & y_m & y_m^2 & \cdots & y_m^{n-1} \end{bmatrix} \tag{4.3.9}$$

This is known as a *Vandermonde matrix*, and the basic result on its rank is as follows:

4.3.2. The $m \times n$ matrix B of (4.3.9) *has rank n if and only if $m \geq n$ and at least n of the points $y_1, \ldots, y_m$ are distinct.*

PROOF. Let r be the number of the y_j that are distinct, and suppose that $r < n$. Then B has only r distinct rows, and rank$(B) \leq r < n$. Conversely, let $r \geq n$ and suppose that rank$(B) < n$. Then the columns of B are linearly dependent, and there is a nonzero vector $\mathbf{x}$ such that $B\mathbf{x} = 0$. This is equivalent to

$$x_1 + x_2 y_i + \cdots + x_n y_i^{n-1} = 0, \qquad i = 1, \ldots, m$$

so that the polynomial $p(y) = x_1 + x_2 y + \cdots + x_n y^{n-1}$ of degree $n - 1$ has $r \geq n$ distinct roots. Since this is impossible, we must have rank$(B) = n$. □

As an immediate corollary of 4.3.1 and 4.3.2, we then have the basic mathematical result for least squares polynomial approximation.

4.3.3. LEAST SQUARES POLYNOMIAL APPROXIMATION. *The problem* (4.3.8) *has a unique solution for given* $z_1, \ldots, z_m$ and $y_1, \ldots, y_m$ *if and only if* $m \geq n$ *and at least* n *of the* y_i *are distinct.*

We give a simple example. Let $m = 4$, $n = 3$ and

$$y_i = i, \quad i = 1, 2, 3, 4, \qquad z_1 = 1, \quad z_2 = 3, \quad z_3 = 1, \quad z_4 = -1$$

The matrix B is

$$B = \begin{bmatrix} 1 & 1 & 1 \\ 1 & 2 & 4 \\ 1 & 3 & 9 \\ 1 & 4 & 16 \end{bmatrix}$$

and 4.3.2 ensures that rank$(B) = 3$. Hence, the least squares problem has a unique solution. The normal equations, $B^T B\mathbf{x} = B^T \mathbf{z}$, are

$$\begin{bmatrix} 4 & 10 & 30 \\ 10 & 30 & 100 \\ 30 & 100 & 354 \end{bmatrix} \begin{bmatrix} x_1 \\ x_2 \\ x_3 \end{bmatrix} = \begin{bmatrix} 4 \\ 6 \\ 6 \end{bmatrix}$$

The solution of this system is $x_1 = -2$, $x_2 = 21/5$, $x_3 = -1$, so that the quadratic that is the unique solution of the least squares problem is

$$p(y) = -2 + \tfrac{21}{5} y - y^2$$

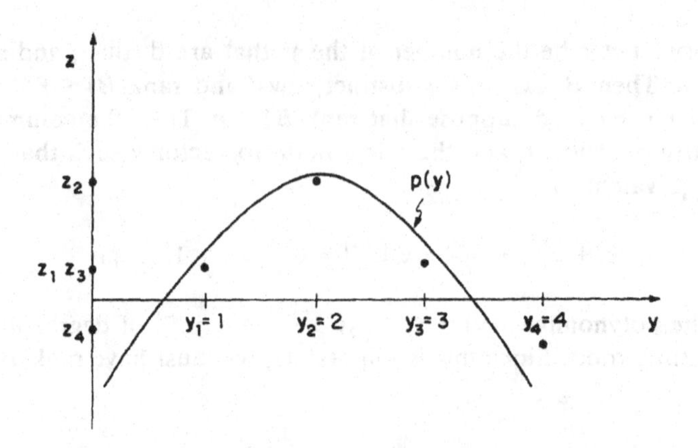

Figure 4.2. Least squares quadratic.

This is illustrated in Figure 4.2.

As a second corollary of 4.3.1, we consider again the linear regression problem (4.3.2). In this case $\phi_i(\mathbf{y}) = y_i$ and the matrix B of (4.3.5) is

$$B = \begin{bmatrix} y_1^{(1)} & \cdots & y_n^{(1)} \\ \vdots & & \vdots \\ y_1^{(m)} & \cdots & y_n^{(m)} \end{bmatrix} \tag{4.3.10}$$

4.3.4. LINEAR REGRESSION. *The function f of* (4.3.2) *has a unique minimizer if and only if the $m \times n$ matrix B of* (4.3.10) *has rank equal to n.*

The interpretation of the condition $\text{rank}(B) = n$ in 4.3.4 is that, first, there must be at least as many observations as unknowns; that is, $m \geq n$. Second, these points $\mathbf{y}^{(i)}$ at which these observations are made must span R^n. In the special case $m = n$ this is the condition that B be nonsingular, and the normal equations (4.3.7) reduce to just $B\mathbf{x} = \mathbf{z}$. The solution then gives the exact interpolating plane through the given data points.

Least Squares and Orthogonal Projection

We next interpret the least squares problem in a more geometric way. Recall from Section 2.4 that if $\mathbf{b}_1, \ldots, \mathbf{b}_n$ are linearly independent vectors in R^m, then the orthogonal projection $\mathbf{p}$ of a vector $\mathbf{z}$ in R^m onto $\text{span}(\mathbf{b}_1, \ldots, \mathbf{b}_n)$ is given by

$$\mathbf{p} = B\mathbf{x}, \qquad \mathbf{x} = (B^TB)^{-1}B^T\mathbf{z}$$

where B is the $m \times n$ matrix with columns $\mathbf{b}_1, \ldots, \mathbf{b}_n$. As we have just seen, the vector $\mathbf{x}$, which gives the coordinates $x_1, \ldots, x_n$ of the orthogonal projection $\mathbf{p}$ in terms of the vectors $\mathbf{b}_1, \ldots, \mathbf{b}_n$, is the solution of the least squares problem (4.3.6). Now note that the quadratic function f of (4.3.6) is just

$$f(\mathbf{x}) = \|B\mathbf{x} - \mathbf{z}\|_2^2,$$

the square of the l_2 norm of $B\mathbf{x} - \mathbf{z}$. Hence, the least squares problem of minimizing f can be interpreted as minimizing the Euclidian distance between a given vector $\mathbf{z}$ and $\text{span}(\mathbf{b}_1, \ldots, \mathbf{b}_n) = \text{range}(B)$. Thus, the orthogonal projection of $\mathbf{z}$ onto range(B) gives the point for which this distance is minimum. This is illustrated in Figure 4.3, in which $\hat{\mathbf{x}}$ is the minimizer of f.

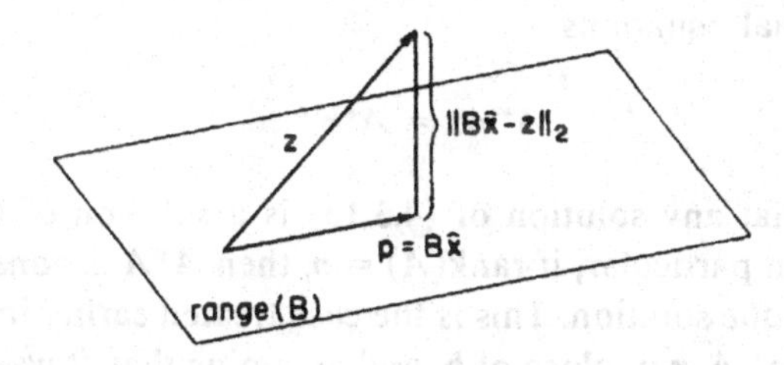

Figure 4.3. Relation of least squares solution to orthogonal projection.

Least Squares Solutions of Linear Systems

We now take a somewhat different approach to the least squares problem by considering the linear system

$$A\mathbf{x} = \mathbf{b} \tag{4.3.11}$$

where A is assumed to be an $m \times n$ real or complex matrix. As was discussed in Section 2.3, there are then three possibilities:

1. The system (4.3.11) has a unique solution;
2. The system (4.3.11) has no solution;
3. The system (4.3.11) has infinitely many solutions.

Our interest here is only in the last two possibilities. Consider first case 2. Here it is reasonable to attempt to do the best we can in the sense of solving

$$\min_{\mathbf{x}} \|A\mathbf{x} - \mathbf{b}\|_2 \tag{4.3.12}$$

In principle we can use any norm in (4.3.12), but we shall restrict our attention here to the l_2 norm. Let $\mathbf{b} = \mathbf{b}_1 + \mathbf{b}_2$ where $\mathbf{b}_1 \in \text{range}(A)$ and $\mathbf{b}_2 \in \text{range}(A)^\perp$. Then, for any $\mathbf{x}$, $A\mathbf{x} - \mathbf{b}_1 \in \text{range}(A)$ so $A\mathbf{x} - \mathbf{b}_1$ and $\mathbf{b}_2$ are orthogonal. Therefore (see Exercise 4.3-5),

$$\|A\mathbf{x} - \mathbf{b}\|_2^2 = \|A\mathbf{x} - \mathbf{b}_1\|_2^2 + \|\mathbf{b}_2\|_2^2 \tag{4.3.13}$$

It is clear from (4.3.13) that $\|A\mathbf{x} - \mathbf{b}\|_2$ is minimized whenever $\mathbf{x}$ satisfies the system

$$A\mathbf{x} = \mathbf{b}_1 \tag{4.3.14}$$

Now, multiply (4.3.14) by A^*. Since, by Theorem 2.4.7, $\text{range}(A)^\perp = \text{null}(A^*)$ we have that $A^*\mathbf{b} = A^*\mathbf{b}_1$ so that any solution of (4.3.12) also satisfies the normal equations

$$A^*A\mathbf{x} = A^*\mathbf{b} \tag{4.3.15}$$

(It is also true that any solution of (4.3.15) is a solution of (4.3.14); see Exercise 4.3-7.) In particular, if $\text{rank}(A) = n$, then A^*A is nonsingular and (4.3.15) has a unique solution. This is the case treated earlier in this section using B in place of A, $\mathbf{z}$ in place of $\mathbf{b}$, and assuming that B was real. (Note that in this present approach we have obtained the normal equations without differentiating $\|A\mathbf{x} - \mathbf{b}\|_2^2$.) In case $\text{rank}(A) < n$, then (4.3.14) and (4.3.15) have infinitely many solutions. Hence, we are again led to case 3.

Suppose now that the system (4.3.11) has infinitely many solutions. We seek the "best" of these solutions, and our criterion for "best" will be the "solution of minimum norm." Thus, we define a *minimum 2-norm solution* of (4.3.11) as a vector $\mathbf{x}_p$ that satisfies

$$\|\mathbf{x}_p\|_2 = \min\{\|\mathbf{x}\|_2 : A\mathbf{x} = \mathbf{b}\} \tag{4.3.16}$$

Again, in principle, we could use any norm in (4.3.16), but we will restrict our attention to the l_2 norm. The basic result is then the following:

4.3.5. If A is an $m \times n$ matrix and the system (4.3.11) *has solutions, then there is a unique minimum 2-norm solution $\mathbf{x}_p$, and every solution of* (4.3.11) *can be written in the form*

$$\mathbf{x} = \mathbf{x}_p + \mathbf{x}_N \tag{4.3.17}$$

where $\mathbf{x}_N \in \text{null}(A)$.

PROOF. By Theorem 2.4.9, $A\mathbf{x} = \mathbf{b}$ has a unique solution $\hat{\mathbf{x}} \in \text{null}(A)^\perp$, and every solution of $A\mathbf{x} = \mathbf{b}$ is of the form $\hat{\mathbf{x}} + \mathbf{x}_N$. Since $\hat{\mathbf{x}}^*\mathbf{x}_N = 0$ for any $\mathbf{x}_N \in \text{null}(A)$, it follows that (Exercise 4.3 – 5)

$$\|\hat{\mathbf{x}} + \mathbf{x}_N\|_2^2 = \|\hat{\mathbf{x}}\|_2^2 + \|\mathbf{x}_N\|_2^2$$

Therefore, $\hat{\mathbf{x}}$ is also the unique vector $\mathbf{x}_p$ that satisfies (4.3.16). □

We next show how $\mathbf{x}_p$ can be represented in terms of the singular value decomposition of A. Recall (Theorem 3.3.7) that the singular value decomposition of an $m \times n$ matrix A is

$$A = PDQ, \qquad D = \begin{bmatrix} D_1 & 0 \\ 0 & 0 \end{bmatrix}, \quad D_1 = \text{diag}(\sigma_1, \ldots, \sigma_r) \tag{4.3.18}$$

where $\sigma_1, \ldots, \sigma_r$ are the nonzero singular values of A, the matrices P and Q are unitary, and D is an $m \times n$ matrix that is 0 except for D_1. Define

$$A^+ = Q^*D^+P^*, \qquad D^+ = \begin{bmatrix} D_1^{-1} & 0 \\ 0 & 0 \end{bmatrix} \tag{4.3.19}$$

Then we will show that

$$\mathbf{x}_p = A^+\mathbf{b} \tag{4.3.20}$$

Since the l_2 norm of a vector remains unchanged under multiplication by unitary matrices and the matrix D is real even if A is complex, we have

$$\begin{aligned} \|A\mathbf{x} - \mathbf{b}\|_2^2 &= \|PDQ\mathbf{x} - \mathbf{b}\|_2^2 = \|DQ\mathbf{x} - P^*\mathbf{b}\|_2^2 \\ &= \|D\mathbf{y} - \mathbf{c}\|_2^2 \end{aligned} \tag{4.3.21}$$

where we have set $\mathbf{y} = Q\mathbf{x}$ and $\mathbf{c} = P^*\mathbf{b}$. If $\mathbf{y}$ and $\mathbf{c}$ are partitioned as D is, then the minimizers of (4.3.21) in the $\mathbf{y}$ variables satisfy the normal equations

$$\begin{bmatrix} D_1^2 & 0 \\ 0 & 0 \end{bmatrix} \begin{bmatrix} \mathbf{y}_1 \\ \mathbf{y}_2 \end{bmatrix} = D^T \begin{bmatrix} \mathbf{c}_1 \\ \mathbf{c}_2 \end{bmatrix} = \begin{bmatrix} D_1\mathbf{c}_1 \\ 0 \end{bmatrix}$$

Therefore, the minimizers are

$$\mathbf{y} = \begin{bmatrix} D_1^{-1}\mathbf{c}_1 \\ \mathbf{y}_2 \end{bmatrix} = D^+P^*\mathbf{b} + \mathbf{w}$$

where $\mathbf{w}^* = (0, \mathbf{y}_2^*)$ and $\mathbf{y}_2$ is arbitrary. Since $\mathbf{w}$ and $D^+P^*\mathbf{b}$ are orthogonal, it follows as in the proof of 4.3.5, that the minimum 2-norm solution, $\mathbf{y}_p$, is $D^+P^*\mathbf{b}$. Thus, since Q^* is unitary and norm preserving, we have

$$\mathbf{x}_p = Q^*\mathbf{y}_p = Q^*D^+P^*\mathbf{b} = A^+\mathbf{b}$$

which is (4.3.20).

Equation (4.3.20) gives the minimum norm solution of $A\mathbf{x} = \mathbf{b}$ in the case that there are infinitely many solutions. Now suppose that $A\mathbf{x} = \mathbf{b}$ has no solution, but A^*A is singular, so that the normal equations

$$A^*A\mathbf{x} = A^*\mathbf{b} \tag{4.3.22}$$

have infinitely many solutions. Then the minimum 2-norm solution of (4.3.22) is

$$\mathbf{x}_p = (A^*A)^+A^*\mathbf{b} \tag{4.3.23}$$

If, again, $A = PDQ$ is a singular value decomposition of A, then

$$A^*A = Q^*D^TP^*PDQ = Q^*D^TDQ \tag{4.3.24}$$

is a singular value decomposition of A^*A. Hence, $(A^*A)^+ = Q^*(D^TD)^+Q$, so that

$$(A^*A)^+A^* = Q^*(D^TD)^+QQ^*D^TP^* = Q^*D^+P^* = A^+ \tag{4.3.25}$$

and (4.3.23) reduces to (4.3.20). That is, the minimum 2-norm solution of (4.3.22) is also given by $A^+\mathbf{b}$.

Finally, we note that if A^*A is nonsingular, then the least squares problem has the unique solution given by (4.3.15); by necessity this is then the minimum 2-norm solution, which suggests that, in this case,

$$A^+ = (A^*A)^{-1}A^* \tag{4.3.26}$$

To verify (4.3.26), we note that because A^*A is nonsingular, the matrix D^TD is also nonsingular. Then, using Eq. (4.3.24)

$$(A^*A)^+ = Q^*(D^TD)^{-1}Q = (A^*A)^{-1}$$

and (4.3.25) reduces to (4.3.26).

We summarize the above in the following basic theorem:

4.3.6. MINIMUM 2-NORM LEAST SQUARES THEOREM. *Let A be an $m \times n$ matrix. Then*

(a) *If $\mathbf{Ax} = \mathbf{b}$ has infinitely many solutions, the minimum 2-norm solution is $A^+\mathbf{b}$;*
(b) *If $\mathbf{Ax} = \mathbf{b}$ has no solutions, the minimum 2-norm solution of $A^*A\mathbf{x} = A^*\mathbf{b}$ is $A^+\mathbf{b}$;*
(c) *If A^*A is nonsingular, $A^+ = (A^*A)^{-1}A^*$;*
(d) *If A is nonsingular, $A^+ = A^{-1}$.*

Generalized Inverses

The matrix A^+ is called a *generalized inverse* or *pseudoinverse.* As noted in 4.3.6, $A^+ = A^{-1}$ if A is nonsingular, and there are various other properties that might be expected of a generalized inverse. One of the most famous sets of such properties is that of the following theorem.

4.3.7. MOORE–PENROSE CONDITIONS. *Let A be an $m \times n$ matrix with singular value decomposition $A = PDQ$. Then $A^+ = Q^*D^+P^*$ satisfies*

$$\begin{aligned} &\text{(a)} \quad AA^+A = A \\ &\text{(b)} \quad A^+AA^+ = A^+ \\ &\text{(c)} \quad (AA^+)^* = AA^+ \\ &\text{(d)} \quad (A^+A)^* = A^+A \end{aligned} \tag{4.3.27}$$

PROOF. For (a) note first that $DD^+D = D$. Therefore, since P and Q are unitary, we have

$$AA^+A = PDQQ^*D^+P^*PDQ = PDD^+DQ = A$$

The relation (b) is proved similarly (Exercise 4.3-10). For (c)

$$AA^+ = PDQQ^*D^+P^* = PDD^+P^*$$

so that

$$(AA^+)^* = P(D^+)^TD^TP^*$$

Thus, AA^+ is Hermitian if the real matrix DD^+ is symmetric. But it is easy to see that both DD^+ and $(D^+)^T D^T$ are equal to $\mathrm{diag}(I_r, 0)$, where I_r is the $r \times r$ identity and r is the number of nonzero singular values of A. The last part, (d), is proved in a similar way (Exercise 4.3-10). □

The conditions (4.3.27) are called the *Moore-Penrose conditions*, and it may be shown that A^+ is the unique $n \times m$ matrix that satisfies all of these conditions. Hence, A^+ is sometimes called the Moore-Penrose inverse. Note that if A is nonsingular, then $A^+ = A^{-1}$ trivially satisfies these conditions. Note also that if we multiply (a) on the right by A^+, we obtain $(AA^+)^2 = AA^+$. Hence, by (c), AA^+ is a Hermitian idempotent matrix, and, therefore, by 2.4.6, it is an orthogonal projector onto $\mathrm{range}(AA^+)$. Using the fact (Exercise 2.3-10) that for any matrix product AB, $\mathrm{range}(AB) \subset \mathrm{range}(A)$, we have from (a) that

$$\mathrm{range}(A) = \mathrm{range}(AA^+A) \subset \mathrm{range}(AA^+) \subset \mathrm{range}(A)$$

Thus, $\mathrm{range}(AA^+) = \mathrm{range}(A)$, so that AA^+ is an orthogonal projector onto $\mathrm{range}(A)$. Similarly, one can show that A^+A is an orthogonal projector onto $\mathrm{range}(A^+) = \mathrm{range}(A^*)$ (see Exercise 4.3-11). These two conditions—AA^+ is an orthogonal projector onto $\mathrm{range}(A)$, and A^+A is an orthogonal projector onto $\mathrm{range}(A^*)$—are another way of characterizing A^+.

We also note that A^+ reduces to the right and left inverses of Section 2.3 (see 2.3.10) in case A has full rank. In particular, if $m \geq n$ and $\mathrm{rank}(A) = n$, then $\mathrm{rank}(D) = n$, and D and D^+ have the form

$$D = \begin{bmatrix} D_1 \\ 0 \end{bmatrix}, \qquad D^+ = [D_1^{-1} \quad 0]$$

where D_1 is $n \times n$, and nonsingular. Hence,

$$A^+A = Q^*D^+P^*PDQ = Q^*D^+DQ = I_n$$

where I_n is the $n \times n$ identity. Thus, A^+ is a left inverse of A. Similarly, if $n \geq m$ and $\mathrm{rank}(A) = m$, then D and D^+ have the form

$$D = [D_1 \quad 0], \qquad D^+ = \begin{bmatrix} D_1^{-1} \\ 0 \end{bmatrix}$$

where now D_1 is $m \times m$. Thus,

$$AA^+ = PDQQ^*D^+P^* = PDD^+P^* = I_m$$

and A^+ is a right inverse of A. As noted previously, if rank$(A) = n$, then $A^+ = (A^*A)^{-1}A^*$, which is another representation of a left inverse.

Summary

The main results of this section are as follows:

- The formulation of the general linear least squares problem as a quadratic minimization problem with coefficient matrix B^TB
- The existence of a unique solution of the least squares problem if and only if B has full rank
- The special cases of linear regression and polynomial approximation
- The interpretation of the least squares problem in terms of orthogonal projections
- Results on solutions of linear systems in terms of a generalized inverse A^+. These are summarized by Figure 4.4.

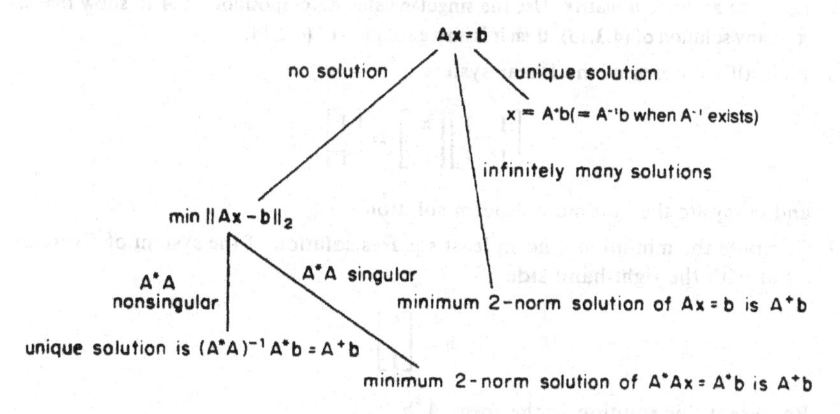

Figure 4.4. Solutions of linear systems.

Exercises 4.3

1. Specialize (4.3.4) by taking $\phi_i(y) = e^{\alpha_i y}$, for known constants α_i. This is the exponential approximation problem.
2. Verify that (4.3.4) may be written as (4.3.6).
3. Find the least squares approximation (4.3.1) for the data

z	y_1	y_2
1	1	1
2	1	2
3	2	1

4. Find the least squares polynomial $y = a_0 + a_1 x$ for the data

y	x
2	1
1	2
3	3

5. Pythagorean Theorem. Let x and y be orthogonal vectors in C^n. Show that

$$\|\mathbf{x} + \mathbf{y}\|_2^2 = \|\mathbf{x}\|_2^2 + \|\mathbf{y}\|_2^2$$

and then extend this result to an arbitrary inner product space.

6. If $B = (b_{ij})$ is a real $m \times n$ matrix, show that

$$\operatorname{Tr}(B^T B) = \operatorname{Tr}(BB^T) = \sum_{i,j} b_{ij}^2$$

7. Let A be an $m \times n$ matrix. Use the singular value decomposition of A to show that if $\hat{\mathbf{x}}$ is any solution of (4.3.15), then it is also a solution of (4.3.14).
8. Find all solutions of the linear system

$$\begin{bmatrix} 1 & 1 \\ 1 & 1 \end{bmatrix} \begin{bmatrix} x_1 \\ x_2 \end{bmatrix} = \begin{bmatrix} 1 \\ 1 \end{bmatrix}$$

and compute the minimum 2-norm solution.

9. Compute the minimum 2-norm least squares solution of the system of Exercise 8 but with the right-hand side

$$\mathbf{b} = \begin{bmatrix} 2 \\ 1 \end{bmatrix}.$$

Represent this solution in the form $A^+\mathbf{b}$.

10. Show that (4.3.27) (b) and (d) are valid.
11. Show that range(A^+) = range(A^*), and then show that A^+A is an orthogonal projector onto range(A^*).

Review Questions—Chapter 4

Answer whether the following statements are true or false and justify your assertions.

1. Let A be a real $n \times n$ matrix and assume that the symmetric part, A_s, of A is positive definite. Then, for a given vector $\mathbf{b}$ and scalar c, $\mathbf{x}^T A\mathbf{x} - 2\mathbf{b}^T\mathbf{x} - c$ and $\mathbf{x}^T A_s \mathbf{x} - 2\mathbf{b}^T\mathbf{x} - c$ will, in general, have different minimizers.
2. If A is a 3×3 real symmetric matrix with $\operatorname{In}(A) = (2, 1, 0)$, then $\mathbf{x}^T A\mathbf{x} - 2(1, 1, 1)\mathbf{x} - 2$ has no minimum.

3. If A is a real symmetric positive definite matrix, then the set of solutions of $\mathbf{x}^T A\mathbf{x} = 1$ is an ellipsoid.
4. If A is an indefinite real symmetric matrix, then we can always find a nonsingular matrix P such that $P^T AP$ is positive definite.
5. If A is a real symmetric positive definite matrix with eigenvalues $\lambda_1, \ldots, \lambda_n$, then the solution set of $\mathbf{x}^T A\mathbf{x} = 2$ is an ellipsoid whose axes have lengths $2\lambda_i$.
6. If A is the matrix of Problem 5, then the eigenvectors of A give the directions of the axes of the ellipsoid.
7. If A is the matrix of Problem 5, then the set of solutions of $\mathbf{x}^T A\mathbf{x} - 2\mathbf{b}^T\mathbf{x} = 0$ is an ellipsoid centered at $A\mathbf{b}$.
8. If A is the matrix of Problem 5, then $(\mathbf{x}^T A\mathbf{x})^{1/2}$ defines a norm whose unit sphere is an ellipsoid.
9. If A is the matrix of Problem 5, then there is a nonsingular matrix P such that in the variables $\mathbf{y} = P\mathbf{x}$, the equation $\mathbf{x}^T A\mathbf{x} = 1$ becomes $\mathbf{y}^T\mathbf{y} = 1$.
10. If A is the matrix of Problem 5, then the function of $\mathbf{x}^T A\mathbf{x} - 2\mathbf{b}^T\mathbf{x} - c$ has a unique minimizer for any given $\mathbf{b}$ and c.
11. If A is a real symmetric positive semidefinite matrix, then $\mathbf{x}^T A\mathbf{x} - 2\mathbf{b}^T\mathbf{x} + c$ has a unique minimizer provided that $\mathbf{b} \in \text{null}(A)$.
12. The algebraically largest eigenvalue of a real symmetric matrix A is always the maximum value of the Rayleigh quotient of A.
13. If A is the matrix of Problem 5, then the solution set of $\mathbf{x}^T A^{-1}\mathbf{x} = 1$ is an ellipsoid.
14. If A is a real symmetric matrix, then any eigenvalue of A can be represented as the minimum of the Rayleigh quotient of A over a suitable subspace.
15. If $y_i = i$, $i = 1, \ldots, n$, and z_i, $i = 1, \ldots, n$, are given numbers, then there is a unique polynomial p of degree n such that $p(y_i) = z_i$, $i = 1, \ldots, n$.
16. If B is a real $m \times n$ matrix and $\mathbf{z}$ is a given real vector, then $\mathbf{x}^T B^T B\mathbf{x} - 2\mathbf{x}^T B^T\mathbf{z} + \mathbf{z}^T\mathbf{z}$ always has minimizers.
17. The normal equations for a least squares problem always have a unique solution.
18. The linear regression problem

$$\underset{\mathbf{x}\in R^n}{\text{minimize}} \sum_{i=1}^{m} (z^{(i)} - \mathbf{x}^T\mathbf{y}^{(i)})^2$$

has a unique solution if and only if $\text{span}(\mathbf{y}^{(1)}, \ldots, \mathbf{y}^{(m)}) = R^n$.
19. If A is an $m \times n$ matrix, the minimum 2-norm solution of $A\mathbf{x} = \mathbf{b}$ is always equal to the unique solution in $\text{null}(A)^\perp$.
20. If A is an $m \times n$ matrix of rank n, then $A^+ = (A^*A)^{-1}A^*$.
21. $A^+ = A^{-1}$ whenever A is nonsingular.
22. $A^+A = AA^+$.
23. A^+A is always Hermitian.
24. If A is $m \times n$ and $\text{rank}(A) = n$, then A^+ is a left inverse of A.

References and Extensions: Chapter 4

1. For more on the geometry of quadratic forms, see Mirsky [1982]. For extensions of quadratic forms to infinite-dimensional spaces, see Gross [1979].
2. There are numerous books on optimization. For treatments stressing the computational aspects of unconstrained and constrained problems, see Fletcher [1980, 1981] and Dennis and Schnabel [1983]. The solution of unconstrained quadratic problems is a key part of several methods for more general optimization problems.
3. For the role of the Rayleigh quotient in computing eigenvalues of Hermitian matrices, see, for example, Parlett [1980].
4. There are numerous inequalities involving positive definite matrices. For example, $|\det A| \leq a_{11} \cdots a_{nn}$ or $|\det A|^{1/n} + |\det B|^{1/n} \leq |\det(A + B)|^{1/n}$. For the proofs of these and a number of other inequalities, see, for example, Mirsky [1982]. There are also several inequalities involving the Schur product of positive definite matrices: for example, $\det(A \circ B) \geq \det A \det B$. For additional such results, see, for example, Basilevsky [1983].
5. There are many good books covering the theory, application, and computational aspects of least squares problems. See Golub and van Loan [1983], Stewart [1973], Lawson and Hanson [1974], and the review paper by Heath [1984]. For additional statistical interpretations and applications, see Seber [1977] and Basilevsky [1983].
6. Another approach to the least squares solution of $A\mathbf{x} = \mathbf{b}$ can be based on the QR factorization of Section 2.4. As noted there, if the QR process is applied to an $m \times n$ matrix A, with $m > n$, then

$$A = QR, \qquad R = \begin{bmatrix} R_1 \\ 0 \end{bmatrix}$$

where R_1 is an $n \times n$ upper triangular matrix. Since

$$\|A\mathbf{x} - \mathbf{b}\|_2 = \|Q(R\mathbf{x} - Q^*\mathbf{b})\|_2 = \|R\mathbf{x} - Q^*\mathbf{b}\|_2$$

it follows that minimizing $\|R\mathbf{x} - Q^*\mathbf{b}\|_2$ is equivalent to minimizing $\|A\mathbf{x} - \mathbf{b}\|_2$. If we write

$$Q^*\mathbf{b} = \mathbf{d} = \begin{bmatrix} \mathbf{d}_1 \\ \mathbf{d}_2 \end{bmatrix}$$

where $\mathbf{d}_1$ is an n-vector and $\mathbf{d}_2$ is $m - n$, then

$$\|R\mathbf{x} - Q^*\mathbf{b}\|_2^2 = \left\| \begin{bmatrix} R_1\mathbf{x} \\ 0 \end{bmatrix} - \begin{bmatrix} \mathbf{d}_1 \\ \mathbf{d}_2 \end{bmatrix} \right\|_2^2 = \|R_1\mathbf{x} - \mathbf{d}_1\|_2^2 + \|\mathbf{d}_2\|_2^2$$

and it is clear that minimizing $\|A\mathbf{x} - \mathbf{b}\|_2$ is equivalent to minimizing $\|R_1\mathbf{x} - \mathbf{d}_1\|_2$. In the case that R_1 is nonsingular, the least squares solution is given by

$$\mathbf{x} = R_1^{-1}\mathbf{d}_1$$

More generally, if R_1 is singular, then the minimum 2-norm solution is given by

$$\mathbf{x} = R_1^+ \mathbf{d}_1$$

For further discussion of this approach and its computational merits, see Golub and van Loan [1983].

7. Much further information on the theory, applications, and history of generalized inverses may be found in Campbell and Meyer [1979].

5

Differential and Difference Equations

In this chapter we treat various questions about ordinary differential and difference equations. We first define and give various properties of the exponential of a matrix, which allows us to express the solution of a system of differential equations in a concise way. The Jordan form is the main tool that allows us to obtain the basic properties of a matrix exponential, and through these properties we are able to express the solution of a system of differential equations $\dot{\mathbf{x}} = A\mathbf{x}$ with constant coefficients in terms of the eigensystem of A. Higher-order equations can be reduced to a first-order system and, thus, treated in the same way. In Section 5.2 we ascertain the stability of solutions when the initial condition is changed, and in Section 5.3 we obtain corresponding results for difference equations. These stability results for difference equations can be interpreted also as convergence theorems for certain iterative methods. Finally, in Section 5.4, we treat Lyapunov's criterion for stability as well as several related results.

5.1. Differential Equations and Matrix Exponentials

Let A be a given $n \times n$ matrix and consider the system of ordinary differential equations with constant coefficients and given initial condition

$$\dot{\mathbf{x}}(t) = A\mathbf{x}(t), \qquad \mathbf{x}(0) = \mathbf{x}_0 \tag{5.1.1}$$

Here, $\mathbf{x}(t)$ is an n-vector whose components $x_i(t)$ are functions of an independent variable t, and $\dot{\mathbf{x}}$ denotes the vector whose components are the derivatives dx_i/dt.

We try a solution of (5.1.1) of the form $\mathbf{x}(t) = e^{\lambda t}\mathbf{p}$, where $\mathbf{p}$ is a fixed vector, independent of t. If this is to be a solution of the differential equation, then we must have

$$\dot{\mathbf{x}}(t) = \lambda e^{\lambda t}\mathbf{p} = A(e^{\lambda t}\mathbf{p}) = e^{\lambda t}A\mathbf{p}$$

Since $e^{\lambda t} \neq 0$, it follows that λ and $\mathbf{p}$ must satisfy $A\mathbf{p} = \lambda\mathbf{p}$; that is, if λ is an eigenvalue of A and $\mathbf{p}$ is a corresponding eigenvector, then $e^{\lambda t}\mathbf{p}$ is a solution of the differential equation. If $\mathbf{x}_1(t)$ and $\mathbf{x}_2(t)$ are any two solutions of the differential equation, then so is any linear combination because

$$\begin{aligned}\frac{d}{dt}[c_1\mathbf{x}_1(t) + c_2\mathbf{x}_2(t)] &= c_1\dot{\mathbf{x}}_1(t) + c_2\dot{\mathbf{x}}_2(t) = c_1A\mathbf{x}_1(t) + c_2A\mathbf{x}_2(t)\\ &= A[c_1\mathbf{x}_1(t) + c_2\mathbf{x}_2(t)]\end{aligned}$$

This is sometimes called the principle of *superposition* of solutions and just reflects the fact that d/dt and A are linear operators. In particular, if A has n linearly independent eigenvectors $\mathbf{p}_1, \ldots, \mathbf{p}_n$ with corresponding eigenvalues $\lambda_1, \ldots, \lambda_n$, then

$$\mathbf{x}(t) = \sum_{i=1}^{n} c_i e^{\lambda_i t}\mathbf{p}_i \tag{5.1.2}$$

is a solution. The constants c_i may be chosen to satisfy the initial condition of (5.1.1) by evaluating the general solution (5.1.2) at $t = 0$:

$$\mathbf{x}_0 = \mathbf{x}(0) = \sum_{i=1}^{n} c_i\mathbf{p}_i = P\mathbf{c}$$

Here, P is the matrix with columns $\mathbf{p}_i$ and is nonsingular by the assumed linear independence of the eigenvectors. Thus, $\mathbf{c} = P^{-1}\mathbf{x}_0$.

In case the matrix A does not have n linearly independent eigenvectors, then we cannot necessarily obtain solutions of (5.1.1) in terms of only the eigenvectors. The generalized eigenvectors of A must be brought into consideration. In order to do this, it is convenient to take a somewhat different approach and develop the solution in terms of matrix exponentials.

Matrix Exponentials

If $n = 1$, the solution to (5.1.1) is

$$\mathbf{x}(t) = e^{At}\mathbf{x}_0 \tag{5.1.3}$$

and we will show that (5.1.3) also holds for arbitrary n. In order to do this, we first need to define the matrix exponential e^{At}.

If $n = 1$, we know that

$$e^{At} = \sum_{k=0}^{\infty} \frac{(At)^k}{k!} \tag{5.1.4}$$

where the series converges for all real or complex values of A. For a given $n \times n$ matrix A, we will define e^{At} by (5.1.4), and we need to show that this series converges. Let

$$S_m = \sum_{k=0}^{m} \frac{(At)^k}{k!} \tag{5.1.5}$$

be the mth partial sum of the series (5.1.4). Then S_m converges to a limit matrix S if

$$\|S_m - S\| \to 0 \qquad \text{as } m \to \infty \tag{5.1.6}$$

in some norm. As we saw in Section 2.5, if (5.1.6) holds in some norm, it holds in any norm and is equivalent to the convergence of elements of S_m to the corresponding elements of S. If (5.1.6) holds, then we define the limit matrix S to be e^{At}, that is, e^{At} is defined by the exponential series (5.1.4) whenever this series converges.

In order to show the convergence, we note that for any $p < m$

$$\|S_m - S_p\| = \left\| \sum_{k=p+1}^{m} \frac{(At)^k}{k!} \right\| \leq \sum_{k=p+1}^{m} \frac{t^k}{k!} \|A\|^k \tag{5.1.7}$$

Since the scalar series for $e^{\|A\|t}$ converges for any A, any t, and any norm, (5.1.7) shows that $\{S_m\}$ is a Cauchy sequence and, hence, has a limit matrix S. Thus, the series (5.1.4) converges for any $n \times n$ matrix A and any t. The matrix e^A is, of course, the special case when $t = 1$.

A basic property of the scalar exponential is that $e^{A+B} = e^A e^B$. This is not, in general, true for matrices, but it does hold if A and B commute.

5.1.1. If A and B are $n \times n$ matrices such that $AB = BA$, then

$$e^{A+B} = e^A e^B = e^B e^A \tag{5.1.8}$$

PROOF. By the series expansions (5.1.4) we have

$$e^{A+B} = I + (A+B) + \tfrac{1}{2}(A+B)^2 + \cdots \tag{5.1.9}$$

$$\begin{aligned} e^A e^B &= (I + A + \tfrac{1}{2}A^2 + \cdots)(I + B + \tfrac{1}{2}B^2 + \cdots) \\ &= I + (A+B) + \tfrac{1}{2}(A^2 + 2AB + B^2) + \cdots \end{aligned} \tag{5.1.10}$$

The first two terms in (5.1.9) and (5.1.10) are identical. For the third,

$$(A+B)^2 = (A+B)(A+B) = A^2 + BA + AB + B^2 = A^2 + 2AB + B^2$$

since $AB = BA$. Thus, the quadratic terms in (5.1.9) and (5.1.10) are identical. Continuing in this fashion, we can show that all terms in the two series must be the same (Exercise 5.1.1). The second equality in (5.1.8) follows immediately from the commutativity of A and B by equating terms in the series for $e^A e^B$ and $e^B e^A$. □

Solution of the Differential Equations

We next wish to show that

$$\frac{d}{dt}(e^{At}) = A\,e^{At} \tag{5.1.11}$$

If (5.1.11) holds and if $\mathbf{x}(t) = e^{At}\mathbf{x}_0$, then $\dot{\mathbf{x}}(t) = A\,e^{At}\mathbf{x}_0 = A\mathbf{x}(t)$, so that $e^{At}\mathbf{x}_0$ satisfies the differential equation (5.1.1).

If we formally differentiate the series (5.1.4) term by term, we obtain

$$\frac{d}{dt}(e^{At}) = A + A^2 t + \frac{A^3 t^2}{2} + \cdots = A\left(I + At + \frac{A^2 t^2}{2} + \cdots\right) = A\,e^{At} \tag{5.1.12}$$

In order for this differentiation of the infinite series to be valid, it is sufficient that the series (5.1.12) converge uniformly—a fact that follows from the uniform convergence of the scalar exponential series.

Alternatively, we can use the definition of the derivative:

$$\frac{d}{dt}(e^{At}) = \lim_{\Delta t \to 0} \frac{(e^{A(t+\Delta t)} - e^{At})}{\Delta t} \tag{5.1.13}$$

Since At and $A\Delta t$ trivially commute, by 5.1.1 we have

$$e^{A(t+\Delta t)} = e^{At}\,e^{A\Delta t} = e^{A\Delta t}\,e^{At}$$

Hence,

$$e^{A(t+\Delta t)} - e^{At} = (e^{A\Delta t} - I)e^{At} = [A\Delta t + \tfrac{1}{2}(A\Delta t)^2 + \cdots]e^{At}$$

and therefore

$$\left\| \frac{e^{A(t+\Delta t)} - e^{At}}{\Delta t} - A\,e^{At} \right\| = \left\| \left(\frac{1}{2}A^2\Delta t + \frac{1}{6}A^3\Delta t^2 + \cdots\right) e^{At} \right\|$$

$$\leq \Delta t\|A\|^2\|e^{At}\|\left(\frac{1}{2}+\frac{1}{6}\|A\|\Delta t+\frac{1}{4!}\|A\|^2\Delta t^2+\cdots\right)$$

$$\leq \Delta t\|A\|^2\|e^{At}\|\left(1+\frac{1}{2}\|A\|\Delta t+\cdots\right)$$

$$= \Delta t\|A\|^2\|e^{At}\|e^{\|A\|\Delta t}$$

For fixed t the right-hand side of this inequality tends to zero as $\Delta t \to 0$. This shows that the limit in (5.1.13) exists and is given by (5.1.11).

Equation (5.1.11) also shows that e^{At} solves the matrix differential equation

$$\dot{X}(t) = AX(t), \qquad X(0) = I \tag{5.1.14}$$

where $X(t)$ is an $n \times n$ matrix. The solution e^{At} is called a *fundamental matrix* for (5.1.14).

Representation of the Exponential by the Jordan Form

We next wish to ascertain certain properties of the matrix exponential that arise from the Jordan form of A. Suppose first that A is the diagonal matrix $D = \text{diag}(\lambda_1, \ldots, \lambda_n)$. Then

$$e^{Dt} \equiv \sum_{k=0}^{\infty} \frac{D^k t^k}{k!} = \text{diag}\left[\sum \frac{(\lambda_1 t)^k}{k!}, \sum \frac{(\lambda_2 t)^k}{k!}, \ldots, \sum \frac{(\lambda_n t)^k}{k!}\right]$$
$$= \text{diag}(e^{\lambda_1 t}, e^{\lambda_2 t}, \ldots, e^{\lambda_n t}) \tag{5.1.15}$$

since each of the scalar series in (5.1.15) converges to the corresponding exponential. Next, suppose that A is similar to a diagonal matrix: $A = PDP^{-1}$ for some nonsingular matrix P. Since

$$A^k = PDP^{-1}PDP^{-1}\cdots PDP^{-1} = PD^kP^{-1} \tag{5.1.16}$$

we have from (5.1.5) that

$$S_m = P\left(\sum_{k=0}^{m} \frac{D^k t^k}{k!}\right)P^{-1} \tag{5.1.17}$$

The summation in (5.1.17) converges to e^{Dt}, and the multiplication by P and P^{-1} does not affect this convergence (see Exercise 5.1-2). Thus,

$$e^{At} = \lim_{m\to\infty} S_m = Pe^{Dt}P^{-1} = P\,\text{diag}(e^{\lambda_1 t}, \ldots, e^{\lambda_n t})P^{-1} \tag{5.1.18}$$

which also shows that $e^{At} = e^{tPDP^{-1}} = Pe^{Dt}P^{-1}$. Thus, $\mathbf{x}(t) = Pe^{Dt}P^{-1}\mathbf{x}_0$ is the solution of (5.1.1), and if we set $c = P^{-1}\mathbf{x}_0$, then

$$\mathbf{x}(t) = \sum_{i=1}^{n} c_i e^{\lambda_i t}\mathbf{p}_i$$

This was what we obtained previously in (5.1.2) if A has n linearly independent eigenvectors.

For the general case let $J = \text{diag}(J_1, \ldots, J_p)$ be the Jordan form of A, and $A = PJP^{-1}$. Then $At = P(Jt)P^{-1}$, and calculations analogous to those of (5.1.16)-(5.1.18) show that $(At)^k = P(Jt)^k P^{-1}$ and

$$\begin{aligned} e^{At} &= e^{P(Jt)P^{-1}} = I + P(Jt)P^{-1} + \tfrac{1}{2}[P(Jt)P^{-1}]^2 + \cdots \\ &= P[I + Jt + \tfrac{1}{2}(Jt)^2 + \cdots]P^{-1} = Pe^{Jt}P^{-1} \end{aligned} \tag{5.1.19}$$

Clearly,

$$e^{Jt} = \text{diag}(e^{J_1 t}, \ldots, e^{J_p t})$$

and it remains to ascertain the form of the $e^{J_i t}$.

Let $\hat{J}$ be an $r \times r$ Jordan block with eigenvalue λ. Then we can write $\hat{J}t$ as

$$\hat{J}t = \lambda tI + Et$$

where E is the $r \times r$ matrix

$$E = \begin{bmatrix} 0 & 1 & & \\ & \ddots & \ddots & \\ & & \ddots & 1 \\ & & & 0 \end{bmatrix} \tag{5.1.20}$$

It is easy to show (Exercise 5.1-3) that

$$E^k = \begin{bmatrix} & 1 & & \\ & & \ddots & \\ & & & 1 \\ & & & \end{bmatrix}, \quad k < r, \qquad E^k = 0, \quad k \geq r. \tag{5.1.21}$$

where E^k has 1's on the kth superdiagonal and is zero elsewhere. Since $E^k = 0$ for $k \geq r$, the series for e^{Et} terminates with the rth term and is just the polynomial

$$e^{Et} = I + Et + \cdots + \frac{(Et)^{r-1}}{(r-1)!}$$

Consequently, since λtI and Et trivially commute, by 5.1.1 we have that

$$e^{\hat{J}t} = e^{\lambda tI + Et} = e^{\lambda tI} e^{Et} = e^{\lambda t}\left(I + Et + \cdots + \frac{(Et)^{r-1}}{(r-1)!}\right)$$

$$= e^{\lambda t}\begin{bmatrix} 1 & t & \frac{t^2}{2} & \cdots & \frac{t^{r-1}}{(r-1)!} \\ & \ddots & \ddots & \ddots & \vdots \\ & & & & \end{bmatrix} \tag{5.1.22}$$

where the diagonal lines in (5.1.22) indicate that the element in the first row is repeated down the corresponding diagonal.

We now summarize the above discussion in the following theorem.

5.1.2. Let $A = PJP^{-1}$ be an $n \times n$ matrix with Jordan form $J = \mathrm{diag}(J_1, \ldots, J_p)$, where J_i is an $r_i \times r_i$ Jordan block with eigenvalue λ_i. Then the fundamental matrix solution, e^{At}, of (5.1.14) *may be represented as*

$$e^{At} = Pe^{Jt}P^{-1} = P\,\mathrm{diag}(e^{J_1t}, e^{J_2t}, \ldots, e^{J_pt})P^{-1} \tag{5.1.23}$$

where

$$e^{J_it} = e^{\lambda_it}\begin{bmatrix} 1 & t & \frac{t^2}{2} & \cdots & \frac{t^{r_i-1}}{(r_i-1)!} \\ & \ddots & \ddots & \ddots & \vdots \\ & & & & \end{bmatrix} \tag{5.1.24}$$

Note that in the special case $t = 1$, we obtain

$$e^A = P\,\mathrm{diag}(e^{J_1}, \ldots, e^{J_p})P^{-1}, \qquad e^{J_i} = e^{\lambda_i}\begin{bmatrix} 1 & 1 & \cdots & \frac{1}{(r_i-1)!} \\ & \ddots & \ddots & \vdots \\ & & & \end{bmatrix} \tag{5.1.25}$$

Note also that when J is diagonal, (5.1.23) reduces to the previously obtained representation (5.1.18).

We next use Theorem 5.1.2 to write the solution $\mathbf{x}(t) = e^{At}\mathbf{x}_0$ of (5.1.1) in terms of the eigensystem of A. If we again set $\mathbf{c} = P^{-1}\mathbf{x}_0$, we have

$$\mathbf{x}(t) = Pe^{Jt}\mathbf{c} = (P_1, \ldots, P_p)\begin{bmatrix} e^{J_1t}\mathbf{c}_1 \\ \vdots \\ e^{J_pt}\mathbf{c}_p \end{bmatrix} = \sum_{i=1}^{p} P_i e^{J_it}\mathbf{c}_i$$

where we have partitioned P into $n \times r_i$ submatrices corresponding to the Jordan blocks of J, and similarly for $\mathbf{c}$. Then, using (5.1.24),

$$P_1 e^{J_1 t}\mathbf{c}_1 = e^{\lambda_1 t}(\mathbf{p}_1, \ldots, \mathbf{p}_{r_1}) \begin{bmatrix} c_1 + c_2 t + \cdots + c_{r_1}\dfrac{t^{r_1-1}}{(r_1-1)!} \\ c_2 + c_3 t + \cdots + c_{r_1}\dfrac{t^{r_1-2}}{(r_1-2)!} \\ \vdots \\ c_{r_1} \end{bmatrix}$$

$$= e^{\lambda_1 t}\left\{\left[c_1 + c_2 t + \cdots + c_{r_1}\frac{t^{r_1-1}}{(r_1-1)!}\right]\mathbf{p}_1 + \left[c_2 + c_3 t + \cdots + c_{r_1}\frac{t^{r_1-2}}{(r_1-2)!}\right]\mathbf{p}_2 + \cdots + c_{r_1}\mathbf{p}_{r_1}\right\}$$

Analogous expressions hold for $P_i e^{J_i t}\mathbf{c}_i$, and we have

$$\mathbf{x}(t) = \sum_{k=1}^{p} e^{\lambda_k t}\left\{\left[c_{q_{k-1}+1} + \cdots + c_{q_k}\frac{t^{r_k-1}}{(r_k-1)!}\right]\mathbf{p}_{q_{k-1}+1} + \left[c_{q_{k-1}+2} + \cdots + c_{q_k}\frac{t^{r_k-2}}{(r_k-2)!}\right]\mathbf{p}_{q_{k-1}+2} + \cdots + c_{q_k}\mathbf{p}_{q_k}\right\} \tag{5.1.26}$$

where $q_0 = 0$ and $q_k = r_1 + \cdots + r_k$, $k = 1, \ldots, p$. This formidable expression represents the solution $\mathbf{x}(t)$ as a linear combination of eigenvectors and generalized eigenvectors with suitable multiples of $e^{\lambda_k t}$ and powers of t. Note that the powers of t appear as a consequence of a nondiagonal canonical form, and that if the Jordan form is diagonal, then (5.1.26) reduces to the linear combination (5.1.2) of exponentials times eigenvectors that was obtained previously.

We give the following example. The matrix

$$A = \frac{1}{2}\begin{bmatrix} 3 & 0 & 0 & 1 \\ 1 & 5 & 1 & 1 \\ 1 & -1 & 3 & 1 \\ 1 & 0 & 0 & 3 \end{bmatrix}$$

has Jordan form J and similarity transformation matrix P given by

$$J = \begin{bmatrix} 1 & 0 & 0 & 0 \\ 0 & 2 & 1 & 0 \\ 0 & 0 & 2 & 1 \\ 0 & 0 & 0 & 2 \end{bmatrix}, \qquad P = \begin{bmatrix} 1 & 0 & 0 & 1 \\ 0 & 1 & 1 & 0 \\ 0 & -1 & 1 & 0 \\ -1 & 0 & 0 & 1 \end{bmatrix}$$

The first column of P is an eigenvector corresponding to the eigenvalue 1. The second column is an eigenvector corresponding to the eigenvalue 2, and the last two columns are generalized eigenvectors. In (5.1.26), $r_1 = 1$ and $r_2 = 3$, so that (5.1.26) becomes

$$\mathbf{x}(t) = e^{\lambda_1 t} c_1 \mathbf{p}_1 + e^{\lambda_2 t}\left(c_2 + c_3 t + c_4 \frac{t^2}{2}\right)\mathbf{p}_2 + e^{\lambda_2 t}(c_3 + c_4 t)\mathbf{p}_3 + e^{\lambda_2 t} c_4 \mathbf{p}_4$$

$$= c_1 e^t \begin{bmatrix} 1 \\ 0 \\ 0 \\ -1 \end{bmatrix} + \left(c_2 + c_3 t + c_4 \frac{t^2}{2}\right) e^{2t} \begin{bmatrix} 0 \\ 1 \\ -1 \\ 0 \end{bmatrix} + (c_3 + c_4 t)\, e^{2t} \begin{bmatrix} 0 \\ 1 \\ 1 \\ 0 \end{bmatrix}$$

$$+ c_4 e^{2t} \begin{bmatrix} 1 \\ 0 \\ 0 \\ 1 \end{bmatrix} \tag{5.1.27}$$

This expresses the general solution of $\dot{\mathbf{x}} = A\mathbf{x}$ in terms of the four arbitrary constants c_1, c_2, c_3, c_4. In order to choose the constants to satisfy a given initial condition, say $\mathbf{x}_0 = (1, 2, 1, 2)^T$, we must solve the linear system $P\mathbf{c} = \mathbf{x}_0$. This gives $c_1 = -\frac{1}{2}$, $c_2 = \frac{1}{2}$, $c_3 = \frac{3}{2}$, $c_4 = \frac{3}{2}$, so that (5.1.27) becomes

$$\mathbf{x}(t) = \frac{1}{2} e^t \begin{bmatrix} 1 \\ 0 \\ 0 \\ -1 \end{bmatrix} + \frac{1}{2}\left(1 + 3t + \frac{3}{2} t^2\right) e^{2t} \begin{bmatrix} 0 \\ 1 \\ -1 \\ 0 \end{bmatrix} + \frac{3}{2}(1 + t)\, e^{2t} \begin{bmatrix} 0 \\ 1 \\ 1 \\ 0 \end{bmatrix}$$

$$+ \frac{3}{2} e^{2t} \begin{bmatrix} 1 \\ 0 \\ 0 \\ 1 \end{bmatrix}$$

Higher-Order Equations

We next consider the application of the previous results to an nth-order homogeneous equation with constant coefficients:

$$x^{(n)}(t) + a_{n-1}x^{(n-1)}(t) + \cdots + a_1 x'(t) + a_0 x(t) = 0 \tag{5.1.28}$$

Here, $x^{(i)}$ denotes the ith derivative of x. Recall from elementary differential equations that the general solution of (5.1.28) is given by

$$x(t) = \sum_{i=1}^{p} e^{\lambda_i t}(c_{i1} + c_{i2}t + \cdots + c_{ir_i}t^{r_i-1}) \tag{5.1.29}$$

where $\lambda_1, \ldots, \lambda_p$ are the distinct roots with multiplicities $r_1, \ldots, r_p$ of the polynomial

$$p(\lambda) = \lambda^n + a_{n-1}\lambda^{n-1} + \cdots + a_1\lambda + a_0 \tag{5.1.30}$$

We will see how (5.1.29) is a consequence of the theory we have just developed.

We first put the nth-order equation (5.1.28) into the form of a system of first-order equations by introducing the derivatives of x as new variables:

$$x_1(t) = x(t), \qquad x_k(t) = x^{(k-1)}(t), \quad k = 2, \ldots, n \tag{5.1.31}$$

Then

$$\dot{x}_k(t) = x_{k+1}(t), \qquad k = 1, \ldots, n-1 \tag{5.1.32}$$

and (5.1.28) becomes

$$\dot{x}_n(t) + a_{n-1}x_n + \cdots + a_1x_2 + a_0x_1 = 0 \tag{5.1.33}$$

Equations (5.1.32) and (5.1.33) are n first-order differential equations in the unknowns $x_1, \ldots, x_n$. Written in matrix-vector form, this system becomes

$$\dot{\mathbf{x}} = A\mathbf{x}, \qquad A = \begin{bmatrix} 0 & 1 & 0 & \cdots & 0 \\ 0 & 0 & 1 & 0 & 0 \\ & & & \cdot & 0 \\ & & & \cdot & 1 \\ -a_0 & -a_1 & & \cdots & -a_{n-1} \end{bmatrix} \tag{5.1.34}$$

The characteristic polynomial of this matrix is given by

$$\det(\lambda I - A) = \lambda^n + a_{n-1}\lambda^{n-1} + \cdots + a_0 \tag{5.1.35}$$

as is easily verified by expansion of the determinant (see Exercise 1.4-8). Note that this is the same as the characteristic polynomial (5.1.30) of (5.1.28).

As discussed in Section 3.2, the multiplicity of an eigenvalue of a matrix does not, in general, determine the block structure in the Jordan form. However, for the matrix A of (5.1.34), the multiplicity of each eigenvalue does indeed determine the Jordan form. In particular, if λ is an eigenvalue of multiplicity r, then there is an associated Jordan block of size $r \times r$.

5.1.3. *Let the* $n \times n$ *matrix* A *of* (5.1.34) *have distinct eigenvalues* $\lambda_1, \ldots, \lambda_p$, *where* λ_i *is of multiplicity* r_i. *Then the Jordan canonical form of* A *is*

$$J = \mathrm{diag}(J_1, \ldots, J_p)$$

where J_i *is an* $r_i \times r_i$ *Jordan block associated with* λ_i.

PROOF. For any eigenvalue λ_i, $\mathrm{rank}(A - \lambda_i I) = n - 1$ because the first $n - 1$ rows of $A - \lambda_i I$ are always linearly independent (Exercise 5.1-5). Since rank is preserved under similarity transformations (Theorem 2.3.2), it follows that $\mathrm{rank}(J - \lambda_i I) = n - 1$. This precludes λ_i from being associated with more than one Jordan block. For example, if both J_i and J_k were Jordan blocks associated with λ_i, then $J_i - \lambda_i I$ and $J_k - \lambda_i I$ are both singular, and $\mathrm{rank}(J - \lambda_i I) \leq n - 2$, which would be a contradiction. □

The solution of the system (5.1.34) can now be written, for an arbitrary initial value $\mathbf{x}_0$, as

$$\mathbf{x}(t) = P\,\mathrm{diag}(e^{J_1 t}, \ldots, e^{J_p t})P^{-1}\mathbf{x}_0 \tag{5.1.36}$$

where the eigenvalues of the Jordan blocks J_i are distinct. In particular, if $\mathbf{p}$ denotes the first row of P and $\mathbf{q} = P^{-1}\mathbf{x}_0$, then x_1, the first component of (5.1.36), which by (5.1.31) is the solution of the original nth-order equation (5.1.28), is given by

$$x(t) = x_1(t) = \mathbf{p}\,\mathrm{diag}(e^{J_1 t}, \ldots, e^{J_p t})\mathbf{q} = \sum_{i=1}^{p} \mathbf{p}_i e^{J_i t}\mathbf{q}_i \tag{5.1.37}$$

where $\mathbf{p}$ and $\mathbf{q}$ have been partitioned into subvectors of lengths commensurate with the dimensions of the $e^{J_i t}$. Then, using (5.1.24) and combining terms, we conclude that

$$x(t) = \sum_{i=1}^{p} e^{\lambda_i t}(c_{i1} + c_{i2}t + \cdots + c_{ir_i}t^{r_i - 1}) \tag{5.1.38}$$

where the constants c_{ij} are combinations of the components of **p** and **q**. One can verify directly (Exercise 5.1-7) that (5.1.38) is a solution of (5.1.28) for arbitrary constants c_{ij}, and these constants can be determined, in principle, by the initial condition $\mathbf{x}_0$ of (5.1.36) through the calculation that led to (5.1.38). An important special case of (5.1.38) is when there are n distinct eigenvalues so that the solution is simply a linear combination of exponentials: $x(t) = \sum_{i=1}^{n} c_i e^{\lambda_i t}$.

The Polar Decomposition and Cayley Transforms

We end this section with a rather different use of the matrix exponential. If α is any complex number, it may be written in polar coordinates as $re^{i\theta}$, where r and θ are nonnegative real numbers. Surprisingly, we can also express matrices in terms of "polar coordinates" in the following sense.

5.1.4. POLAR DECOMPOSITION. *Any real or complex $n \times n$ matrix A may be written in the form $A = H_1 e^{iH_2}$, where H_1 and H_2 are Hermitian positive semidefinite matrices.*

PROOF. By 3.3.7 the singular value decomposition of A is $A = PDQ$, where P and Q are unitary and D is diagonal and positive semidefinite. If we define $H_1 = PDP^*$ and $U = PQ$, then H_1 is Hermitian positive semidefinite and U is unitary. Thus, $A = H_1 U$ is the product of a Hermitian positive semidefinite matrix and a unitary matrix. We then need to show that U can be expressed as e^{iH_2} for some Hermitian positive semidefinite matrix H_2. Since U is normal, by 3.1.12 there is a unitary matrix V and a diagonal matrix $D_1 = \text{diag}(\lambda_1, \ldots, \lambda_n)$ such that $U = VD_1V^*$. Since U is unitary, its eigenvalues λ_j are all of absolute value 1 (Exercise 5.1-9), and therefore we can find nonnegative real numbers θ_j such that $\lambda_j = e^{i\theta_j}$, $j = 1, \ldots, n$. With $D_2 = \text{diag}(\theta_1, \ldots, \theta_n)$, we then have

$$U = Ve^{iD_2}V^* = e^{iVD_2V^*} = e^{iH_2}$$

where $H_2 = VD_2V^*$ is Hermitian and positive semidefinite. □

We note that the first part of the proof of 5.1.4 showed that A can be written as a product of a Hermitian positive semidefinite matrix and a unitary matrix; this complements Exercise 3.1-15. If A is nonsingular, then H_1 is nonsingular also. If A is real, then both H_1 and H_2 may be taken as real. The geometric interpretation of 5.1.4 is that any $n \times n$ matrix generates a linear transformation that can be viewed as being a unitary transformation followed by a transformation that performs contractions or dilations in the directions of the eigenvectors of H_1.

Somewhat related to the polar decomposition of a matrix are the so-called *Cayley transforms.* For a real scalar variable x,

$$z = (1 + ix)(1 - ix)^{-1} \tag{5.1.39}$$

defines a mapping of the real line onto the unit circle in the complex plane. The inverse of this mapping,

$$x = i(1 - z)(1 + z)^{-1}, \tag{5.1.40}$$

defines a mapping of the unit circle in the complex plane onto the real line. Both (5.1.39) and (5.1.40) have natural analogs for matrices. Let H be an $n \times n$ Hermitian matrix and define

$$U = (I + iH)(I - iH)^{-1} \tag{5.1.41}$$

The inverse in this expression is well defined, since iH has only imaginary eigenvalues, and, hence, no eigenvalues of $I - iH$ can be zero. Then

$$\begin{aligned} U^*U &= ((I - iH)^{-1})^*(I + iH)^*(I + iH)(I - iH)^{-1} \\ &= (I + iH)^{-1}(I - iH)(I + iH)(I - iH)^{-1} = I \end{aligned}$$

since $I + iH$ and $I - iH$ commute. Hence, U is unitary, and (5.1.41) represents a mapping, corresponding to (5.1.39), of Hermitian matrices onto unitary matrices. If we solve (5.1.41) for H, we obtain (Exercise 5.1-10)

$$H = i(I - U)(I + U)^{-1} \tag{5.1.42}$$

which is a mapping of unitary matrices into Hermitian matrices and corresponds to (5.1.40). Note that the inverse in (5.1.42) is well defined if U has no eigenvalue equal to -1. If U does have an eigenvalue equal to -1, this corresponds to the singularity at $z = -1$ in (5.1.40).

Summary

The main results of this section are as follows:

- The exponential, e^A, of an $n \times n$ matrix is defined by the exponential power series;
- The solution of $\dot{\mathbf{x}} = A\mathbf{x}$, $\mathbf{x}(0) = \mathbf{x}_0$ can be represented as $e^{At}\mathbf{x}_0$;
- If $J = P^{-1}AP$ is the Jordan form of A, then $e^{At} = Pe^{Jt}P^{-1}$;
- Higher-order differential equations can be solved as first-order systems, and the corresponding Jordan form has only a single Jordan block associated with each distinct eigenvalue;

- Any $n \times n$ matrix can be written as a product of a Hermitian positive semidefinite matrix and a unitary matrix; the unitary matrix can in turn be written as a matrix exponential.

Exercises 5.1

1. Let A and B be $n \times n$ matrices such that $AB = BA$. Show that the binomial formula

$$(A+B)^k = A^k + kA^{k-1}B + \cdots + \binom{k}{j} A^{k-j}B^j + \cdots + B^k$$

holds for any positive integer k, where the $\binom{k}{j}$ are the binomial coefficients. (*Hint*: Use induction.) Use this result to show that the series (5.1.9) and (5.1.10) are identical.

2. Suppose that $\{x_i^{(k)}\}$, $i = 1, \ldots, n$, are n sequences of real or complex numbers with limits $x_1, \ldots, x_n$. Show that

$$\sum_{i=1}^{n} c_i x_i^{(k)} \to \sum_{i=1}^{n} c_i x_i \qquad \text{as } k \to \infty$$

Extend this to show that if $\{A_k\}$ is a sequence of $m \times n$ matrices such that $A_k \to A$ as $k \to \infty$, then for any $m \times m$ matrix P and $n \times n$ matrix Q, $PA_kQ \to PAQ$ as $k \to \infty$.

3. Let E be the $r \times r$ matrix of (5.1.20). Show that (5.1.21) holds.

4. For each of the following matrices, compute P and J such that $A = PJP^{-1}$, where J is the Jordan canonical form. Then compute e^{At} from (5.1.23) and (5.1.24). Finally, compute the solution (5.1.3) of the differential equation (5.1.1).

(a) $\begin{bmatrix} 2 & 1 \\ 1 & 2 \end{bmatrix}$

(b) $\begin{bmatrix} 1 & -2 \\ 2 & 1 \end{bmatrix}$

(c) $\begin{bmatrix} 3 & -1 \\ 1 & 1 \end{bmatrix}$

5. Verify that the first $n-1$ rows of $A - \lambda I$ are linearly independent for any eigenvalue λ of the matrix A of (5.1.34).

6. Do the detailed computation to obtain (5.1.38) from (5.1.37).

7. Verify directly that (5.1.38) is a solution of (5.1.28) for arbitrary constants c_{ij}.

8. Obtain the general solution of the second-order differential equation $\ddot{x}(t) - 2\dot{x}(t) + x(t) = 0$ by converting it to a first-order system $\dot{\mathbf{x}} = A\mathbf{x}$ and solving this system explicitly in terms of the Jordan form of A.
9. Show that the eigenvalues of a unitary or orthogonal matrix are all of absolute value one.
10. Solve (5.1.41) for H to obtain (5.1.42).
11. For any $n \times n$ matrix A, show that e^A is nonsingular. Show also that $\det e^A = e^{\mathrm{Tr}(A)}$.
12. Find a Hermitian matrix H such that

$$\begin{bmatrix} 0 & 1 \\ 1 & 0 \end{bmatrix} = e^{iH}$$

13. Find the polar decomposition of

$$\text{(a)} \begin{bmatrix} 1 & 1 \\ 1 & 1 \end{bmatrix}$$

$$\text{(b)} \begin{bmatrix} 1 & 1 \\ 0 & 0 \end{bmatrix}$$

14. Find the Cayley transform (5.1.41) for the matrix (a) of Exercise 4, and for the matrix of Exercise 12.
15. If A is an $n \times n$ matrix with eigenvalues $\lambda_1, \ldots, \lambda_n$ and corresponding eigenvectors $\mathbf{x}_1, \ldots, \mathbf{x}_n$, show that e^{λ_i} and $\mathbf{x}_i$ are eigenvalues and eigenvectors of e^A.
16. Let S be a real $n \times n$ skew-symmetric matrix. Show that $I + S$ is nonsingular and $(I - S)(I + S)^{-1}$ is orthogonal.
17. Let A be a real $n \times n$ matrix. Ascertain if the Jordan canonical form of the following matrices is necessarily diagonal: (a) e^{iA^TA}, (b) $(A^T - A)^5$.
18. Let

$$A = \begin{bmatrix} 2 & 1 & & \\ & 2 & 1 & \\ & & 2 & \\ & & & 3 \end{bmatrix}.$$

Find e^{At} and the Jordan form of e^A.

5.2. Stability

An important property of solutions of differential equations is stability. By this we mean, intuitively, that if $\mathbf{x}(t)$ is a solution and the differential equation or the initial condition is changed slightly, giving rise to another solution $\hat{\mathbf{x}}(t)$, then $\mathbf{x}(t)$ and $\hat{\mathbf{x}}(t)$ remain "close together" as $t \to \infty$. This question of stability can be addressed for rather general, and nonlinear, differential equations, but we will examine in this section only changes in the initial condition for the linear system with constant coefficients

$$\dot{\mathbf{x}}(t) = A\mathbf{x}(t), \qquad \mathbf{x}(0) = \mathbf{x}_0 \tag{5.2.1}$$

where A is a given $n \times n$ matrix.

Let $\hat{\mathbf{x}}(t)$ be a solution of (5.2.1) with a different initial condition $\hat{\mathbf{x}}_0$. Then $\mathbf{y}(t) = \hat{\mathbf{x}}(t) - \mathbf{x}(t)$ satisfies

$$\dot{\mathbf{y}}(t) = A\mathbf{y}, \qquad \mathbf{y}(0) = \hat{\mathbf{x}}_0 - \mathbf{x}_0$$

Hence, to study changes in the solution of (5.2.1) caused by changes in the initial condition, it suffices to consider changes to the *trivial solution* $\mathbf{x} \equiv 0$ of (5.2.1), corresponding to the initial condition $\mathbf{x}_0 = 0$.

5.2.1. DEFINITION: *Stability.* The trivial solution of (5.2.1) is *stable* if, given $\varepsilon > 0$, there is a $\delta > 0$ such that

$$\|\mathbf{x}(t)\| \leq \varepsilon \qquad \text{for } t \in (0, \infty) \text{ whenever } \|\mathbf{x}_0\| \leq \delta \tag{5.2.2}$$

where $\mathbf{x}(t)$ is the solution of (5.2.1) with initial condition $\mathbf{x}_0$, and $\|\ \|$ is any norm on R^n (or C^n). The trivial solution is *unstable* if it is not stable. The trivial solution is *asymptotically stable* if

$$\mathbf{x}(t) \to 0 \qquad \text{as } t \to \infty \text{ for any } \mathbf{x}_0 \tag{5.2.3}$$

Definition 5.2.1 states that the trivial solution is stable if, when $\mathbf{x}_0$ is small, the corresponding solution $\mathbf{x}(t)$ remains small as $t \to \infty$. The choice of the norm is arbitrary, but the constants δ and ε will, in general, depend on the norm used. Thus, if (5.2.2) holds for a given norm, it may not hold for another norm $\|\ \|'$ with the same constants ε and δ. However, by the Norm Equivalence Theorem, 2.5.4, we can obtain other constants ε' and δ' so that (5.2.2) will hold for $\|\ \|'$. Thus, the stability of a solution is independent of the norm used (Exercise 5.2-1). We note that the terms "stable" and "neutrally stable" are sometimes used for what we have called "asymptotically stable" and "stable."

Consider, as a first example, (5.2.1) for $n = 1$ and A the complex number $\alpha + i\beta$. The solution is

$$x(t) = e^{(\alpha+i\beta)t}x_0 = e^{\alpha t}e^{i\beta t}x_0$$

Recall that $e^{i\beta t} = \cos \beta t + i \sin \beta t$ and $|e^{i\beta t}| = 1$. Thus, $|x(t)| = e^{\alpha t}|x_0|$, and we have the three cases:

1. $\alpha > 0$. Then $|x(t)| \to \infty$ as $t \to \infty$ if $x_0 \neq 0$, and the trivial solution is unstable.
2. $\alpha = 0$. Then $|x(t)| = |x_0|$, and the trivial solution is stable but not asymptotically stable.
3. $\alpha < 0$. Then $x(t) \to 0$ as $t \to \infty$ for all x_0, and the trivial solution is stable and asymptotically stable.

Next, consider (5.2.1) for $n = 2$ and

$$A = \begin{bmatrix} \lambda & 1 \\ 0 & \lambda \end{bmatrix}, \qquad \lambda = \alpha + i\beta \tag{5.2.4}$$

Here the Jordan canonical form of A is just A itself, and from the last section we have

$$\mathbf{x}(t) = e^{At}\mathbf{x}_0 = e^{\lambda t}\begin{bmatrix} 1 & t \\ 0 & 1 \end{bmatrix}\mathbf{x}_0$$

In terms of the components of $\mathbf{x}$, we have

$$x_1(t) = e^{(\alpha+i\beta)t}(x_1^0 + tx_2^0)$$
$$x_2(t) = e^{(\alpha+i\beta)t}x_2^0$$

where x_1^0 and x_2^0 are the components of $\mathbf{x}_0$. It is clear that $x_2(t)$ behaves just as in the $n = 1$ case previously discussed, and the same is true for the first part of $x_1(t)$. The additional factor now appearing in the second term of $x_1(t)$ is the multiplier t, which implies that $|x_1(t)| \to \infty$ as $t \to \infty$ even when $\alpha = 0$. Thus, we have only two cases:

1. $\alpha \geq 0$. Then $|x_1(t)| \to \infty$ as $t \to \infty$ if $x_2^0 \neq 0$, and the trivial solution is unstable.
2. $\alpha < 0$. Then $|x_i(t)| \to 0$ as $t \to \infty$, $i = 1, 2$, and the trivial solution is stable and asymptotically stable.

In order to conclude the second statement, we have used the basic result from the calculus that if $\alpha < 0$, then $te^{\alpha t} \to 0$ as $t \to \infty$. This is a special case of

$$t^k e^{\alpha t} \to 0 \qquad \text{as } t \to \infty, \text{ if } \alpha < 0 \tag{5.2.5}$$

for any k; its proof is left to Exercise 5.2-2.

The previous example indicates the role of the eigenvalues and Jordan form of A in concluding stability. We now embody the required properties of A in the following definition and then give the general result. Here, Re λ denotes the real part of λ.

5.2.2. DEFINITION: An $n \times n$ matrix A with eigenvalues $\lambda_1, \ldots, \lambda_n$ is *negative stable* if Re $\lambda_i < 0$, $i = 1, \ldots, n$. It is *weakly negative stable* if Re $\lambda_i \leq 0$, $i = 1, \ldots, n$, and whenever Re $\lambda_i = 0$, all associated Jordan blocks of λ_i are 1×1.

We give some examples of negative stable and weakly negative stable matrices. Consider the matrices

$$\begin{bmatrix} -1+i & & \\ & -1-i & \\ & & -2 \end{bmatrix} \quad \begin{bmatrix} -i & & \\ & i & \\ & & 0 \end{bmatrix} \quad \begin{bmatrix} 0 & 1 & \\ & 0 & \\ & & i \end{bmatrix}$$

(a) (b) (c)

The matrix (a) is negative stable because it has eigenvalues $-1 \pm i$ and -2, all of whose real parts are negative. The real parts of the eigenvalues of (b) are all zero, and each eigenvalue has a 1×1 Jordan block; hence, it is weakly negative stable. The eigenvalues of (c) also have real parts all equal to zero, but now there is a 2×2 Jordan block associated with the zero eigenvalue. Therefore, (c) is not weakly negative stable.

We now give the basic stability result.

5.2.3. STABILITY THEOREM. *If A is a real or a complex $n \times n$ matrix, then the trivial solution of Equation* (5.2.1) *is stable if and only if A is weakly negative stable. It is asymptotically stable if and only if A is negative stable.*

PROOF. We first consider asymptotic stability. If we write the solution of (5.2.1) as

$$\mathbf{x}(t) = e^{At}\mathbf{x}_0 \tag{5.2.6}$$

and $\mathbf{x}_0 = \mathbf{e}_j$, the jth unit vector, then $\mathbf{x}(t)$ is the jth column of e^{At}. Thus, $\mathbf{x}(t) \to 0$ as $t \to \infty$ for all $\mathbf{x}_0$ if and only if $e^{At} \to 0$ as $t \to \infty$. If $A = PJP^{-1}$, where J is the Jordan form of A, then, by 5.1.2, $e^{At} = Pe^{Jt}P^{-1}$, and it is clear that $e^{At} \to 0$ as $t \to \infty$ if and only if $e^{Jt} \to 0$ as $t \to \infty$. For this, it is necessary and sufficient that $e^{J_jt} \to 0$ as $t \to \infty$ for each block J_j. By (5.1.22)

$$e^{J_jt} = e^{\lambda_jt}\begin{bmatrix} 1 & t & \cdots & \frac{t^2}{2} & \cdots & \frac{t^{r-1}}{(r-1)!} \\ & \ddots & \ddots & & & \vdots \\ & & & & & \end{bmatrix} \tag{5.2.7}$$

If Re $\lambda_j \geq 0$, then $|e^{\lambda_j t}| \geq 1$, and it is clear that $e^{J_j t}$ cannot tend to zero as $t \to \infty$. On the other hand, if Re $\lambda_j < 0$ and $\lambda_j = \alpha + i\beta$, then for any integer k

$$|e^{\lambda_j t} t^k| = e^{\alpha t} t^k \to 0 \qquad \text{as } t \to \infty$$

by (5.2.5). Thus, all terms in (5.2.7) tend to zero as $t \to \infty$, and this completes the proof for asymptotic stability.

Next, consider stability. Again, if Re $\lambda_j > 0$, $e^{J_j t}$ does not remain bounded, and the trivial solution is not stable. If Re $\lambda_j = 0$, and λ_j is associated with a nondiagonal Jordan block, then, from (5.2.7), it is clear that the terms t^k again cause $e^{J_j t}$ to become unbounded as $t \to \infty$, and the trivial solution is not stable. On the other hand, if Re $\lambda_j = 0$ and all Jordan blocks associated with λ_j are 1×1, then there are no polynomial terms in t, and the elements of $e^{J_j t}$ are bounded. Thus, if A is weakly negative stable, then all elements of e^{Jt} either tend to zero or are bounded as $t \to \infty$. If $\|e^{Jt}\| \leq \mu$ for $0 \leq t < \infty$, then

$$\|\mathbf{x}(t)\| \leq \|P\| \mu \|P^{-1}\| \|\mathbf{x}_0\|, \qquad 0 \leq t < \infty$$

and $\|\mathbf{x}(t)\|$ may be kept as small as we like by limiting the size of $\|\mathbf{x}_0\|$.

□

It is important to note that even if the trivial solution is asymptotically stable, a solution for a nonzero initial condition can grow considerably before decreasing to zero. A simple example is the following:

$$\begin{bmatrix} \dot{x}_1 \\ \dot{x}_2 \end{bmatrix} = \begin{bmatrix} -1 & \alpha \\ 0 & -1 \end{bmatrix} \begin{bmatrix} x_1 \\ x_2 \end{bmatrix}, \qquad x_1(0) = 0,\ x_2(0) = \beta \qquad (5.2.8)$$

Since the eigenvalues of the coefficient matrix are -1, the trivial solution is asymptotically stable, so that the solution of (5.2.8) tends to zero as $t \to \infty$ for any values of α and β. However, as is readily verified, the solution of (5.2.8) is

$$x_1(t) = \alpha\beta t\, e^{-t}, \qquad x_2(t) = \beta\, e^{-t}$$

and the component x_1 grows until $t = 1$, where its magnitude is $\alpha\beta/e$. Thus, the 2-norm of the solution is at least $\alpha\beta/e$ at $t = 1$, compared to β at $t = 0$, and the growth can be arbitrarily large by choice of α. We will return to this example from another point of view in Section 5.4, along with additional results on stability.

Higher-Order Equations

We apply Theorem 5.2.3 to a system of the form

$$\dot{\mathbf{x}} = A\mathbf{x}, \qquad A = \begin{bmatrix} 0 & 1 & 0 & & \cdots & 0 \\ 0 & 0 & 1 & 0 & \cdots & 0 \\ & & & & & \vdots \\ & & & & & 0 \\ & & & & & 1 \\ -a_0 & -a_1 & & & \cdots & -a_{n-1} \end{bmatrix} \tag{5.2.9}$$

As we saw in the previous section, (5.2.9) arises from converting an nth-order equation to a first-order system. The result for asymptotic stability of solutions of (5.2.8) is just that Re $\lambda_i < 0$, $i = 1, \ldots, n$, where the λ_i are the eigenvalues of A. For stability, Theorem 5.1.3 showed that whenever the matrix A of (5.2.9) has an eigenvalue of multiplicity r, it has an associated $r \times r$ Jordan block. Hence, whenever Re $\lambda_i = 0$, λ_i must be a simple eigenvalue if A is to be weakly negative stable. Therefore, we have the following result.

5.2.4. The trivial solution of the system (5.2.9) *is stable if and only if the eigenvalues of A satisfy* Re $\lambda_i \leq 0$, *and whenever* Re $\lambda_i = 0$, λ_i *is simple.*

Exercises 5.2

1. Assume that (5.2.2) holds. Use (2.5.10) to show that if $\| \ \|'$ is any other norm on R^n (or C^n), then there are constants ε' and δ' so that

$$\|\mathbf{x}(t)\|' \leq \varepsilon' \qquad \text{for } t \in (0, \infty) \text{ whenever } \|\mathbf{x}_0\| \leq \delta'$$

2. Verify (5.2.5) for any k. (Hint: Define $f(t) = \log(t^k e^{\alpha t})$ and show that $f(t) \to -\infty$ as $t \to \infty$.)
3. Ascertain if the trivial solution of $\dot{\mathbf{x}} = A\mathbf{x}$ is asymptotically stable for each of the matrices

(a) $\begin{bmatrix} -\frac{1}{2} & 1 \\ 0 & -\frac{1}{2} \end{bmatrix}$

(b) $\begin{bmatrix} 0 & 1 \\ 0 & 0 \end{bmatrix}$

(c) $\begin{bmatrix} \frac{1}{2} & 1 \\ 0 & -\frac{1}{2} \end{bmatrix}$

If the trivial solution is not asymptotically stable, ascertain if it is stable.

4. For each of the two matrices below, determine for what values of the real parameter α the trivial solution of $\dot{\mathbf{x}}(t) = A\mathbf{x}(t)$ is stable and asymptotically stable:

$$\text{(a)} \begin{bmatrix} 0 & \alpha \\ -1 & 0 \end{bmatrix}$$

$$\text{(b)} \begin{bmatrix} -2 & \alpha \\ -1 & -2 \end{bmatrix}$$

5.3. Difference Equations and Iterative Methods

Corresponding to the nth-order differential equation (5.1.28) is the *nth-order linear difference equation with constant coefficients.*

$$y_{n+k} + a_{n-1}y_{n+k-1} + \cdots + a_1 y_{k+1} + a_0 y_k = c, \qquad k = 0, 1, \ldots \tag{5.3.1}$$

where $a_0, \ldots, a_{n-1}$ and c are given constants, and $y_0, \ldots, y_{n-1}$ are given initial conditions. If $c = 0$, then the difference equation is *homogeneous.* Similarly, in analogy to the differential equation $\dot{\mathbf{x}} = A\mathbf{x}$, if B is a given $n \times n$ matrix and $\mathbf{d}$ a given vector, then

$$\mathbf{x}^{(k+1)} = B\mathbf{x}^{(k)} + \mathbf{d}, \qquad k = 0, 1, 2, \ldots \tag{5.3.2}$$

is a *system of first-order linear difference equations with constant coefficients* and with initial condition $\mathbf{x}^{(0)}$. If $\mathbf{d} = 0$, the difference equation is *homogeneous* and the *trivial solution* is $\mathbf{x}^{(k)} = 0$, $k = 0, 1, \ldots$. Note that the superscripts in (5.3.2) are indices and do not denote differentiation.

As with differential equations, the nth-order difference equation (5.3.1) may be written as a first-order system by setting

$$\mathbf{x}^{(k)} = \begin{bmatrix} y_k \\ \vdots \\ y_{n+k-1} \end{bmatrix}, \quad B = \begin{bmatrix} 0 & 1 & & & \\ & & 1 & & \\ & & & \ddots & \\ & & & & 1 \\ -a_0 & -a_1 & & & -a_{n-1} \end{bmatrix}, \quad \mathbf{d} = \begin{bmatrix} 0 \\ \vdots \\ 0 \\ c \end{bmatrix} \tag{5.3.3}$$

It is easy to verify (Exercise 5.3-1) that (5.3.2) and (5.3.3) are equivalent to (5.3.1).

The solution of (5.3.2) may be written in the form

$$\mathbf{x}^{(k)} = B^k\mathbf{x}^{(0)} + \sum_{j=0}^{k-1} B^j\mathbf{d}, \qquad k = 1, 2, \ldots \tag{5.3.4}$$

as is easy to verify (Exercise 5.3-2). If $\mathbf{d} = 0$, this reduces to

$$\mathbf{x}^{(k)} = B^k \mathbf{x}^{(0)}, \qquad k = 1, 2, \ldots \tag{5.3.5}$$

and the matrix B^k plays a role for difference equations analogous to the role of e^{At} for differential equations.

We can define stability and asymptotic stability for solutions of difference equations in a way similar to differential equations.

5.3.1. DEFINITION: *Stability*. The trivial solution of the difference equation (5.3.2) with $\mathbf{d} = 0$ is *stable* if, given $\varepsilon > 0$, there is a $\delta > 0$ so that

$$\|\mathbf{x}^{(k)}\| \leq \varepsilon, \quad k = 1, 2, \ldots, \qquad \text{if } \|\mathbf{x}^{(0)}\| \leq \delta \tag{5.3.6}$$

The trivial solution is *asymptotically stable if*

$$\mathbf{x}^{(k)} \to 0 \quad \text{as } k \to \infty \text{ for any } \mathbf{x}^{(0)} \tag{5.3.7}$$

We next give a result on stability of solution of difference equations which corresponds to Theorem 5.2.3 for differential equations. We first need the following definition.

5.3.2. DEFINITION. An $n \times n$ matrix B is *spectral radius diagonable* if it is similar to a matrix of the form

$$\begin{bmatrix} D & 0 \\ 0 & C \end{bmatrix}$$

where D is diagonal with all diagonal elements equal to the spectral radius $\rho(B)$ in absolute value, and $\rho(C) < \rho(B)$.

An equivalent way of stating 5.3.2 is that B has no eigenvalues of absolute value equal to $\rho(B)$ that are associated with nondiagonal Jordan blocks. Some simple examples are the following:

$$\begin{bmatrix} 1 & & \\ & 1 & \\ & & -1 \end{bmatrix} \quad \begin{bmatrix} 1 & 1 & \\ & 1 & \\ & & -1 \end{bmatrix} \quad \begin{bmatrix} 1 & 1 & \\ & 1 & \\ & & 2 \end{bmatrix}$$

(a) (b) (c)

In cases (a) and (b) the eigenvalues are -1 and 1 with multiplicity two. In case (a) the eigenvalues 1 are each associated with 1×1 Jordan blocks, and this matrix is spectral radius diagonable. In case (b), however, the multiple eigenvalue is associated with a 2×2 Jordan block, and the matrix is not spectral radius diagonable. In case (c) there is still the 2×2 Jordan block associated with the eigenvalue 1, but now the spectral radius is 2, and this matrix is spectral radius diagonable.

5.3.3. STABILITY THEOREM FOR DIFFERENCE EQUATIONS. *The trivial solution of the difference equation* (5.3.2) *with* $\mathbf{d} = 0$ *is asymptotically stable if and only if* $\rho(B) < 1$. *It is stable if and only if either* $\rho(B) < 1$, *or* $\rho(B) = 1$ *and B is spectral radius diagonable.*

PROOF. The solution of the homogeneous equation is given by (5.3.5). For $\mathbf{x}^{(0)} = \mathbf{e}_j$, $\mathbf{x}^{(k)}$ is the jth column of B^k, and it is necessary and sufficient for asymptotic stability that $B^k \to 0$ as $k \to \infty$. Let $B = PJP^{-1}$, where J is the Jordan form of B. Then $B^k = PJ^kP^{-1}$, and $B^k \to 0$ as $k \to \infty$ if and only if $J^k \to 0$ as $k \to \infty$, and the latter is true if and only if powers of each Jordan block tend to zero. Let J_i be an $r \times r$ Jordan block. If $r = 2$, it is easy to verify that

$$J_i^k = \begin{bmatrix} \lambda_i^k & k\lambda_i^{k-1} \\ 0 & \lambda_i^k \end{bmatrix}$$

In general (Exercise 5.3-3),

$$J_i^k = \begin{bmatrix} \lambda_i^k & k\lambda_i^{k-1} & \cdots & \binom{k}{j}\lambda_i^{k-j} & \cdots & \binom{k}{r-1}\lambda_i^{k-r+1} \\ & \ddots & \ddots & & \ddots & \vdots \\ & & & & & \end{bmatrix} \tag{5.3.8}$$

where the lines indicate that the entries in each diagonal are the same. Clearly, if $|\lambda_i| \geq 1$, J_i^k does not tend to zero. If $|\lambda_i| < 1$, we need to show that the elements of J_i^k tend to zero, and for this it is sufficient to show that

$$k^p|\lambda_i|^k \to 0 \qquad \text{as } k \to \infty \tag{5.3.9}$$

for any positive integer p. Since $k^p|\lambda_i|^k = k^p e^{(\ln|\lambda_i|)k}$, this follows from (5.2.5). This completes the proof for asymptotic stability. For stability it is clear that J_i^k becomes unbounded if $|\lambda_i| > 1$, or if $|\lambda_i| = 1$ and J_i is not 1×1. On the other hand, if $|\lambda_i| = 1$ and J_i is 1×1, then J_i^k remains bounded. Hence, if $\rho(B) = 1$ and, whenever $|\lambda_i| = 1$, it is associated with only 1×1 Jordan blocks, then J^k remains bounded. □

Theorem 5.3.3 is really a result on the behavior of powers of a matrix, and because of its importance, we state this separately.

5.3.4. Let B be an $n \times n$ matrix. Then $B^k \to 0$ as $k \to \infty$ if and only if $\rho(B) < 1$. B^k remains bounded as $k \to \infty$ if and only if $\rho(B) < 1$, or $\rho(B) = 1$ and B is spectral radius diagonable.

We note that the assumption that B is spectral radius diagonable corresponds to weak negative stability of the matrix in the context of differential equations.

Theorem 5.3.4 is closely related to a basic result on matrix norms. Recall, from Section 2.5, that $\rho(B) \leq \|B\|$ in any norm, but equality does not necessarily hold. Indeed, for a given norm there are matrices B such that $\rho(B)$ and $\|B\|$ may be arbitrarily far apart (Exercise 5.3-5). On the other hand, the following theorem shows that for given B we can always find some norm such that $\|B\|$ is arbitrarily close to $\rho(B)$. The theorem also gives necessary and sufficient conditions for equality to be achieved.

5.3.5. Let B be an $n \times n$ real or complex matrix. Then for any $\varepsilon > 0$ there is a norm on R^n (or C^n) such that

$$\|B\| \leq \rho(B) + \varepsilon \tag{5.3.10}$$

Moreover, there is a norm such that $\rho(B) = \|B\|$ if and only if B is spectral radius diagonable.

PROOF. For given $\varepsilon > 0$ there is, by 3.2.12, a nonsingular matrix P such that $B = P\hat{J}P^{-1}$, where $\hat{J}$ is the Jordan form of B with the off-diagonal 1's replaced by ε. With $Q = P^{-1}$ we then have

$$\|QBQ^{-1}\|_2 = \|\hat{J}\|_2 = \max_{\|\mathbf{x}\|_2=1} \|\hat{J}\mathbf{x}\|_2$$

If ε_i is ε or 0, corresponding to the off-diagonal positions of $\hat{J}$, and $\lambda_1, \ldots, \lambda_n$ are the eigenvalues of B, then the Cauchy-Schwarz inequality gives

$$\begin{aligned}\|\hat{J}\mathbf{x}\|_2^2 &= \sum_{i=1}^{n-1} (|\lambda_i|^2|x_i|^2 + \varepsilon_i\lambda_i\bar{x}_{i+1}x_i + \varepsilon_i\bar{\lambda}_i\bar{x}_i x_{i+1} + \varepsilon_i^2|x_{i+1}|^2) + |\lambda_n|^2|x_n|^2 \\ &\leq [\rho(B)^2 + 2\rho(B)\varepsilon + \varepsilon^2]\|\mathbf{x}\|_2^2 = [\rho(B) + \varepsilon]^2\|\mathbf{x}\|_2^2\end{aligned} \tag{5.3.11}$$

so that $\|\hat{J}\|_2 \leq \rho(B) + \varepsilon$. Now, by 2.5.3, we can define a norm on C^n by $\|\mathbf{x}\| = \|Q\mathbf{x}\|_2$, and by 2.5.7, $\|B\| = \|QBQ^{-1}\|_2 = \|\hat{J}\|_2 \leq \rho(B) + \varepsilon$.

For the second part, choose $\varepsilon > 0$ so that $|\lambda| + \varepsilon < \rho(B)$ for every eigenvalue λ of B such that $|\lambda| < \rho(B)$. Since B is spectral radius diagonable, the matrix $\hat{J}$ has no off-diagonal ε's in those rows with eigenvalues equal in absolute value to $\rho(B)$. We may assume that these eigenvalues of modulus

$\rho(B)$ are the first p, and we let μ be the maximum of the absolute values of the other eigenvalues. Then, again using the Cauchy-Schwarz inequality, (5.3.11) becomes

$$\|\hat{J}\mathbf{x}\|_2^2 \leq \rho(B)^2 \sum_{i=1}^{p} |x_i|^2 + \mu^2 \sum_{i=p+1}^{n} |x_i|^2 + 2\varepsilon\mu \sum_{i=p+1}^{n} |x_i|^2 + \varepsilon^2 \sum_{i=p+1}^{n} |x_i|^2$$

$$= \rho(B)^2 \sum_{i=1}^{p} |x_i|^2 + (\mu + \varepsilon)^2 \sum_{i=p+1}^{n} |x_i|^2 \leq \rho(B)^2 \|\mathbf{x}\|_2^2$$

Therefore, $\|\hat{J}\|_2 \leq \rho(B)$, so that $\|B\| \leq \rho(B)$. But any norm is at least as large as $\rho(B)$, so $\|B\| = \rho(B)$.

For the converse, assume that $\|B\| = \rho(B)$ in some norm, but that there is an $m \times m$ Jordan block J_i, with $m \geq 2$, associated with an eigenvalue λ_i such that $|\lambda_i| = \rho(B)$. Then it suffices to show that it is not possible to have a norm such that $\|J_i\| = |\lambda_i|$. Clearly, we may assume that $\lambda_i \neq 0$, for otherwise we would have $\|B\| = 0$, which implies $B = 0$. Assume that $\|\hat{J}_i\| = 1$, where $\hat{J}_i = \lambda_i^{-1} J_i$. A direct computation shows that $\hat{J}_i^k \mathbf{e}_1 = (k/\lambda_i, 1, 0, \ldots, 0)^T$, so that $\|\hat{J}_i^k \mathbf{e}_1\| \to \infty$ as $k \to \infty$. But this contradicts $\|\hat{J}_i\| = 1$. □

We note that a somewhat simpler proof can be given using the ∞-norm rather than the 2-norm (Exercise 5.3-7). However, we will need later the following equivalent way of stating 5.3.5 in the 2-norm.

5.3.6. Let B be an $n \times n$ matrix. Then for any $\varepsilon > 0$ there is a nonsingular matrix Q such that $\|QBQ^{-1}\|_2 \leq \rho(B) + \varepsilon$. Moreover, there is a Q such that $\|QBQ^{-1}\|_2 = \rho(B)$ if and only if B is spectral radius diagonable.

Iterative Methods

Consider the system of linear equations

$$A\mathbf{x} = \mathbf{b} \tag{5.3.12}$$

where we assume that the diagonal elements of A are nonzero. Thus, $D = \operatorname{diag}(a_{11}, \ldots, a_{nn})$ is nonsingular, and we define

$$B = I - D^{-1}A, \qquad \mathbf{d} = D^{-1}\mathbf{b} \tag{5.3.13}$$

Then we generate a sequence of vectors $\mathbf{x}^{(k)}$ by

$$\mathbf{x}^{(k+1)} = B\mathbf{x}^{(k)} + \mathbf{d}, \qquad k = 0, 1, \ldots \tag{5.3.14}$$

which is the *Jacobi iterative method* for approximating a solution of (5.3.12). Similarly, if L is the strictly lower triangular part of A (i.e., the main diagonal of L is zero) and U is the strictly upper triangular part, so that $A = D + L + U$, then with

$$B = -(D+L)^{-1}U, \qquad \mathbf{d} = (D+L)^{-1}\mathbf{b} \tag{5.3.15}$$

(5.3.14) defines the *Gauss–Seidel iterative method.* More generally, (5.3.14) describes a large number of other possible iterative methods (called *linear stationary one-step methods*) through the choice of B. Iterative methods are widely used for the numerical solution of (5.3.12) in situations where methods such as Gaussian elimination are not economical because of the size and structure of A.

Assume that A is nonsingular and $\hat{\mathbf{x}}$ is the unique solution of (5.3.12). Then it is easy to verify (Exercise 5.3-9) that $\hat{\mathbf{x}}$ satisfies

$$\hat{\mathbf{x}} = B\hat{\mathbf{x}} + \mathbf{d} \tag{5.3.16}$$

for the Jacobi and Gauss–Seidel iterations. More generally, the iterative method (5.3.14) is said to be *consistent* with (5.3.12) if (5.3.16) holds. The error at the kth step of the iterative method is $\mathbf{e}^{(k)} = \mathbf{x}^{(k)} - \hat{\mathbf{x}}$, and for any consistent iterative method we can subtract (5.3.16) from (5.3.14) to obtain the error equation

$$\mathbf{e}^{(k+1)} = B\mathbf{e}^{(k)}, \qquad k = 0, 1, \ldots \tag{5.3.17}$$

We may sometimes be able to choose the initial approximation $\mathbf{x}^{(0)}$ close to $\hat{\mathbf{x}}$, but, in general, we will have no control over the direction of the initial error. Therefore, for an iterative method to be successful, we shall require that $\mathbf{x}^{(k)} \to \hat{\mathbf{x}}$ as $k \to \infty$ for all $\mathbf{x}^{(0)}$, which is equivalent to $\mathbf{e}^{(k)} \to 0$ as $k \to \infty$ for all $\mathbf{e}^{(0)}$. Hence, we see that convergence of the iterative method (5.3.14) is precisely equivalent to asymptotic stability of the trivial solution of (5.3.17), viewed as a difference equation. Therefore, as an immediate corollary of 5.3.3 we have the next result.

5.3.7. Convergence of Iterative Methods. *If the system* (5.3.12) *has a unique solution* $\hat{\mathbf{x}}$ *and the iterative method* (5.3.14) *is consistent with* (5.3.12), *then the iterates of* (5.3.14) *converge to* $\hat{\mathbf{x}}$ *for any* $\mathbf{x}^{(0)}$ *if and only if* $\rho(\mathbf{B}) < 1$.

By means of this fundamental theorem, the analysis of convergence of the iterative method (5.3.14) depends on ascertaining whether $\rho(B) < 1$. We shall see in Section 6.1 some particular sufficient conditions for this.

Geometric Series

The solution of the difference equation (5.3.14) was given by (5.3.4) as

$$\mathbf{x}^{(k)} = B^k\mathbf{x}^{(0)} + \sum_{j=0}^{k-1} B^j\mathbf{d}, \qquad k = 1, 2, \ldots$$

If $\rho(B) < 1$, then $B^k\mathbf{x}^{(0)} \to 0$ and $\mathbf{x}^{(k)} \to \hat{\mathbf{x}}$ as $k \to \infty$. Hence, we expect that

$$\hat{\mathbf{x}} = (I - B)^{-1}\mathbf{d} = \left(\sum_{j=0}^{\infty} B^j\right)\mathbf{d}$$

This is verified by the following basic result, which is the matrix analogue of the geometric series for scalars.

5.3.8. NEUMANN EXPANSION. *If B is an $n \times n$ matrix with $\rho(B) < 1$, then $(I - B)^{-1}$ exists and*

$$(I - B)^{-1} = \lim_{k\to\infty} \sum_{i=0}^{k} B^i = \sum_{i=0}^{\infty} B^i \tag{5.3.18}$$

PROOF. Since $\rho(B) < 1$, B can have no eigenvalue equal to 1, and, hence, $I - B$ is nonsingular. To prove (5.3.18), we multiply the identity

$$(I - B)(I + B + \cdots + B^{k-1}) = I - B^k$$

by $(I - B)^{-1}$ to obtain

$$I + B + \cdots + B^{k-1} = (I - B)^{-1} - (I - B)^{-1}B^k \tag{5.3.19}$$

Since $\rho(B) < 1$, $B^k \to 0$ as $k \to \infty$, and, therefore, the last term of (5.3.19) tends to zero as $k \to \infty$. □

As an application of the previous result, we give the following perturbation theorem.

5.3.9. Let A be a nonsingular $n \times n$ matrix and E an $n \times n$ matrix such that in some norm $\alpha = \|A^{-1}E\| < 1$. Then $A + E$ is nonsingular, and

$$\|(A + E)^{-1} - A^{-1}\| \leq \frac{\alpha}{1 - \alpha}\|A^{-1}\| \tag{5.3.20}$$

PROOF. Set $B = -A^{-1}E$. Then $\alpha = \|B\| < 1$, so that $\rho(B) < 1$ and (5.3.18) holds. Hence,

$$\|(I - B)^{-1}\| \leq \sum_{k=0}^{\infty} \|B\|^k = \sum_{k=0}^{\infty} \alpha^k = \frac{1}{1-\alpha} \tag{5.3.21}$$

But

$$A + E = A(I + A^{-1}E) = A(I - B)$$

so that $A + E$, as the product of nonsingular matrices, is nonsingular. Moreover,

$$(A + E)^{-1} - A^{-1} = [(I - B)^{-1} - I]A^{-1} = B(I - B)^{-1}A^{-1}$$

Thus, the result follows from (5.3.21). □

Theorem 5.3.9 is illustrative of many results in numerical analysis and other areas in which one is forced to deal with inexact data. For example, the matrix $A + E$ may be the computer representation of an exact matrix A whose entries, such as $\frac{1}{3}$, do not have a finite binary fractional representation. Then (5.3.20) estimates the difference between the exact inverse, A^{-1}, and the inverse of the approximation $A + E$.

Exercises 5.3

1. Verify that (5.3.2) and (5.3.3) are equivalent to (5.3.1).
2. Verify that (5.3.4) is the solution of (5.3.2).
3. Let $\hat{J}$ be an $r \times r$ Jordan block with eigenvalue λ, and write $\hat{J}$ as $\lambda I + E$, where E is the matrix of (5.1.20). Recall that E^k satisfies (5.1.21). Use the binomial expansion

$$(\lambda I + E)^k = \lambda^k I + k\lambda^{k-1}E + \cdots + \lambda^{k-j}\binom{k}{j}E^j + \cdots + E^k$$

to verify (5.3.8).

4. Ascertain stability and asymptotic stability of the trivial solution of the difference equation $\mathbf{x}^{(k+1)} = B\mathbf{x}^{(k)}$, $k = 0, 1, \ldots$, for each of the following matrices:

$$\text{(a)} \quad B = \begin{bmatrix} 2 & 1 \\ 1 & 2 \end{bmatrix}$$

(b) $B = \frac{1}{4}\begin{bmatrix} 2 & 1 \\ 1 & 2 \end{bmatrix}$

(c) $B = \begin{bmatrix} \frac{1}{2} & 1 \\ 0 & \frac{1}{2} \end{bmatrix}$

(d) $B = \begin{bmatrix} 1 & 1 \\ 0 & 1 \end{bmatrix}$

5. Let

$$B = \begin{bmatrix} 0 & \alpha \\ 0 & 0 \end{bmatrix}.$$

Conclude that $\rho(B)$ and $\|B\|_\infty$ may be arbitrarily far apart by choice of α.

6. Let

$$A = \begin{bmatrix} 1 & 2 \\ 0 & 1 \end{bmatrix}.$$

Find a norm on R^2 such that $\|A\| \leq 1.1$. Can you find a norm for which $\|A\| = 1$?

7. Prove 5.3.5 by using the ∞-norm rather than the 2-norm.
8. Show that the trivial solution of (5.3.1) with the matrix B given by (5.3.3) is stable if and only if $\rho(B) \leq 1$, and whenever $|\lambda| = 1$, λ is a simple eigenvalue. (Hint: Review 5.2.4.)
9. Verify that (5.3.16) holds for B and $\mathbf{d}$ defined by (5.3.13) or (5.3.15).

5.4. Lyapunov's Theorem and Related Results

In the last two sections we have seen that the condition for asymptotic stability of solutions of differential equations is that the coefficient matrix is negative stable, whereas for difference equations the condition is that $\rho(B) < 1$. There are several criteria for ensuring that these conditions are satisfied without actually computing the eigenvalues of the matrix. In this section we wish to develop one such criterion, Lyapunov's Theorem, as well as a number of related results.

We begin with the following result for the difference equation:

$$\mathbf{x}^{(k+1)} = B\mathbf{x}^{(k)}, \qquad k = 0, 1, \ldots \tag{5.4.1}$$

5.4.1. STEIN'S THEOREM. *Let B be an $n \times n$ matrix. Then*

(a) *The trivial solution of (5.4.1) is asymptotically stable if and only if there is a Hermitian positive definite matrix H such that $H - B^*HB$ is positive definite.*
(b) *The trivial solution of (5.4.1) is stable if and only if there is a Hermitian positive definite matrix H such that $H - B^*HB$ is positive semidefinite.*

PROOF. (a) If the trivial solution of (5.4.1) is asymptotically stable, then, by 5.3.3, $\rho(B) < 1$. Therefore, by 5.3.6, there is a nonsingular matrix Q such that

$$\|QBQ^{-1}\|_2 < 1 \tag{5.4.2}$$

Set $H = Q^*Q$. Then H is Hermitian positive definite, and for any $\mathbf{x} \neq 0$

$$\begin{aligned} \mathbf{x}^*(H - B^*HB)\mathbf{x} &= \mathbf{x}^*Q^*Q\mathbf{x} - \mathbf{x}^*B^*Q^*QB\mathbf{x} \\ &= \|Q\mathbf{x}\|_2^2 - \|QB\mathbf{x}\|_2^2 = \|Q\mathbf{x}\|_2^2 - \|QBQ^{-1}Q\mathbf{x}\|_2^2 \\ &\geq (1 - \|QBQ^{-1}\|_2^2)\|Q\mathbf{x}\|_2^2 > 0 \end{aligned} \tag{5.4.3}$$

by (5.4.2). Thus, $H - B^*HB$ is positive definite. Conversely, if there is a Hermitian positive definite matrix H, so that $H - B^*HB$ is positive definite, then in the norm $\|\mathbf{x}\|_H = (\mathbf{x}^*H\mathbf{x})^{1/2}$ we have

$$\|B\mathbf{x}\|_H^2 = \mathbf{x}^*B^*HB\mathbf{x} < \mathbf{x}^*H\mathbf{x} = \|\mathbf{x}\|_H^2 \tag{5.4.4}$$

Hence, $\|B\|_H < 1$, so that $\rho(B) < 1$. Thus, by 5.3.3, the trivial solution of (5.4.1) is asymptotically stable.

(b) If the trivial solution is stable, then, by 5.3.3, either $\rho(B) < 1$, with which we have already dealt, or $\rho(B) = 1$ and B is spectral radius diagonable. In the latter case, by 5.3.6, there is a nonsingular Q, so that $\|QBQ^{-1}\|_2 = 1$. With $H = Q^*Q$, (5.4.3) becomes

$$\mathbf{x}^*(H - B^*HB)\mathbf{x} \geq 0$$

so that $H - B^*HB$ is positive semidefinite. Conversely, if $H - B^*HB$ is positive semidefinite, then (5.4.4) becomes

$$\|B\mathbf{x}\|_H^2 \leq \|\mathbf{x}\|_H^2$$

so that $\|B\|_H \leq 1$. Thus, B^k is bounded as $k \to \infty$, and the trivial solution is stable. □

For future use we restate Stein's Theorem in the following equivalent way.

5.4.2. Let B be an $n \times n$ matrix. Then $\rho(B) < 1$ if and only if there is a Hermitian positive definite matrix H such that $H - B^*HB$ is positive definite. $\rho(B) = 1$ and B is spectral radius diagonable if and only if there is a Hermitian positive definite matrix H such that $H - B^*HB$ is positive semidefinite.

For a differential equation

$$\dot{\mathbf{x}} = A\mathbf{x}, \qquad \mathbf{x}(0) = \mathbf{x}_0 \tag{5.4.5}$$

we next give the stability result corresponding to Stein's Theorem.

5.4.3. LYAPUNOV'S THEOREM. *Let A be an $n \times n$ matrix. Then*

(a) *The trivial solution of (5.4.5) is asymptotically stable if and only if there is a Hermitian positive definite matrix H such that $A^*H + HA$ is negative definite.*
(b) *The trivial solution of (5.4.5) is stable if and only if there is a Hermitian matrix H such that $A^*H + HA$ is negative semidefinite.*

We note that we can replace $A^*H + HA$ by $AH + HA^*$ in 5.4.3, and either form can be used as desired. This follows from the fact that if $C = AH + HA^*$ is negative (semi)definite, then, with $\hat{H} = H^{-1}$,

$$\hat{C} = H^{-1}CH^{-1} = H^{-1}A + A^*H^{-1} = \hat{H}A + A^*\hat{H}$$

and $\hat{C}$ is negative (semi)definite. Thus, there is a Hermitian positive definite H such that $AH + HA^*$ is negative (semi)definite if and only if there is a Hermitian positive definite $\hat{H}$ such that $\hat{H}A + A^*\hat{H}$ is negative (semi)definite, and we may use either form as the context dictates.

PROOF OF 5.4.3. (a) If the trivial solution of (5.4.5) is asymptotically stable, then, by 5.2.3, A is negative stable. If A has eigenvalues λ_i, then $I - A$ has eigenvalues $1 - \lambda_i$, and since $\text{Re}\,\lambda_i < 0$, it follows that $I - A$ is nonsingular. Let

$$B = (I - A)^{-1}(I + A) \tag{5.4.6}$$

Then it is easy to show (Exercise 5.4-2) that the eigenvalues of B are

$$\mu_i = \frac{1+\lambda_i}{1-\lambda_i}, \qquad i = 1, \ldots, n \tag{5.4.7}$$

and since $\text{Re}\,\lambda_i < 0$, it follows that $\rho(B) < 1$ (see Exercise 5.4-2). Thus, by 5.4.2, there is a positive definite Hermitian H so that $H - B^*HB$ is positive definite. But, by (5.4.6), and noting that $(I-A)^{-1}$ and $I + A$ commute (Exercise 5.4-1),

$$\begin{aligned} H - B^*HB &= H - (I + A^*)(I - A^*)^{-1}H(I - A)^{-1}(I + A) \\ &= (I - A^*)^{-1}[(I - A^*)H(I - A) \\ &\quad - (I + A^*)H(I + A)](I - A)^{-1} \\ &= -2((I - A)^{-1})^*(A^*H + HA)(I - A)^{-1} \end{aligned} \tag{5.4.8}$$

which shows that $A^*H + HA$ is negative definite. Conversely, if there is a Hermitian positive definite H such that $A^*H + HA$ is negative definite, then (5.4.8) shows that $H - B^*HB$ is positive definite, so that, by 5.4.2, $\rho(B) < 1$. Thus, by (5.4.7), it follows (Exercise 5.4-2) that $\text{Re}\,\lambda_i < 0$, $i = 1, \ldots, n$, so that, by 5.2.3, the trivial solution of (5.4.5) is asymptotically stable.

(b) If the trivial solution of (5.4.5) is stable, then, by 5.2.3, A is weakly negative stable. Thus, the eigenvalues (5.4.7) of B satisfy $|\mu_i| \leq 1$ (Exercise 5.4-2). Moreover, if $\text{Re}\,\lambda_i = 0$, then λ_i is associated only with diagonal Jordan blocks. Since B and A have the same eigenvectors and the same structure in their Jordan forms (Exercise 5.4-3), it follows that any μ_i, such that $|\mu_i| = 1$, is associated only with diagonal Jordan blocks of B. Thus, by 5.3.3 and 5.4.1, there is a Hermitian positive definite H such that $H - B^*HB$ is positive semidefinite; (5.4.8) then shows that $A^*H + HA$ is negative semidefinite. Conversely, if $A^*H + HA$ is negative semidefinite, (5.4.8) shows that $H - B^*HB$ is positive semidefinite. Thus, by 5.4.1 and 5.3.3, $\rho(B) \leq 1$, and B is spectral radius diagonable if $\rho(B) = 1$. Therefore, the eigenvalues of A satisfy $\text{Re}\,\lambda_i \leq 0$, and if $\text{Re}\,\lambda_i = 0$, then λ_i is associated with only 1×1 Jordan blocks. Hence, by 5.2.3, the trivial solution of (5.4.5) is stable. □

We will give a geometric interpretation of Lyapunov's Theorem. Let $\mathbf{x}(t)$ be a solution of $\dot{\mathbf{x}} = A\mathbf{x}$. Then

$$\begin{aligned} \frac{d}{dt}\|\mathbf{x}(t)\|_2^2 &= \frac{d}{dt}[\mathbf{x}^*(t)\mathbf{x}(t)] = \dot{\mathbf{x}}^*(t)\mathbf{x}(t) + \mathbf{x}^*(t)\dot{\mathbf{x}}(t) \\ &= \mathbf{x}^*(t)A^*\mathbf{x}(t) + \mathbf{x}^*(t)A\mathbf{x}(t) = \mathbf{x}^*(t)(A^* + A)\mathbf{x}(t) \end{aligned} \tag{5.4.9}$$

Now suppose that the Hermitian matrix $A^* + A$ is negative definite with eigenvalues $\lambda_1 \leq \cdots \leq \lambda_n < 0$. Then, by 3.1.9, $\mathbf{x}^*(A^* + A)\mathbf{x} \leq \lambda_n \mathbf{x}^*\mathbf{x}$ for all $\mathbf{x}$, so that

$$\frac{d}{dt}\|\mathbf{x}(t)\|_2^2 \leq \lambda_n \mathbf{x}^*(t)\mathbf{x}(t) = \lambda_n \|\mathbf{x}(t)\|_2^2$$

This differential inequality for $\|\mathbf{x}(t)\|_2^2$ implies (Exercise 5.4-4) that

$$\|\mathbf{x}(t)\|_2^2 \leq e^{\lambda_n t}\|\mathbf{x}(0)\|_2^2$$

so that the 2-norm of any solution decreases exponentially at least as fast as $e^{\lambda_n t}$.

On the other hand, if $A^* + A$ is indefinite, then $\lambda_n > 0$. Thus, if $\mathbf{x}_0$ is an eigenvector of $A^* + A$ corresponding to λ_n, then we have from (5.4.9) that the derivative of $\|\mathbf{x}(t)\|_2^2$ at zero is positive:

$$\frac{d}{dt}\|\mathbf{x}(0)\|_2^2 = \mathbf{x}_0^*(A^* + A)\mathbf{x}_0 = \lambda_n \mathbf{x}_0^*\mathbf{x}_0 > 0$$

Therefore, the 2-norm of $\mathbf{x}(t)$ initially increases, and this can happen even if A is negative stable, so that all solutions tend to zero. A simple example of this was given in Section 5.2, in which

$$A = \begin{bmatrix} -1 & \alpha \\ 0 & -1 \end{bmatrix}, \qquad A^* + A = \begin{bmatrix} -2 & \alpha \\ \alpha & -2 \end{bmatrix} \tag{5.4.10}$$

Here, A has eigenvalues -1, so that A is negative stable for all α. But $A^* + A$ has eigenvalues $-2 \pm \alpha$, so that if $\alpha > 2$, then $A^* + A$ is not negative definite. In particular, for example, if $\alpha = 4$, then $A^* + A$ has eigenvalues -6 and $+2$, and $\mathbf{x}_0 = (1, 1)^T$ is an eigenvector corresponding to the eigenvalue $+2$. If this $\mathbf{x}_0$ is the initial condition for $\dot{\mathbf{x}} = A\mathbf{x}$, then the 2-norm of the solution will grow initially before beginning to decrease.

On the other hand, Lyapunov's Theorem, 5.4.3, tells us that if A is negative stable we can always find an inner product norm, but not necessarily the l_2 norm, in which every solution of $\dot{\mathbf{x}} = A\mathbf{x}$ is decreasing for all time. Let H be the positive definite hermitian matrix of 5.4.3 for which $A^*H + HA$ is negative definite, and define the inner product and norm

$$(\mathbf{x}, \mathbf{y})_H = \mathbf{x}^* H \mathbf{y}, \qquad \|\mathbf{x}\|_H = (\mathbf{x}, \mathbf{x})_H^{1/2} \tag{5.4.11}$$

If $\mathbf{x}(t)$ is any solution of $\dot{\mathbf{x}} = A\mathbf{x}$, then in the norm (5.4.11)

$$\frac{d}{dt}\|\mathbf{x}(t)\|_H^2 = \frac{d}{dt}[\mathbf{x}(t)H\mathbf{x}(t)] = \dot{\mathbf{x}}(t)^* H\mathbf{x}(t) + \mathbf{x}(t)H\dot{\mathbf{x}}(t)$$
$$= [A\mathbf{x}^*(t)]^* H\mathbf{x}(t) + \mathbf{x}^*(t)HA\mathbf{x}(t)$$
$$= \mathbf{x}^*(t)(A^*H + HA)\mathbf{x}(t) < 0 \tag{5.4.12}$$

so that $\|\mathbf{x}(t)\|_H$ is decreasing for all t. If A is only weakly negative stable, then, by 5.4.3, $A^*H + HA$ is negative semidefinite, so that (5.4.12) becomes

$$\frac{d}{dt}\|\mathbf{x}(t)\|_H^2 \leq 0$$

and every solution is nonincreasing. Thus, a corollary of Lyapunov's Theorem is the following.

5.4.4. *If the $n \times n$ matrix A is (weakly) negative stable, then there is an inner product norm* (5.4.11) *on C^n in which $\|\mathbf{x}(t)\|_H$ is a decreasing (nonincreasing) function of t for every solution $\mathbf{x}(t)$ of $\dot{\mathbf{x}} = A\mathbf{x}$.*

An interesting special case of 5.4.4 is when there is an H such that $A^*H + HA = 0$, in which case

$$\frac{d}{dt}\|\mathbf{x}(t)\|_H^2 = 0$$

for any solution and all t. A special case of this is when A is skew-Hermitian, so that $A^* + A = 0$, and then

$$\frac{d}{dt}\|\mathbf{x}(t)\|_2^2 = 0$$

Definiteness of Non-Hermitian Matrices

The definiteness properties of the matrix $A^*H + HA$ can be phrased in another way to provide a useful generalization of definiteness properties of Hermitian matrices.

5.4.5. DEFINITION: *Definiteness.* Let (,) be an inner product on C^n. Then the $n \times n$ matrix A is

(a) *positive definite* with respect to (,) if $\text{Re}(\mathbf{x}, A\mathbf{x}) > 0$ for all $\mathbf{x} \neq 0$, and *negative definite* if $\text{Re}(\mathbf{x}, A\mathbf{x}) < 0$.
(b) *positive semidefinite* with respect to (,) if $\text{Re}(\mathbf{x}, A\mathbf{x}) \geq 0$ for all $\mathbf{x}$, and *negative semidefinite* if $\text{Re}(\mathbf{x}, A\mathbf{x}) \leq 0$.

Note that this definition applies to matrices that are not necessarily Hermitian. However, if A is Hermitian and $(\ ,\ )$ is the usual inner product on C^n, then 5.4.5 reduces to the usual definition of definiteness. A matrix A that is positive definite in the sense of 5.4.5 is sometimes also called *coercive* or *accretive.*

Now let H be Hermitian and positive definite, and let the inner product be defined by (5.4.11). Then

$$HA = \tfrac{1}{2}(A^*H + HA) + \tfrac{1}{2}(HA - A^*H)$$

is the decomposition of HA into its Hermitian and skew-Hermitian parts. Thus,

$$\text{Re}(\mathbf{x}, A\mathbf{x})_H = \text{Re}(\mathbf{x}^*HA\mathbf{x}) = \tfrac{1}{2}\mathbf{x}^*(A^*H + HA)\mathbf{x}$$

and we have the following characterizations.

5.4.6. Let H be an $n \times n$ Hermitian positive definite matrix, and let $(\ ,\ )_H$ be the inner product (5.4.11). Then A is positive (negative) (semi)definite in this inner product if and only if the Hermitian matrix $A^*H + HA$ is positive (negative) (semi)definite.

An equivalent way of stating Lyapunov's Theorem, 5.4.3, is then the following.

5.4.7. Let A be an $n \times n$ matrix. Then A is (weakly) negative stable if and only if there is a Hermitian positive definite matrix H such that A is negative (semi)definite in the inner product defined by H.

In the usual inner product $\mathbf{x}^*\mathbf{y}$ on C^n, Definition 5.4.5 becomes as follows: A is positive definite if $\text{Re}\, \mathbf{x}^*A\mathbf{x} > 0$, and by 5.4.6 this is equivalent to $A^* + A$ being positive definite. This, of course, reduces to the usual definition if A is Hermitian.

Still another way of stating the above results is in terms of the following concept.

5.4.8. Definition: *Numerical Range.* Let A be an $n \times n$ matrix. Then the *numerical range* of A with respect to the inner product $(\ ,\ )$ on C^n is $\{(\mathbf{x}, A\mathbf{x}): (\mathbf{x}, \mathbf{x}) = 1, \mathbf{x} \in C^n\}$.

That is, the numerical range is the set of all complex numbers that $(\mathbf{x}, A\mathbf{x})$, with $(\mathbf{x}, \mathbf{x}) = 1$, can take on. The numerical range is also called the *field of values* of A with respect to the inner product $(\ ,\ )$. If $(\ ,\ )$ is the usual inner product on C^n, then $\{\mathbf{x}^*A\mathbf{x}: \mathbf{x}^*\mathbf{x} = 1, \mathbf{x} \in C^n\}$ is the *field of values* or *numerical range* of A.

It is clear that the numerical range of A with respect to (,) is in the left half complex plane if and only if A is negative definite with respect to (,). Thus, still another statement of Lyapunov's Theorem is

5.4.9. Let A be an $n \times n$ matrix. Then A is negative stable if and only if there is a Hermitian positive definite matrix H such that the numerical range of A with respect to the inner product defined by H is in the left half of the complex plane.

Clearly, the numerical range with respect to the usual inner product contains all eigenvalues of A, because if λ is an eigenvalue and $\mathbf{x}$ a corresponding eigenvector with $\mathbf{x}^*\mathbf{x} = 1$, then $\mathbf{x}^*A\mathbf{x} = \lambda\mathbf{x}^*\mathbf{x} = \lambda$. The numerical range, in turn, is contained in a rectangle in the complex plane that is determined by the eigenvalues of the Hermitian and skew-Hermitian parts of A. We first note that if S is a skew-Hermitian matrix, then $-iS$ is Hermitian. Therefore, by 3.1.9,

$$\mu_1\mathbf{x}^*\mathbf{x}^* \leq \mathbf{x}^*(-iS)\mathbf{x} \leq \mu_n\mathbf{x}^*\mathbf{x}$$

where μ_1 and μ_n are the smallest and largest eigenvalues of $-iS$.

5.4.10. Let $A = A_1 + A_2$ be an $n \times n$ matrix, where A_1 and A_2 are the Hermitian and skew-Hermitian parts of A, respectively. Let γ_1 and γ_n be the smallest and largest eigenvalues of A_1, and μ_1 and μ_n the smallest and largest eigenvalues of $-iA_2$. Then the numerical range of A, and hence all eigenvalues of A, lies in the rectangle

$$R = [\gamma_1, \gamma_n] \times [i\mu_1, i\mu_n] \tag{5.4.13}$$

in the complex plane.

PROOF. Let λ be in the numerical range of A. Then there is an $\mathbf{x}$, with $\mathbf{x}^*\mathbf{x} = 1$, so that $\lambda = \mathbf{x}^*A\mathbf{x}$. Then $\bar{\lambda} = \mathbf{x}^*A^*\mathbf{x}$ and

$$\operatorname{Re}\lambda = \frac{\lambda + \bar{\lambda}}{2} = \frac{1}{2}\mathbf{x}^*(A + A^*)\mathbf{x} = \mathbf{x}^*A_1\mathbf{x}$$

Thus, by 3.1.9, $\gamma_1 \leq \operatorname{Re}\lambda \leq \gamma_n$. Similarly, for the imaginary part,

$$\operatorname{Im}\lambda = \frac{-i(\lambda - \bar{\lambda})}{2} = \frac{-i}{2}\mathbf{x}^*(A - A^*)\mathbf{x} = \mathbf{x}^*(-iA_2)\mathbf{x}$$

so that $\mu_1 \leq \text{Im}\,\lambda \leq \mu_n$. □

The rectangle (5.4.13) in the complex plane is illustrated by Figure 5.1.

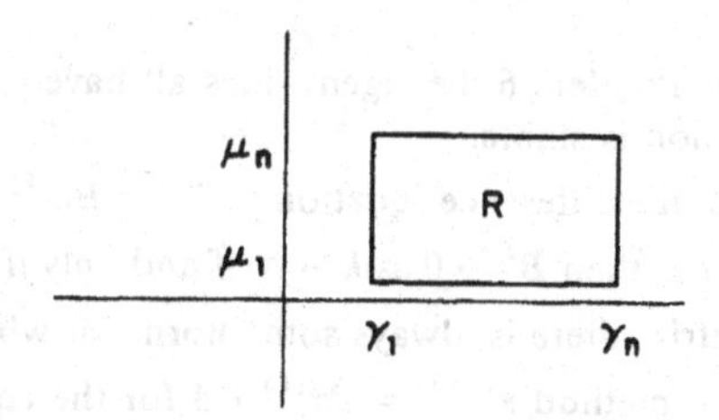

Figure 5.1. Rectangle containing the numerical range.

Exercises 5.4

1. Show that $(I - A)^{-1}(I + A) = (I + A)(I - A)^{-1}$.
2. If A has eigenvalues $\lambda_1, \ldots, \lambda_n$, show that the eigenvalues of $(I - A)^{-1}(I + A)$ are $\mu_i = (1 + \lambda_i)/(1 - \lambda_i)$, $i = 1, \ldots, n$. Conclude that if $\text{Re}\,\lambda_i < 0$, then $|\mu_i| < 1$, and if $\text{Re}\,\lambda_i \leq 0$, then $|\mu_i| \leq 1$. Conversely, conclude that if $|\mu_i| < 1$, then $\text{Re}\,\lambda_i < 0$, and if $|\mu_i| \leq 1$, then $\text{Re}\,\lambda_i \leq 0$.
3. Show that $(I - A)^{-1}(I + A)$ and A have the same eigenvectors and the same structure in their Jordan forms.
4. Let σ be a nonnegative differentiable function such that $d\sigma/dt \leq \lambda\sigma$ for $t \geq 0$. Show that $\sigma(t) \leq e^{\lambda t}\sigma(0)$.
5. Show that if A is normal and negative stable, then $A + A^*$ is negative definite.
6. If A is skew-Hermitian, show that $\|e^{At}\|_2 = 1$ for all $t > 0$.

Review Questions—Chapter 5

Answer whether the following statements are true or false and justify your assertions.

1. The exponential of an $n \times n$ matrix A is defined only if A is nonsingular.
2. If A is an $n \times n$ matrix, then e^A is similar to a diagonal matrix if and only if A is similar to a diagonal matrix.
3. If an nth-order differential equation with constant coefficients is converted to a first-order system, then the coefficient matrix is always similar to a diagonal matrix.
4. The polar decomposition of an $n \times n$ matrix A is the same as the singular value decomposition if A is Hermitian.
5. If A is an $n \times n$ matrix all of whose eigenvalues have zero real part, then A is weakly negative stable.

6. If an nth-order differential equation with constant coefficients is converted to a first-order system and the eigenvalues of the coefficient matrix all have negative real part, then the trivial solution of the differential equation is asymptotically stable.
7. If in the situation of Problem 6 the eigenvalues all have nonpositive real part, then the trivial solution is stable.
8. The trivial solution of the difference equation $\mathbf{x}^{(k+1)} = B\mathbf{x}^{(k)}$ is stable if $\rho(B) \leq 1$.
9. If B is an $n \times n$ matrix, then $B^k \to 0$ as $k \to \infty$ if and only if $\|B\|_2^k \to 0$ as $k \to \infty$.
10. If B is an $n \times n$ matrix, there is always some norm for which $\|B\| = \rho(B)$.
11. A consistent iterative method $\mathbf{x}^{(k+1)} = B\mathbf{x}^{(k)} + \mathbf{d}$ for the equation $A\mathbf{x} = \mathbf{b}$ with A nonsingular may be viewed as a difference equation, and convergence of the iterates is equivalent to asymptotic stability of the trivial solution of the homogeneous difference equation.
12. The geometric series $\sum_{i=0}^{k} B^i$ converges as $k \to \infty$, provided B has no eigenvalue equal to 1.
13. If the eigenvalues of A all have positive real part, then there is an inner product in which $(\mathbf{x}, A\mathbf{x}) > 0$ for all $\mathbf{x} \neq 0$.

References and Extensions: Chapter 5

1. The basic theory and applications of linear differential equations are covered in a large number of books at various levels. A classical reference is Coddington and Levinson [1955]. See also Lukes [1982].
2. Other matrix functions can be defined, as was done for e^A, by means of power series. If $f(\lambda) = \sum_{k=0}^{\infty} c_k\lambda^k$ is a power series that is convergent for all complex λ, then we can define $f(A) = \sum_{k=0}^{\infty} c_k A^k$ for any $n \times n$ matrix A. More generally, $f(A)$ may be defined by the power series if the spectrum of A lies in the domain of convergence of the power series. By means of this power series definition, we can define such matrix functions as sin A, cos A, cosh A, log A, and so forth. For a full discussion, see, for example, Gantmacher [1959].
3. For further information on the polar decomposition and Cayley transforms as well as related results, see Gantmacher [1959].
4. The designation "negative stable" for a matrix all of whose eigenvalues have negative real part has become fairly common, although a variety of other terms may be found in the literature. The designation "weakly negative stable" is much less commonly used.
5. The designation "spectral radius diagonable" is nonstandard, and other terms may be found in the literature. For example, Householder [1964] calls such matrices "of class M."
6. Higher-order systems of the form

$$A_0\mathbf{x}^{(m)}(t) + A_1\mathbf{x}^{m-1}(t) + \cdots + A_{m-1}\mathbf{x}(t) + A_m\mathbf{x}(t) = 0$$

where $\mathbf{x}(t)$ is an n-vector and $A_0, \ldots, A_m$ are given $n \times n$ matrices, may also be converted to a first-order system of n^2 equations, provided that A_0 is nonsingular. The situation is more difficult if A_0 is singular; see Campbell [1980] and Gohberg et al. [1982]. The same situation is true for higher-order systems of difference equations.

7. Further discussion of the relation between iterative methods and difference equations, including nonlinear problems, is given in Ortega [1973].
8. There is a vast literature on iterative methods and their analysis. See, for example, Hageman and Young [1981], Young [1971], and Varga [1962]. For nonlinear problems see Ortega and Rheinboldt [1970].
9. Further results on the Lyapunov theory of Section 5.4 may be found, for example, in Bellman [1960] and LaSalle and Lefschetz [1961]. In particular, if A is negative stable and C is Hermitian and positive definite, then the integral

$$H = \int_0^\infty e^{At} C e^{A^*t}\, dt$$

exists, is Hermitian positive definite, and satisfies $AH + HA^* = -C$. This gives an "explicit" representation of the matrix H of Theorem 5.4.3.

10. There are several other criteria for ascertaining if a matrix A is negative stable. See Gantmacher [1959] for several of these.
11. Another approach to the Lyapunov theorem, 5.4.3, is given by Ostrowski and Schneider [1962]. They extend the concept of inertia, given in Section 3.1 for Hermitian matrices, to an arbitrary $n \times n$ matrix A by defining $\text{In}(A) = (i, j, k)$, where i, j, k are the number of eigenvalues with positive, negative, and zero real parts, respectively. They then prove that for a given A there is a Hermitian matrix H such that AH is positive definite if and only if A has no purely imaginary eigenvalues. Such an H satisfies $\text{In}(H) = \text{In}(A)$. Theorem 5.4.3 is an easy consequence of this result, because if A is negative stable it has no purely imaginary eigenvalue; hence, H exists with $\text{In}(H) = \text{In}(A) = (0, n, 0)$. Thus, $-H$ is positive definite, so that $-AH$ is negative definite. Conversely, if there is a negative definite Hermitian H such that AH is positive definite, then $\text{In}(H) = \text{In}(A)$ implies that A is negative stable.

6

Other Topics

We collect in this final chapter a number of different topics. In Section 6.1 we deal with matrices that have nonnegative or positive entries, or whose inverses have this property. Such matrices arise in a number of application areas, and there are many beautiful and striking results concerning them. In Section 6.2 we treat various extensions of the eigenvalue problem, including the so-called generalized eigenvalue problem and higher-order problems. In Section 6.3 we consider some very special, but important, types of matrices, including Kronecker products and circulants. Finally, in Section 6.4, we deal with matrix equations and commutativity of matrices.

6.1. Nonnegative Matrices and Related Results

A real n-vector $\mathbf{x}$, or a real $m \times n$ matrix A, is *nonnegative* if all components are nonnegative; that is, if $x_i \geqslant 0$, $i = 1, \ldots, n$, or $a_{ij} \geqslant 0$, $i = 1, \ldots, m$, $j = 1, \ldots, n$. The vector $\mathbf{x}$ or matrix A is *positive* if all components are positive: $x_i > 0$, $i = 1, \ldots, n$ or $a_{ij} > 0$, $i = 1, \ldots, m$, $j = 1, \ldots, n$. We will use the notation $\mathbf{x} \geqslant 0$, $A \geqslant 0$, $\mathbf{x} > 0$, $A > 0$ to denote nonnegative or positive vectors and matrices. We will also use the notation $\mathbf{x} \geqslant \mathbf{y}$ or $A \geqslant B$ to mean $\mathbf{x} - \mathbf{y} \geqslant 0$ or $A - B \geqslant 0$.

Nonnegative or positive matrices arise in a variety of application areas, including numerical analysis, probability theory, economics, and operations research. They also arise by taking the absolute values of the elements of a real or complex matrix. Here, $|\mathbf{x}|$ or $|A|$ denotes the vector or matrix of absolute values of the elements of $\mathbf{x}$ or A; clearly, $|\mathbf{x}| \geqslant 0$ and $|A| \geqslant 0$.

A particularly important type of $n \times n$ nonnegative matrix A occurs when all columns sum to one: $\sum_{i=1}^{n} a_{ij} = 1$, $j = 1, \ldots, n$. Such a matrix is called *stochastic* or *Markov*, and it arises in problems such as described in the following example of a Markov process.

Let $x_i^{(k)}$ be the probability that a system is in state i at time k, and let a_{ij} be the probability that a system in state j at time k will be in state i at time $k+1$. Then

$$x_i^{(k+1)} = \sum_{j=1}^{n} a_{ij} x_j^{(k)}, \qquad i = 1, \ldots, n \tag{6.1.1}$$

or

$$\mathbf{x}^{(k+1)} = A\mathbf{x}^{(k)}, \qquad \mathbf{x}^{(0)} \text{ given, } k = 0, 1, \ldots \tag{6.1.2}$$

Assuming that the system must be in one of the n allowable states at each time, it is clear that we must have

$$\sum_{i=1}^{n} x_i^{(k)} = 1, \quad k = 0, 1, \ldots; \qquad \sum_{i=1}^{n} a_{ij} = 1, \quad j = 1, \ldots, n \tag{6.1.3}$$

Thus A is a stochastic matrix. Moreover, if $A > 0$, we have the following remarkable result.

6.1.1. *If A is a positive stochastic matrix, then there is a unique positive vector $\hat{\mathbf{x}}$, with $\sum_{i=1}^{n} \hat{x}_i = 1$, such that for any nonnegative vector $\mathbf{x}^0$ with $\sum_{i=1}^{n} x_i^0 = 1$, the sequence of* (6.1.2) *satisfies*

$$\mathbf{x}^{(k)} \to \hat{\mathbf{x}} \qquad \text{as } k \to \infty \tag{6.1.4}$$

The proof of 6.1.1 depends on a basic theorem about the spectral radius $\rho(A)$ of positive matrices.

6.1.2. PERRON'S THEOREM. *Let A be a positive $n \times n$ matrix. Then $\rho(A)$ is a simple eigenvalue of A, and all other eigenvalues are less than $\rho(A)$ in modulus. Moreover, an eigenvector associated with $\rho(A)$ may be taken to be positive.*

In general, of course, a nonsymmetric real matrix may have all complex eigenvalues, but Perron's Theorem guarantees that for a positive matrix its largest eigenvalue in absolute value is always real. Perron's Theorem will be a corollary of the more general theorem 6.1.4, but we will use it now to prove 6.1.1.

PROOF OF 6.1.1. Since A is stochastic, we have $\|A\|_1 = 1$, and thus $\rho(A) \leq 1$. If $\mathbf{e}$ is the vector all of whose components are 1, then $A^T\mathbf{e} = \mathbf{e}$, again because A is stochastic. Hence, A^T has an eigenvalue equal to 1, and so does A, since A^T and A have the same eigenvalues. Thus, $\rho(A) = 1$, and, by Perron's Theorem, 1 is a simple eigenvalue and all other eigenvalues are less than 1 in absolute value. It follows that the Jordan form of A must be of the form

$$J = \begin{bmatrix} 1 & \\ & \hat{J} \end{bmatrix}$$

where $\rho(\hat{J}) < 1$. Then, by 5.3.4, $\hat{J}^k \to 0$ as $k \to \infty$ so that

$$J^k \to \text{diag}(1, 0, \ldots, 0) \qquad \text{as } k \to \infty$$

Therefore, if $A = PJP^{-1}$, we have

$$\mathbf{x}^{(k)} = A^k\mathbf{x}^{(0)} = PJ^kP^{-1}\mathbf{x}^{(0)} \to (\mathbf{p}_1, 0, \ldots, 0)\mathbf{q} = q_1\mathbf{p}_1 \qquad \text{as } k \to \infty \quad (6.1.5)$$

where $\mathbf{p}_1$ is the first column of P and q_1 is the first component of $\mathbf{q} = P^{-1}\mathbf{x}^{(0)}$. By Perron's Theorem the eigenvector $\mathbf{p}_1$ may be assumed to be positive. By (6.1.3) the sum of the components of each $\mathbf{x}^{(k)}$ equals 1, and this property must persist in the limit. Thus $\sum_{i=1}^n \hat{x}_i = 1$, so that $q_1 > 0$ and $\hat{\mathbf{x}} = q_1\mathbf{p}_1 > 0$. Since 1 is a simple eigenvalue, all associated eigenvectors are multiples of $\mathbf{p}_1$, and thus $\hat{\mathbf{x}}$ is unique since $\sum \hat{x}_i = 1$. □

Irreducible Matrices

Perron's Theorem extends to nonnegative matrices that have the property of irreducibility. An $n \times n$ matrix A is *reducible* if there is a permutation matrix P such that

$$PAP^T = \begin{bmatrix} A_{11} & A_{12} \\ 0 & A_{22} \end{bmatrix} \qquad (6.1.6)$$

where A_{11} and A_{22} are square submatrices. The matrix S is *irreducible* if it is not reducible.

Clearly, any matrix all of whose elements are nonzero is irreducible; in particular, positive matrices are irreducible. On the other hand, any matrix that has a zero row or column is reducible (Exercise 6.1-5). More generally, a useful approach to ascertaining whether a matrix is irreducible

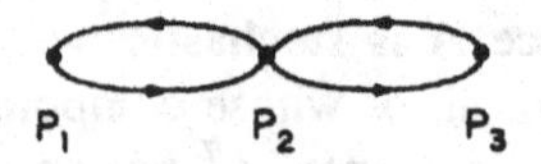

Figure 6.1. Directed graph of (6.1.7).

is through the *directed graph* of a matrix. This is illustrated in Figure 6.1 for the matrix

$$A = \begin{bmatrix} 2 & 1 & 0 \\ 1 & 2 & 1 \\ 0 & 1 & 2 \end{bmatrix} \tag{6.1.7}$$

In general, the directed graph of an $n \times n$ matrix is obtained by connecting n points $P_1, \ldots, P_n$ (on the real line or in the plane) by a *directed link* from P_i to P_j if $a_{ij} \neq 0$. Other examples are given in Figure 6.2 for the matrices

$$A = \begin{bmatrix} 1 & 0 & 1 \\ 1 & 0 & 2 \\ 1 & 0 & 0 \end{bmatrix}, \qquad A = \begin{bmatrix} 0 & 0 & 0 \\ 1 & 1 & 1 \\ 2 & 2 & 2 \end{bmatrix}$$

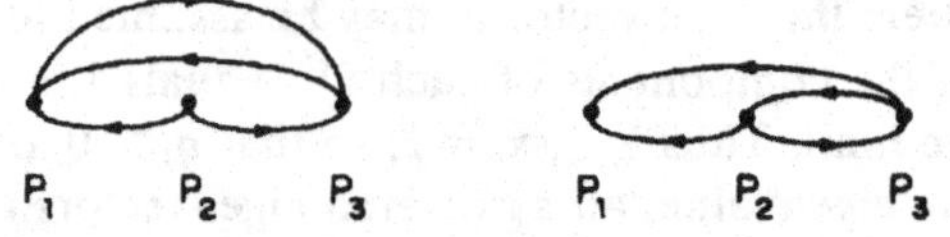

Figure 6.2. Directed graphs.

Note that the definition of a directed graph of a matrix A is sometimes interpreted to mean that if the ith diagonal element of A is nonzero, there would be a path from P_i to itself. This does not affect our use of directed graphs, and we have not indicated such loops in the figures.

A directed graph is *strongly connected* if for any two points P_i and P_j, there is a directed path from P_i to P_j. It is clear that the graph of Figure 6.1 is strongly connected. However, neither of the graphs of Figure 6.2 are strongly connected. In the first, there is no directed path leading to P_2; in the second, there is no directed path from P_1 to either P_2 or P_3.

It is easy to see (Exercise 6.1-6), that the matrix of (6.1.7) is irreducible, while those of Figure 6.2 are reducible. This is a manifestation of the connection between irreducibility and strongly connected graphs that is given by the following result. The proof is left to Exercise 6.1-6.

6.1.3. An $n \times n$ matrix A is irreducible if and only if the directed graph of A is strongly connected.

We now prove the following important theorem, which contains Perron's Theorem 6.1.2 as a special case.

6.1.4. PERRON–FROBENIUS THEOREM. *Let A be an $n \times n$ nonnegative irreducible matrix. Then $\rho(A)$ is a simple eigenvalue of A, and an associated eigenvector may be taken to be positive. Moreover, if A has at least one row with all nonzero elements, then any other eigenvalue λ of A satisfies $|\lambda| < \rho(A)$.*

The proof is based on the following lemma.

6.1.5. *If B is an $n \times n$ irreducible nonnegative matrix with positive diagonal elements, then $B^{n-1} > 0$.*

PROOF. We may assume without loss of generality, since otherwise we could factor out the positive diagonal of B, that $B = I + C$, where $C \geqslant 0$ contains the off-diagonal elements of B and is also irreducible. $B^{n-1} > 0$ is equivalent to $B^{n-1}\mathbf{e}_i > 0$, $i = 1, \ldots, n$, where $\mathbf{e}_i$ is the vector with a 1 in the ith component and 0's elsewhere. Hence, it is sufficient to show that $B^{n-1}\mathbf{x} > 0$ for any nonnegative nonzero vector $\mathbf{x}$. Clearly, $B\mathbf{x} = \mathbf{x} + C\mathbf{x}$ has as many nonzero elements as $\mathbf{x}$ itself. If it has exactly as many, then there is a permutation matrix P such that

$$P\mathbf{x} = \begin{bmatrix} \mathbf{y} \\ 0 \end{bmatrix}, \qquad P(\mathbf{x} + C\mathbf{x}) = \begin{bmatrix} \mathbf{z} \\ 0 \end{bmatrix}$$

where $\mathbf{y}$ and $\mathbf{z}$ are positive and of the same length. Therefore, if $\hat{C} = PCP^T$ is partitioned accordingly, we have

$$\begin{bmatrix} \mathbf{z} \\ 0 \end{bmatrix} = P(\mathbf{x} + C\mathbf{x}) = \begin{bmatrix} \mathbf{y} \\ 0 \end{bmatrix} + \hat{C}\begin{bmatrix} \mathbf{y} \\ 0 \end{bmatrix} = \begin{bmatrix} \mathbf{y} \\ 0 \end{bmatrix} + \begin{bmatrix} \hat{C}_{11} & \hat{C}_{12} \\ \hat{C}_{21} & \hat{C}_{22} \end{bmatrix}\begin{bmatrix} \mathbf{y} \\ 0 \end{bmatrix}$$

Since $\mathbf{y} > 0$, we must have $\hat{C}_{21} = 0$, which contradicts the irreducibility of C. Hence, $B\mathbf{x}$ has more nonzero elements than $\mathbf{x}$, and repeated application of this shows that $B^{n-1}\mathbf{x} > 0$. □

PROOF OF 6.1.4. Let

$$S = \{\lambda > 0\colon A\mathbf{x} \geqslant \lambda\mathbf{x} \text{ for some } \mathbf{x} \geqslant 0 \text{ with } \|\mathbf{x}\|_\infty = 1\}$$

S is not empty, because if $\mathbf{x}$ is the vector all of whose components are 1, we must have $A\mathbf{x} > 0$, or else A would have a zero row and be reducible (Exercise 6.1-5). Thus, there is a $\lambda > 0$ so that $A\mathbf{x} \geqslant \lambda\mathbf{x}$. Moreover, S is bounded, because if $\lambda\mathbf{x} \leqslant A\mathbf{x}$, then

$$\lambda\|\mathbf{x}\|_\infty \leqslant \|A\mathbf{x}\|_\infty \leqslant \|A\|_\infty\|\mathbf{x}\|_\infty$$

so that $\lambda \leq \|A\|_\infty$. Let $\hat{\lambda} = \sup S$, and let $\{\lambda_i\}$ be a sequence converging to $\hat{\lambda}$, with $\{\mathbf{x}^{(i)}\}$ the corresponding vectors. Since $\|\mathbf{x}^{(i)}\|_\infty = 1$, there is a subsequence converging to a vector $\hat{\mathbf{x}}$, with $\|\hat{\mathbf{x}}\|_\infty = 1$, and since $\lambda_i \mathbf{x}^{(i)} \leq A\mathbf{x}^{(i)}$, it follows that

$$\hat{\lambda}\hat{\mathbf{x}} \leq A\hat{\mathbf{x}} \tag{6.1.8}$$

We now wish to show that (6.1.8) is actually an equality. Suppose that $(A - \hat{\lambda}I)\hat{\mathbf{x}} = \mathbf{y} \neq 0$. Since $\mathbf{y} \geq 0$, by 6.1.5, $(I + A)^{n-1}\mathbf{y} > 0$. Set $\mathbf{w} = (I + A)^{n-1}\hat{\mathbf{x}}$. Then, since $(I + A)^{n-1}$ and A commute, we have

$$A\mathbf{w} - \hat{\lambda}\mathbf{w} = (I + A)^{n-1}(A - \hat{\lambda}I)\hat{\mathbf{x}} = (I + A)^{n-1}\mathbf{y} > 0$$

Since, by 6.1.5, $\mathbf{w} > 0$, there is an $\varepsilon > 0$ so that $A\mathbf{w} \geq (\hat{\lambda} + \varepsilon)\mathbf{w}$, and this contradicts the definition of $\hat{\lambda}$. Hence,

$$A\hat{\mathbf{x}} = \hat{\lambda}\hat{\mathbf{x}} \tag{6.1.9}$$

The eigenvector $\hat{\mathbf{x}}$ must be positive since

$$0 < (I + A)^{n-1}\hat{\mathbf{x}} = (1 + \hat{\lambda})^{n-1}\hat{\mathbf{x}} \tag{6.1.10}$$

We next show that $\hat{\lambda} = \rho(A)$ and that $\hat{\lambda}$ is simple. Let λ be any other eigenvalue of A and $\mathbf{x}$ a corresponding eigenvector. Then, since $A \geq 0$,

$$|\lambda||\mathbf{x}| = |\lambda\mathbf{x}| = |A\mathbf{x}| \leq A|\mathbf{x}|$$

so that by the definition of $\hat{\lambda}$ we must have $\hat{\lambda} \geq |\lambda|$. Hence, $\hat{\lambda} = \rho(A)$. Suppose that $\hat{\lambda}$ is a multiple eigenvalue and that $\hat{\mathbf{x}}$ and $\mathbf{z}$ are two linearly independent eigenvectors. Since $\hat{\lambda}$ is real, $\mathbf{z}$ can be taken to be real. Then $\hat{\mathbf{x}} + \alpha\mathbf{z}$ is an eigenvector for all scalars α, and, since $\hat{\mathbf{x}} > 0$, we may increase or decrease α from 0 until at least one component of $\hat{\mathbf{x}} + \alpha\mathbf{z}$ is zero and $\hat{\mathbf{x}} + \alpha\mathbf{z} \geq 0$. Clearly, $\hat{\mathbf{x}} + \alpha\mathbf{z} \neq 0$ if $\hat{\mathbf{x}}$ and $\mathbf{z}$ are linearly independent. But then, as in (6.1.10), we must have $\mathbf{x} + \alpha\mathbf{z} > 0$, which is a contradiction. Thus, if $\hat{\lambda}$ is a multiple eigenvalue, it is associated with a nondiagonal Jordan block, and, by (3.2.8), there is a vector $\mathbf{y}$, which we may assume to be real because $\hat{\lambda}$ is real, such that

$$(A - \hat{\lambda}I)^2\mathbf{y} = 0, \qquad (A - \hat{\lambda}I)\mathbf{y} \neq 0$$

Therefore, $(A - \hat{\lambda}I)\mathbf{y}$ is an eigenvector associated with $\hat{\lambda}$ and must be a multiple of $\hat{\mathbf{x}}$:

$$\hat{\mathbf{x}} = \alpha(A - \hat{\lambda}I)\mathbf{y} = (A - \hat{\lambda}I)\mathbf{z}$$

with $\mathbf{z} = \alpha\mathbf{y}$. Thus, $A\mathbf{z} = \hat{\lambda}\mathbf{z} + \hat{\mathbf{x}} > \hat{\lambda}\mathbf{z}$, so that $A|\mathbf{z}| \geq |A\mathbf{z}| > \hat{\lambda}|\mathbf{z}|$, which contradicts the maximum property of $\hat{\lambda}$.

Finally, we prove that if A has at least one row with all non-zero elements, then there are no other eigenvalues of modulus $\hat{\lambda}$. Suppose there is an eigenvalue λ with $|\lambda| = \hat{\lambda}$, and let $\mathbf{z}$ be an associated eigenvector. Then $\hat{\lambda}|\mathbf{z}| = |\lambda\mathbf{z}| \leq A|\mathbf{z}|$, and the argument that led from (6.1.8) to (6.1.9) shows that $\hat{\lambda}|\mathbf{z}| = A|\mathbf{z}|$. Hence, $|\mathbf{z}|$ is a multiple of $\hat{\mathbf{x}}$, $|\mathbf{z}| = \alpha\hat{\mathbf{x}}$, and we must have

$$|A\mathbf{z}| = |\hat{\lambda}\mathbf{z}| = \hat{\lambda}|\mathbf{z}| = A|\mathbf{z}|$$

In particular, for the kth row of A,

$$\left|\sum_{j=1}^{n} a_{kj}z_j\right| = \sum_{j=1}^{n} a_{kj}|z_j|$$

and if $a_{kj} > 0$, $j = 1, \ldots, n$, this is possible (see Exercise 6.1-4) only if $\mathbf{z}$ is a multiple of a nonnegative nonzero vector: $\mathbf{z} = \beta\mathbf{w}$. Thus, $|\mathbf{z}| = |\beta|\mathbf{w} = \alpha\hat{\mathbf{x}}$, so that $\mathbf{w}$ is a multiple of $\hat{\mathbf{x}}$. Hence, $\mathbf{z}$ is a multiple of $\hat{\mathbf{x}}$, and we must have $\lambda = \hat{\lambda}$. □

We give a simple example of 6.1.4. It is easy to see by 6.1.3 that the matrix

$$A = \begin{bmatrix} 0 & 1 & 0 \\ 0 & 0 & 1 \\ 1 & 0 & 0 \end{bmatrix}$$

is irreducible. Moreover, $\|A\|_\infty = 1$, and 1 is an eigenvalue with eigenvector $(1, 1, 1)^T$. Hence, $\rho(A) = 1$, and, by 6.1.4, 1 is a simple eigenvalue. We cannot conclude, however, that there are no other eigenvalues with absolute value 1. Indeed, the characteristic equation is $\lambda^3 - 1 = 0$, so that the eigenvalues are the cube roots of unity. The other two eigenvalues are $(-1 \pm i\sqrt{3})/2$, each with absolute value 1. (See Exercise 6.1-11 for the $n \times n$ case of this example.)

Nonnegative Inverses

An important type of matrix is one that has a nonnegative inverse. As an example of conditions on the matrix A that ensure that $A^{-1} \geq 0$, we give the following result.

6.1.6. Let A be a real $n \times n$ matrix with positive diagonal elements, and set $D = \mathrm{diag}(a_{11}, \ldots, a_{nn})$. Assume that $B = D - A \geq 0$ and $\rho(D^{-1}B) < 1$. Then A^{-1} exists and $A^{-1} \geq 0$. Moreover, if A is irreducible, then $A^{-1} > 0$.

PROOF. Set $C = D^{-1}B$. Then $C \geqslant 0$ and $A = D(I - C)$. By the Neumann expansion 5.3.8

$$(I - C)^{-1} = \sum_{i=0}^{\infty} C^i \tag{6.1.11}$$

Since every matrix in the summation is nonnegative, we have $(I - C)^{-1} \geqslant 0$, and thus $A^{-1} = (I - C)^{-1}D^{-1} \geqslant 0$. If A is irreducible, then so is C. Hence, by 6.1.5,

$$0 < (I + C)^{n-1} = I + \alpha_1 C + \cdots + \alpha_{n-2}C^{n-2} + C^{n-1}$$

where the α_i are the binomial coefficients. Since $C \geqslant 0$ and the α_i are positive, it follows that $I + C + \cdots + C^{n-1} > 0$. Thus, from (6.1.11), all elements of $(I - C)^{-1}$ are positive. □

Theorem 6.1.6 pertains to an important type of matrix defined as follows.

6.1.7. DEFINITION. *A real $n \times n$ matrix A is an M-matrix if $a_{ij} \leqslant 0$, $i, j = 1, \ldots, n$, $i \neq j$, and $A^{-1} \geqslant 0$.*

We show that if A is an M-matrix, then A must have positive diagonal elements and $\rho(D^{-1}B) < 1$, where, again, $A = D - B$ is the splitting of A into its diagonal and off-diagonal elements. We have

$$I = (D - B)A^{-1} = DA^{-1} - BA^{-1}$$

Since $BA^{-1} \geqslant 0$, this shows that the diagonal elements of D are positive. Let $C = D^{-1}B$. Then $(I - C)^{-1} = A^{-1}D \geqslant 0$, so that by the identity (5.3.19)

$$I + C + \cdots + C^k = (I - C)^{-1} - (I - C)^{-1}C^{k+1} \leqslant (I - C)^{-1}$$

Hence, the left-hand side must converge, and since $C \geqslant 0$, we must have $C^k \to 0$ as $k \to \infty$. Therefore, $\rho(D^{-1}B) < 1$, and on the basis of 6.1.6 we can state the following:

6.1.8. A real $n \times n$ matrix A with nonpositive off-diagonal elements is an M-matrix if and only if A has positive diagonal elements and $\rho(D^{-1}B) < 1$.

We next give an example of a class of M-matrices that arise in various applications. Let A be the tridiagonal matrix

$$A = \begin{bmatrix} a_1 & b_1 & & \\ c_1 & a_2 & \ddots & \\ & \ddots & \ddots & b_{n-1} \\ & & c_{n-1} & a_n \end{bmatrix} \tag{6.1.12}$$

where

$$a_i > 0, \quad i = 1, \ldots, n, \qquad b_i < 0, \quad c_i < 0, \qquad i = 1, \ldots, n-1 \tag{6.1.13}$$

and

$$a_1 + b_1 > 0, \qquad a_n + c_{n-1} > 0, \qquad a_i + b_i + c_{i-1} \geqslant 0, \quad i = 2, \ldots, n-1 \tag{6.1.14}$$

A simple example of a matrix that satisfies these conditions is

$$A = \begin{bmatrix} 2 & -1 & & \\ -1 & \ddots & \ddots & \\ & \ddots & \ddots & -1 \\ & & -1 & 2 \end{bmatrix} \tag{6.1.15}$$

This important matrix arises in a number of applications, including the numerical solution of differential equations.

6.1.9. *If the matrix* (6.1.12) *satisfies* (6.1.13) *and* (6.1.14), *then A is an M-matrix and $A^{-1} > 0$. In particular, this is true for the matrix* (6.1.15).

PROOF. By (6.1.13) the diagonal elements of A are positive, and the off-diagonal elements are nonpositive. Hence, $C = D^{-1}B \geqslant 0$, where $A = D - B$. Then (6.1.14) implies that $\|C\|_\infty \leqslant 1$. If strict equality held in each of the inequalities of (6.1.14), we would have

$$\rho(C) \leqslant \|C\|_\infty < 1$$

and we could conclude from 6.1.6 that $A^{-1} > 0$. Otherwise, we note that the directed graph of the matrix (6.1.12) is as shown in Figure 6.3 and is

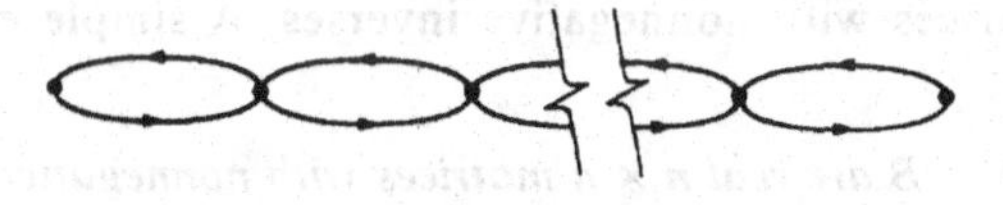

Figure 6.3. Directed graph of (6.1.12).

clearly strongly connected, so that A is irreducible. Then C is also irreducible, and we use the following generally useful result to complete the proof. □

6.1.10. If C is an irreducible $n \times n$ real or complex matrix such that

$$\sum_{j=1}^{n} |c_{ij}| \leq 1, \qquad i = 1, \ldots, n \tag{6.1.16}$$

and strict inequality holds in (6.1.16) *for at least one i, then $\rho(C) < 1$.*

PROOF. Clearly, $\rho(C) \leq \|C\|_\infty \leq 1$ follows from (6.1.16). Suppose there were an eigenvalue λ of C with $|\lambda| = 1$. Let $\mathbf{x}$ be a corresponding eigenvector with $\|\mathbf{x}\|_\infty = 1$, and suppose that all components of $\mathbf{x}$ have absolute value one. Then we would have, from $C\mathbf{x} = \lambda\mathbf{x}$,

$$1 = |\lambda||x_i| = |\lambda x_i| = \left|\sum_{j=1}^{n} c_{ij}x_j\right| \leq \sum_{j=1}^{n} |c_{ij}| \tag{6.1.17}$$

for all i, which would contradict the hypothesis of strict inequality in (6.1.16) for at least one i. Therefore, not all components of $\mathbf{x}$ have the same absolute value, and there is a permutation matrix P such that if $\mathbf{y} = P\mathbf{x}$, then

$$|y_i| < 1, \quad i = 1, \ldots, r, \qquad |y_i| = 1, \quad i = r+1, \ldots, n$$

Let $\hat{C} = PCP^T$. Then the inequalities (6.1.16) also hold for $\hat{C}$ (Exercise 6.1-12). Since $\hat{C}\mathbf{y} = \lambda\mathbf{y}$, it follows as in (6.1.17) that, for any $i \geq r+1$,

$$1 \leq \sum_{j=1}^{n} |\hat{c}_{ij}||y_j| < \sum_{j=1}^{n} |\hat{c}_{ij}| \leq 1$$

unless $\hat{c}_{ij} = 0$ for $j = 1, \ldots, r$ and $i = r+1, \ldots, n$. Thus, $\hat{C}$ is of the form

$$\hat{C} = \begin{bmatrix} \hat{C}_{11} & \hat{C}_{12} \\ 0 & \hat{C}_{22} \end{bmatrix}$$

This contradicts the irreducibility of C, and we conclude that $|\lambda| = 1$ is not possible. □

Comparison Theorems

There are many so-called comparison theorems for nonnegative matrices or matrices with nonnegative inverses. A simple example is the following result:

6.1.11. If $A \geq B$ are real $n \times n$ matrices with nonnegative inverses, then $A^{-1} \leq B^{-1}$.

PROOF. From $A^{-1}(A-B) \geq 0$ we conclude that $A^{-1}B \leq I$. Therefore, $(I - A^{-1}B)B^{-1} \geq 0$, which is equivalent to $A^{-1} \leq B^{-1}$. □

Another typical result involves the spectral radii.

6.1.12. If $|A| \leq B$, then $\rho(A) \leq \rho(B)$.

PROOF. Let $\sigma = \rho(B)$, and for any $\varepsilon > 0$, set $B_1 = (\sigma + \varepsilon)^{-1}B$ and $A_1 = (\sigma+\varepsilon)^{-1}A$. Then $|A_1| \leq B_1$ and $\rho(B_1) < 1$, so that $0 \leq |A_1|^k \leq B_1^k \to 0$ as $k \to \infty$. Thus, by 5.3.4, $\rho(A_1) < 1$, so that $\rho(A) < \sigma + \varepsilon$. Since ε was arbitrary, we must have $\rho(A) \leq \sigma$. □

Still another example involves M-matrices.

6.1.13. Let A be an M-matrix and D a nonnegative diagonal matrix. Then $A + D$ is an M-matrix, and $(A+D)^{-1} \leq A^{-1}$.

PROOF. Let $A = D_1 - B$, where D_1 is the diagonal of A. Then, by 6.1.8, $\rho(D_1^{-1}B) < 1$. Clearly,

$$0 \leq (D + D_1)^{-1}B \leq D_1^{-1}B$$

so that, by 6.1.12,

$$\rho((D+D_1)^{-1}B) \leq \rho(D_1^{-1}B) < 1$$

Thus, by 6.1.6, $(A+D)^{-1} \geq 0$, and, by 6.1.11, $(A+D)^{-1} \leq A^{-1}$. □

As an example of the previous result, consider the matrix A of (6.1.15), and let D be any diagonal matrix with nonnegative diagonal elements. By 6.1.9, A is an M-matrix, and then 6.1.13 shows that $A + D$ is also an M-matrix with $(A+D)^{-1} \leq A^{-1}$.

Diagonal Dominance and Gerschgorin Theorems

We conclude this section with two basic results on inverses and eigenvalues that are rather easy consequences of the previous results in this section.

6.1.14. DEFINITION. An $n \times n$ real or complex matrix A is *diagonally dominant* if

$$|a_{ii}| \geq \sum_{j \neq i} |a_{ij}|, \qquad i = 1, \ldots, n \tag{6.1.18}$$

A is *strictly diagonally dominant* if strict inequality holds in (6.1.18) for all i, and A is *irreducibly diagonally dominant* if it is irreducible and (6.1.18) holds with strict inequality for at least one i.

Diagonal dominance provides a relatively simple criterion for guaranteeing the nonsingularity of a matrix. However, diagonal dominance by itself does not ensure this; for example,

$$A = \begin{bmatrix} 1 & 1 \\ 1 & 1 \end{bmatrix}$$

is diagonally dominant but singular. However, either of the other two conditions of 6.1.14 will suffice.

6.1.15. If the $n \times n$ real or complex matrix A is strictly or irreducibly diagonally dominant, then A is nonsingular.

PROOF. Let $A = D - B$ be the splitting of A into its diagonal and off-diagonal parts. If A is strictly diagonally dominant, then D is nonsingular, and if $C = D^{-1}B$, then $\rho(C) \leqslant \|C\|_\infty < 1$. Thus, $I - C = D^{-1}A$ is nonsingular, so that A is nonsingular. If A is irreducibly diagonally dominant, then again D is nonsingular; otherwise, A would have a zero row, which would contradict the irreducibility assumption. Then $\rho(C) < 1$ by 6.1.10, and the result follows as before. □

Simple examples of irreducibly diagonally dominant matrices are (6.1.15), or (6.1.12) under the assumption that

$$|a_1| > |b_1|, \qquad |a_n| > |c_{n-1}|, \qquad |a_i| \geqslant |b_i| + |c_{i-1}|, \quad i = 1, \ldots, n-1 \tag{6.1.19}$$

with all b_i and c_i nonzero. If strict inequality holds for all i in (6.1.19), then the matrix is strictly diagonally dominant. Further examples are given in the exercises.

We note that the proof of 6.1.15 shows that the Jacobi iterative method (5.3.13-5.1.14), converges. Indeed, the matrix C is just the Jacobi iteration matrix $I - D^{-1}A$, and since we have shown that $\rho(C) < 1$, the convergence follows from 5.3.7. We state this as a corollary.

6.1.16. If the $n \times n$ real or complex matrix A is strictly or irreducibly diagonally dominant, then the Jacobi iteration converges.

Theorem 6.1.15 is related to a way of obtaining estimates on the location of the eigenvalues of a matrix.

6.1.17. GERSCHGORIN'S THEOREM. *Let A be an $n \times n$ real or complex matrix. Let S_i be the disk in the complex plane with center at a_{ii} and radius $r_i = \sum_{j \neq i} |a_{ij}|$. Then all eigenvalues of A lie in the union of the disks $S_1, S_2, \ldots, S_n$.*

PROOF. Let λ be any eigenvalue of A with corresponding eigenvector $\mathbf{x}$, with $\|\mathbf{x}\|_\infty = 1$. From $A\mathbf{x} = \lambda\mathbf{x}$ we have

$$(\lambda - a_{ii})x_i = \sum_{j \neq i} a_{ij}x_j, \qquad i = 1, \ldots, n$$

We assume that $\|\mathbf{x}\|_\infty = 1$, so that $|x_k| = 1$ for some k. Then

$$|\lambda - a_{kk}| = |(\lambda - a_{kk})x_k| \leq \sum_{j \neq k} |a_{kj}||x_j| \leq r_k$$

so that λ is in the disk S_k. □

As an example of the use of Gerschgorin's Theorem, consider the matrix

$$A = \begin{bmatrix} 8 & 1 & & \\ 1 & 10 & 1 & \\ & 1 & 12 & 1 \\ & & 1 & 10 \end{bmatrix}$$

Since $\|A\|_\infty = 14$, we know that all eigenvalues of A are less than or equal to 14 in magnitude. However, 6.1.17 gives a much sharper estimate: All eigenvalues are in the union of the disks illustrated in Figure 6.4.

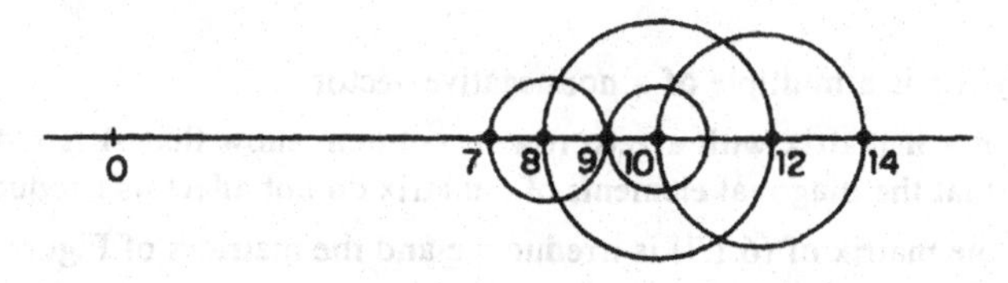

Figure 6.4. Gerschgorin disks.

Moreover, since A is symmetric, its eigenvalues are real and, hence, all lie in the interval [7, 14].

Actually, we can obtain a smaller interval by doing the following. Let $D = \text{diag}(d, 1, 1, 1)$. Then

$$DAD^{-1} = \begin{bmatrix} 8 & d & & \\ d^{-1} & 10 & 1 & \\ & 1 & 12 & 1 \\ & & 1 & 10 \end{bmatrix} \tag{6.1.20}$$

is similar to A and has the same eigenvalues. If we apply Gerschgorin's Theorem to (6.1.20), we obtain $\min\{8 - d, 10 - 1 - d^{-1}\}$ as the lower bound of the eigenvalues. Thus, the lower bound will be as large as possible when $8 - d = 10 - 1 - d^{-1}$ or when $d = (-1 + \sqrt{5})/2 \doteq 0.61$. Hence, all eigenvalues are larger than 7.38. A similar improvement of the upper limit can also be made.

Exercises 6.1

1. Conclude, without any computation, that the eigenvalues of

$$\begin{bmatrix} 2 & 3 \\ 1 & 2 \end{bmatrix}$$

   are real.

2. Compute the eigenvalues and eigenvectors of the matrix of Exercise 1.
3. Suppose that A is an $n \times n$ matrix with positive off-diagonal elements ($a_{ij} > 0$ if $i \neq j$). Show that the eigenvalue of A with largest real part is real and simple, and its associated eigenvector can be taken to be positive.
4. Let $b_i > 0$, $i = 1, \ldots, n$. Show that

$$\left| \sum_{i=1}^{n} b_i z_i \right| = \sum_{i=1}^{n} b_i |z_i|$$

   if and only if $\mathbf{z}$ is a multiple of a nonnegative vector
5. If A is an $n \times n$ matrix with a zero row or column, show that A is reducible. Show also that the diagonal elements of a matrix do not affect its irreducibility.
6. Show that the matrix of (6.1.7) is irreducible and the matrices of Figure 6.2 are reducible. Then prove 6.1.3.
7. Ascertain if the following matrices are irreducible by forming the directed graph of each matrix:

$$\text{(a)} \quad \begin{bmatrix} 1 & 1 & 1 \\ 0 & 2 & 2 \\ 2 & 0 & 1 \end{bmatrix}$$

$$\text{(b)} \quad \begin{bmatrix} 1 & 1 & 1 \\ 0 & 2 & 0 \\ 2 & 1 & 0 \end{bmatrix}$$

(c) $\begin{bmatrix} 1 & 0 & 1 \\ 0 & 1 & 0 \\ 1 & 0 & 1 \end{bmatrix}$

8. Give an example of an irreducible 3×3 matrix A such that no 2×2 principal submatrix of A is irreducible.
9. Suppose that A and B are symmetric matrices with eigenvalues $\lambda_1 \geqslant \cdots \geqslant \lambda_n$ and $\mu_1 \geqslant \cdots \geqslant \mu_n$, respectively. If $A \geqslant B$, is it necessarily true that $\lambda_i \geqslant \mu_i$, $i = 1, \cdots, n$?
10. An $n \times n$ nonnegative matrix A is *primitive* if $A^m > 0$ for some m. Let A be a nonnegative irreducible matrix with at least one positive row. Show that A is primitive. (Hint: Show that if A^k has at least k positive rows then A^{k+1} has at least $k+1$ positive rows.)
11. Show that the $n \times n$ matrix

$$A = \begin{bmatrix} 0 & 1 & & & \\ & 0 & 1 & & \\ & & \ddots & \ddots & \\ & & & & 1 \\ 1 & & & & 0 \end{bmatrix}$$

is irreducible and that the vector $\mathbf{e} = (1, \ldots, 1)^T$ is an eigenvector associated with the eigenvalue $\lambda = 1$. Use the Perron–Frobenius Theorem to conclude that 1 is a simple eigenvalue. What are the other eigenvalues of A?

12. Let the $n \times n$ real or complex matrix C satisfy (6.1.16). Let P be a permutation matrix and $\hat{C} = PCP^T$. Show that

$$\sum_{j=1}^{n} |\hat{c}_{ij}| \leqslant 1, \qquad i = 1, \ldots, n.$$

13. Let

$$A = \begin{bmatrix} 2 & -1 \\ -1 & 2 \end{bmatrix}, \qquad B = \begin{bmatrix} 3 & -1 \\ -1 & 3 \end{bmatrix}$$

Conclude, without any computation, that $A^{-1} \geqslant 0$, $B^{-1} \geqslant 0$, and $A^{-1} \geqslant B^{-1}$.

14. Show that the only nonnegative matrices that have a nonnegative inverse are of the form $A = PD$ or DP, where P is a permutation matrix and D is a nonnegative nonsingular diagonal matrix.
15. Without finding A^{-1}, ascertain whether or not the matrix

$$A = \begin{bmatrix} 3 & -1 & -2 & 0 \\ -1 & 2 & 0 & -1 \\ 0 & -1 & 2 & -1 \\ -1 & 0 & -1 & 3 \end{bmatrix}$$

has a positive inverse. Justify your assertion.

16. Use the trigonometric identity $\sin(\alpha \pm \beta) = \sin\alpha\cos\beta \pm \cos\alpha\sin\beta$ to show that the eigenvalues of the matrix (6.1.15) are $2 - 2\cos[k\pi/(n+1)]$, $k = 1, \ldots, n$, with corresponding eigenvectors

$$\left(\sin\frac{k\pi}{n+1}, \sin\frac{2k\pi}{n+1}, \cdots, \sin\frac{nk\pi}{n+1}\right)^T$$

Also show that these eigenvectors are orthogonal.

17. Let A be the $n \times n$ matrix

$$A = \begin{bmatrix} 3 & 2 & & & \\ 1 & 3 & 2 & & \\ & 1 & 3 & 2 & \\ & & \ddots & \ddots & \ddots \\ & & & & 2 \\ & & & 1 & 3 \end{bmatrix}$$

Is A nonsingular? Prove your assertion.

18. Compute the Gerschgorin disks for the matrix of Problem 17 and obtain bounds for the eigenvalues.

19. Show that there is a diagonal matrix D such that if A is the matrix of Problem 17, then DAD^{-1} is symmetric and, hence, the eigenvalues are real.

20. For the difference equation

$$\mathbf{x}^{(k+1)} = A\mathbf{x}^{(k)} + \mathbf{b}, \qquad k = 0, 1, \ldots$$

with

$$A = \frac{1}{4}\begin{bmatrix} 2 & 2 \\ 2 & 1 \end{bmatrix}, \qquad \mathbf{b} = \begin{bmatrix} 1 \\ 2 \end{bmatrix}$$

show that there is a unique vector $\hat{\mathbf{x}}$ so that $\mathbf{x}^{(k)} \to \hat{\mathbf{x}}$ as $k \to \infty$ for any $\mathbf{x}^{(0)}$, and find $\hat{\mathbf{x}}$.

6.2. Generalized and Higher-Order Eigenvalue Problems

In this section we will treat various extensions of the eigenvalue problem. Consider first the so-called *generalized eigenvalue problem*

$$A\mathbf{x} = \lambda B\mathbf{x} \tag{6.2.1}$$

Here, A and B are $n \times n$ matrices, and we seek scalars λ and corresponding nonzero vectors $\mathbf{x}$ that satisfy (6.2.1). There are several different possibilities to be considered:

1. ***B* is nonsingular**: In this case (6.2.1) is equivalent to the standard eigenvalue problem $C\mathbf{x} = \lambda\mathbf{x}$, where $C = B^{-1}A$.

2. ***B* is singular, but *A* is nonsingular**: In this case we can multiply (6.2.1) by A^{-1} to obtain $C\mathbf{x} = \mu\mathbf{x}$, with $C = A^{-1}B$ and $\mu = \lambda^{-1}$. Again, we have a standard eigenvalue problem but with the following difference: The matrix C is singular (since B is singular) and hence has at least one zero eigenvalue. For these zero eigenvalues μ, the corresponding λ are infinite. Hence, the problem (6.2.1) has "infinite eigenvalues" corresponding to the zero eigenvalues of C.

3. ***A* and *B* are Hermitian, and *B* is positive definite**: This is a special case of 1, and we can again convert the problem to the standard eigenvalue problem. However, the matrix C is not, in general, Hermitian.

Case 3 is very important in applications, and we wish to treat it in more detail. In particular, the next result shows that by a different type of conversion to a standard eigenvalue problem we are able to retain the Hermitian character of the problem.

6.2.1. *Let A and B be n × n Hermitian matrices with B positive definite. Then the eigenvalue problem* (6.2.1) *has n real eigenvalues and n corresponding linearly independent eigenvectors that can be chosen to be orthogonal in the inner product* $(\mathbf{x}, \mathbf{y}) = \mathbf{x}^* B\mathbf{y}$. *Moreover, if A is positive definite, then the eigenvalues are all positive.*

PROOF. Since B is positive definite, it has a Choleski decomposition (see 1.4.12) $B = LL^*$, where L is nonsingular. Then, with the notation $L^{-*} = (L^{-1})^*$, (6.2.1) can be written as

$$L^{-1}AL^{-*}L^*\mathbf{x} = \lambda L^*\mathbf{x}$$

or

$$C\mathbf{y} = \lambda\mathbf{y}, \qquad C = L^{-1}AL^{-*}, \ \mathbf{y} = L^*\mathbf{x} \tag{6.2.2}$$

Therefore, (6.2.2) is the standard eigenvalue problem, and, since C is Hermitian, there are n real eigenvalues and corresponding eigenvectors $\mathbf{y}_1, \ldots, \mathbf{y}_n$, which satisfy $\mathbf{y}_i^*\mathbf{y}_j = \delta_{ij}$. Thus, $\mathbf{x}_i = L^{-*}\mathbf{y}_i$ are eigenvectors of the original problem (6.2.1), and in the inner product defined by B we have

$$\mathbf{x}_i^* B\mathbf{x}_j = \mathbf{y}_i^* L^{-1}BL^{-*}\mathbf{y}_j = \mathbf{y}_i^*\mathbf{y}_j = \delta_{ij}$$

Finally, if A is positive definite, then so is C, and thus the eigenvalues are positive. □

Theorem 6.2.1 shows that the generalized eigenvalue problem (6.2.1), in which A and B are Hermitian and B is positive definite, is really just a standard eigenvalue problem if we work in the inner product defined by B.

We next give a number of results related to 6.2.1.

6.2.2. SIMULTANEOUS REDUCTION TO DIAGONAL FORM. *Let A and B be $n \times n$ Hermitian matrices with B positive definite. Then there is a nonsingular $n \times n$ matrix Q such that Q^*AQ and Q^*BQ are both diagonal.*

PROOF. Again let $B = LL^*$ be the Choleski decomposition, and let C be the matrix of (6.2.2). Then there is a unitary matrix P so that $P^*CP = D$, where $D = \text{diag}(\lambda_1, \ldots, \lambda_n)$, and the λ_i are the eigenvalues of C. With $Q = L^{-*}P$, we have

$$Q^*AQ = P^*L^{-1}AL^{-*}P = P^*CP = D$$

and $Q^*BQ = P^*P = I$. □

Note that 6.2.2 says that there is a change of variable $\mathbf{x} = Q\mathbf{y}$, so that the quadratic form $\mathbf{x}^*(A - \lambda B)\mathbf{x}$ becomes $\sum \lambda_i y_i^2 - \lambda \sum y_i^2$. Of course, in general, this is not a simultaneous similarity transformation, and the λ_i are the eigenvalues of the generalized eigenvalue problem, and not of A.

Next, we recall that if A and B are Hermitian, the product AB is, in general, not Hermitian, and the eigenvalues of AB are not necessarily real. However, if either A or B is also positive definite, then we have the following result.

6.2.3. *If A and B are $n \times n$ Hermitian matrices and if either is positive definite, then AB has real eigenvalues and is diagonalizable. If both A and B are positive definite, then the eigenvalues of AB are positive. Conversely, if AB has positive eigenvalues and either A or B is positive definite, then both are positive definite.*

PROOF. Suppose that A is positive definite. Then the eigenvalue problem $AB\mathbf{x} = \lambda\mathbf{x}$ is equivalent to $B\mathbf{x} = \lambda A^{-1}\mathbf{x}$. Since A^{-1} is also positive definite, we can apply 6.2.1 to conclude that the eigenvalues $\lambda_1, \ldots, \lambda_n$ are real. Moreover, there are n eigenvectors $\mathbf{x}_1, \ldots, \mathbf{x}_n$, which are orthogonal in the inner product defined by A^{-1} and, hence, are linearly independent. If $X = (\mathbf{x}_1, \ldots, \mathbf{x}_n)$, and $D = \text{diag}(\lambda_1, \ldots, \lambda_n)$, then $ABX = XD$, so that AB is similar to a diagonal matrix. Likewise, if B is positive definite, we can make the change of variable $\mathbf{x} = B^{-1}\mathbf{y}$. Then the eigenvalue problem has the form $A\mathbf{y} = \lambda B^{-1}\mathbf{y}$, and the same conclusions follow. If both A and B are positive definite, 6.2.1 ensures that the eigenvalues are positive.

Conversely, suppose that AB has positive eigenvalues and that A is positive definite. By Exercise 3.1-13, A has a positive definite Hermitian square root $A^{1/2}$, and then

$$AB = A^{1/2}A^{1/2}BA^{1/2}A^{-1/2}$$

Thus, $A^{1/2}BA^{1/2}$ is similar to AB and has positive eigenvalues. Therefore, since $A^{1/2}BA^{1/2}$ is congruent to B, B is positive definite. The proof in the case that B is assumed to be positive definite is similar (Exercise 6.2-6). □

We note that the first statement of Theorem 6.2.3 is equivalent to a number of other seemingly disparate statements.

6.2.4. Let A be an $n \times n$ matrix. Then the following are equivalent:

(*a*) *A is the product of two Hermitian matrices, one of which is positive definite;*
(*b*) *A has real eigenvalues and n linearly independent eigenvectors;*
(*c*) *A is similar to a Hermitian matrix;*
(*d*) *There is a Hermitian positive definite matrix S such that SA is Hermitian.*

PROOF. (a) implies (b): This is 6.2.3.

(b) implies (c): Clearly, (b) implies that A is similar to a real diagonal matrix, which is Hermitian.

(c) implies (d): Let $A = PHP^{-1}$, where H is Hermitian, and $S = P^{-*}P^{-1}$. Then $SA = P^{-*}P^{-1}A = P^{-*}HP^{-1}$. Clearly, S is Hermitian positive definite, and $P^{-*}HP^{-1}$ is Hermitian.

(d) implies (a): If $SA = C$ is Hermitian, then $A = S^{-1}C$ and S^{-1} is Hermitian positive definite. □

Higher-Order Eigenvalue Problems

Consider the second-order system of differential equations

$$\ddot{\mathbf{x}}(t) + A_1\dot{\mathbf{x}}(t) + A_0\mathbf{x}(t) = 0 \tag{6.2.3}$$

where $\mathbf{x}(t)$ is an n-vector and A_1 and A_0 are given constant $n \times n$ matrices. If, as in Section 5.1, we try a solution of the form $\mathbf{x}(t) = e^{\lambda t}\mathbf{p}$, we see that λ and $\mathbf{p}$ must satisfy

$$\lambda^2 e^{\lambda t}\mathbf{p} + \lambda e^{\lambda t}A_1\mathbf{p} + e^{\lambda t}A_0\mathbf{p} = 0$$

or, since $e^{\lambda t} \neq 0$,

$$(\lambda^2 I + \lambda A_1 + A_0)\mathbf{p} = 0 \tag{6.2.4}$$

In order for this system of equations to have a nontrivial solution $\mathbf{p}$, the coefficient matrix must be singular, which gives the characteristic equation

$$\det(\lambda^2 I + \lambda A_1 + A_0) = 0 \tag{6.2.5}$$

By the cofactor expansion of this determinant, it is easy to conclude (Exercise 6.2-3) that (6.2.5) is a polynomial equation in λ of degree $2n$. Hence, there are $2n$ roots, which are the eigenvalues of (6.2.4).

An important special case of (6.2.4) is when $A_1 = 0$ and A_0 is Hermitian. Let $\mu_1 \leq \cdots \leq \mu_r < 0 \leq \mu_{r+1} \leq \cdots \leq \mu_n$ be the eigenvalues of A_0 with corresponding eigenvectors $\mathbf{p}_1, \ldots, \mathbf{p}_n$, and let $\omega_j = |\mu_j|^{1/2}$, $j = 1, \ldots, n$. Then the eigenvalues λ of (6.2.4) are $\pm\omega_j$, $j = 1, \ldots, r$, and $\pm i\omega_j$, $j = r + 1, \ldots, n$, and

$$e^{\pm\omega_j t}\mathbf{p}_j, \quad j = 1, \ldots, r, \qquad e^{\pm i\omega_j t}\mathbf{p}_j, \quad j = r + 1, \ldots, n \tag{6.2.6}$$

are solutions of the differential equation (6.2.3). Thus, solutions corresponding to the negative eigenvalues of A_0 give exponentially increasing and decreasing solutions, and those corresponding to positive eigenvalues of A_0 give oscillatory solutions. If A_0 is positive definite, there are only oscillatory solutions.

The problem (6.2.4) can be converted to a standard eigenvalue problem

$$A\mathbf{q} = \lambda\mathbf{q}, \qquad A = \begin{bmatrix} 0 & I \\ -A_0 & -A_1 \end{bmatrix}, \quad \mathbf{q} = \begin{bmatrix} \mathbf{q}_1 \\ \mathbf{q}_2 \end{bmatrix} \tag{6.2.7}$$

for the $2n \times 2n$ matrix A. Equation (6.2.7) is equivalent to the two equations

$$\mathbf{q}_2 = \lambda\mathbf{q}_1, \qquad -A_0\mathbf{q}_1 - A_1\mathbf{q}_2 = \lambda\mathbf{q}_2$$

Substituting the first of these into the second and setting $\mathbf{p} = \mathbf{q}_1$ gives Equation (6.2.4). In principle, then, one can solve (6.2.4) by obtaining the eigenvalues and eigenvectors of the matrix A of (6.2.7). Note that even if A has $2n$ linearly independent eigenvectors, (6.2.4) can have no more than n linearly independent vectors $\mathbf{p}$ since the $\mathbf{p}$'s are n-vectors. However, these linearly independent vectors give rise to twice as many linearly independent solutions of the differential equation (6.2.3). For example, if $\omega_1 \neq 0$, then $e^{-\omega_1 t}\mathbf{p}_1$ and $e^{\omega_1 t}\mathbf{p}_1$ are two linearly independent solutions of (6.2.3).

One can treat in a similar way still higher-order problems. Consider the mth-order differential equation

$$\mathbf{x}^{(m)}(t) + A_{m-1}\mathbf{x}^{(m-1)}(t) + \cdots + A_1\dot{\mathbf{x}}(t) + A_0\mathbf{x}(t) = 0 \tag{6.2.8}$$

where $\mathbf{x}^{(i)}$ denotes the ith derivative of $\mathbf{x}$, and the matrices A_i are $n \times n$. Corresponding to (6.2.8) is the eigenvalue problem

$$(\lambda^m I + \lambda^{m-1}A_{m-1} + \cdots + \lambda A_1 + A_0)\mathbf{p} = 0 \tag{6.2.9}$$

Again, there are nonzero solutions $\mathbf{p}$ to this equation only for those values of λ for which

$$\det(\lambda^m I + \lambda^{m-1}A_{m-1} + \cdots + \lambda A_1 + A_0) = 0 \tag{6.2.10}$$

which is a polynomial of degree nm in λ. Alternatively, we can consider the eigenvalue problem

$$A\mathbf{q} = \lambda\mathbf{q}, \qquad A = \begin{bmatrix} 0 & I & & & \\ & & I & & \\ & & & \ddots & \\ & & & & I \\ -A_0 & & & \cdots & -A_{m-1} \end{bmatrix}, \quad \mathbf{q} = \begin{bmatrix} \mathbf{q}_1 \\ \vdots \\ \mathbf{q}_m \end{bmatrix} \tag{6.2.11}$$

which is equivalent to the equations

$$\mathbf{q}_{i+1} = \lambda\mathbf{q}_i, \qquad i = 1, \ldots, m-1, \tag{6.2.12a}$$

$$-A_0\mathbf{q}_1 - A_1\mathbf{q}_2 - \cdots - A_{m-1}\mathbf{q}_m = \lambda\mathbf{q}_m \tag{6.2.12b}$$

By (6.2.12a), $\mathbf{q}_i = \lambda^{i-1}\mathbf{q}_1$, $i = 2, \ldots, m$, and if we substitute these $\mathbf{q}_i$ into (6.2.12b) and set $\mathbf{p} = \mathbf{q}_1$, then (6.2.12b) reduces to (6.2.9).

Exercises 6.2

1. Compute the eigenvalues and eigenvectors for the problem $A\mathbf{x} = \lambda B\mathbf{x}$ if

$$A = \begin{bmatrix} -1 & 1 \\ 1 & -1 \end{bmatrix}, \qquad B = \begin{bmatrix} 2 & 1 \\ 1 & 2 \end{bmatrix}$$

Show that the eigenvectors are orthogonal in the inner product $(\mathbf{x}, \mathbf{y}) \equiv \mathbf{x}^T B \mathbf{y}$.

2. Use 6.2.4 to conclude that the matrix

$$A = \begin{bmatrix} 1 & 1 \\ 0 & 1 \end{bmatrix}$$

cannot be written as the product of symmetric positive definite matrices.

3. Use the cofactor expansion to conclude that (6.2.5) is a polynomial of degree $2n$.

4. Let

$$A = \begin{bmatrix} \alpha & 1 \\ 1 & -\alpha \end{bmatrix}, \qquad B = \begin{bmatrix} 0 & 1 \\ 1 & 0 \end{bmatrix}$$

Show that the eigenvalues of AB are $1 \pm i\alpha$.

5. Let A and B be the $n \times n$ matrices

$$A = \begin{bmatrix} 4 & -2 & & & \\ -2 & 4 & -2 & & \\ & \ddots & \ddots & \ddots & \\ & & & & -2 \\ & & & -2 & 4 \end{bmatrix}, \qquad B = \operatorname{diag}(1, 2, \ldots, n)$$

Let $C = AB$. If the eigenvalues of C are $\lambda_1, \ldots, \lambda_n$ and $|\lambda_1| \leq \cdots \leq |\lambda_n|$, then show that λ_1 is real, positive, and simple.

6. Let A and B be Hermitian with B positive definite. If AB has positive eigenvalues, show that A is positive definite.

6.3. Some Special Matrices

In this section we discuss some special types of matrices that arise frequently in various applications.

Kronecker Products

Let A be an $n \times p$ real or complex matrix and B an $m \times q$ real or complex matrix. Then the $nm \times pq$ matrix

$$A \otimes B = \begin{bmatrix} a_{11}B & a_{12}B & \cdots & a_{1p}B \\ \vdots & \vdots & & \vdots \\ a_{n1}B & a_{n2}B & \cdots & a_{np}B \end{bmatrix} \tag{6.3.1}$$

is the *Kronecker* (or *tensor*) *product* of A and B. An example is

$$A = \begin{bmatrix} 1 & 2 \\ -1 & 4 \end{bmatrix}, \qquad B = \begin{bmatrix} 2 & 1 \\ 1 & 2 \end{bmatrix} \tag{6.3.2}$$

$$A \otimes B = \begin{bmatrix} 2 & 1 & 4 & 2 \\ 1 & 2 & 2 & 4 \\ -2 & -1 & 8 & 4 \\ -1 & -2 & 4 & 8 \end{bmatrix} \tag{6.3.3}$$

Kronecker products have many interesting properties. For example, if A and B are nonsingular, then $A \otimes B$ is nonsingular and

$$(A \otimes B)^{-1} = A^{-1} \otimes B^{-1} \tag{6.3.4}$$

If A is $n \times n$, B is $m \times m$, and $A^{-1} = (\hat{a}_{ij})$, then (6.3.4) follows from the computation

$$(A^{-1} \otimes B^{-1})(A \otimes B) = (\hat{a}_{ij}B^{-1})(a_{ij}B) = \left(\sum_{k=1}^{n} \hat{a}_{ik}a_{kj}I_m \right) = I_{nm}$$

Some other simple properties of the Kronecker product are given in the exercises.

We next show that Kronecker products have the rather remarkable property that their eigensystems are easily expressed in terms of the eigensystems of A and B.

6.3.1. EIGENSYSTEMS OF KRONECKER PRODUCTS. *Let A be an $n \times n$ matrix with eigenvalues $\lambda_1, \ldots, \lambda_n$, and B an $m \times m$ matrix with eigenvalues $\mu_1, \ldots, \mu_m$. Then the mn eigenvalues of $A \otimes B$, are $\lambda_i\mu_j$, $i = 1, \ldots, n$, $j = 1, \ldots, m$. Moreover, if $\mathbf{x}_1, \ldots, \mathbf{x}_p$ are linearly independent eigenvectors of A corresponding to the eigenvalues $\lambda_1, \ldots, \lambda_p$, and $\mathbf{y}_1, \ldots, \mathbf{y}_q$ are linearly independent eigenvectors of B corresponding to the eigenvalues $\mu_1, \ldots, \mu_q$, then the Kronecker products of the $\mathbf{x}_i$ and $\mathbf{y}_j$,*

$$\mathbf{x}_i \otimes \mathbf{y}_j, \qquad i = 1, \ldots, p, \; j = 1, \ldots, q \tag{6.3.5}$$

are linearly independent eigenvectors of $A \otimes B$ corresponding to the eigenvalues $\lambda_i\mu_j$. In particular, if A and B both have diagonal Jordan canonical forms, then so does $A \otimes B$.

PROOF. Let $A = PJ_AP^{-1}$ and $B = QJ_BQ^{-1}$, where J_A and J_B are the Jordan forms of A and B. By (6.3.4), $P \otimes Q$ is nonsingular and $(P \otimes Q)^{-1} = P^{-1} \otimes Q^{-1}$. Then, with $P^{-1} = (\hat{p}_{ij})$ and $Q^{-1} = (\hat{q}_{ij})$, we have

$$\begin{aligned}(P^{-1} \otimes Q^{-1})(A \otimes B)(P \otimes Q) &= (\hat{p}_{ij}Q^{-1})(a_{ij}B)(p_{ij}Q) \\ &= (\hat{p}_{ij}I_m)(a_{ij}J_B)(p_{ij}I_m) = J_A \otimes J_B \\ &= \begin{bmatrix} \lambda_1 J_B & \varepsilon_1 J_B & & \\ & \lambda_2 J_B & \ddots & \\ & & \ddots & \varepsilon_{n-1} J_B \\ & & & \lambda_n J_B \end{bmatrix} \end{aligned} \tag{6.3.6}$$

where ε_i is 0 or 1 according to J_A. Thus $J_A \otimes J_B$ is similar to $A \otimes B$, and the diagonal elements of $J_A \otimes J_B$ are the mn eigenvalues of $A \otimes B$. Clearly, if J_A is diagonal, then (6.3.6) indicates the Jordan form of $A \otimes B$, and if J_B is also diagonal, then $J_A \otimes J_B$ is diagonal.

We could deduce from (6.3.6) the desired statements about the eigenvectors of $A \otimes B$, but it is instructive to proceed as follows. If $A\mathbf{x} = \lambda\mathbf{x}$ and $B\mathbf{y} = \mu\mathbf{y}$, then

$$\begin{aligned}\begin{bmatrix} a_{11}B & \cdots & a_{1n}B \\ \vdots & & \vdots \\ a_{n1}B & \cdots & a_{nn}B \end{bmatrix} \begin{bmatrix} x_1\mathbf{y} \\ \vdots \\ x_n\mathbf{y} \end{bmatrix} &= \begin{bmatrix} (a_{11}x_1 + \cdots + a_{1n}x_n)B\mathbf{y} \\ \vdots \\ (a_{n1}x_1 + \cdots + a_{nn}x_n)B\mathbf{y} \end{bmatrix} \\ &= \begin{bmatrix} \lambda x_1 \mu \mathbf{y} \\ \vdots \\ \lambda x_n \mu \mathbf{y} \end{bmatrix} = \lambda\mu \begin{bmatrix} x_1\mathbf{y} \\ \vdots \\ x_n\mathbf{y} \end{bmatrix} \end{aligned} \tag{6.3.7}$$

so that $(x_i\mathbf{y})$ is an eigenvector of $A \otimes B$ corresponding to the eigenvalue $\lambda\mu$. Now suppose that $\mathbf{x}_1, \ldots, \mathbf{x}_p$ and $\mathbf{y}_1, \ldots, \mathbf{y}_q$ are linearly independent eigenvectors of A and B, respectively. Then, by (6.3.7), the vectors of (6.3.5) are all eigenvectors of $A \otimes B$, and we wish to show that they are linearly independent. Suppose not. Then

$$0 = \sum_{i=1}^{p} \sum_{j=1}^{q} c_{ij} \begin{bmatrix} x_{i1}\mathbf{y}_j \\ \vdots \\ x_{in}\mathbf{y}_j \end{bmatrix} = \sum_{j=1}^{q} \begin{bmatrix} \left(\sum_{i=1}^{p} c_{ij}x_{i1}\right)\mathbf{y}_j \\ \vdots \\ \left(\sum_{i=1}^{p} c_{ij}x_{in}\right)\mathbf{y}_j \end{bmatrix}$$

Since the $\mathbf{y}_j$ are linearly independent, this can be true only if

$$\sum_{i=1}^{p} c_{ij} x_{ik} = 0, \qquad k = 1, \ldots, n, j = 1, \ldots, q$$

or

$$\sum_{i=1}^{p} c_{ij} \mathbf{x}_i = 0, \qquad j = 1, \ldots, q$$

But this contradicts the linear independence of the $\mathbf{x}_i$. □

We apply the above result to the matrices of (6.3.2) and (6.3.3). The eigensystems of A and B are

$$A: \quad \begin{matrix} \lambda_1 = 2, \\ \lambda_2 = 3, \end{matrix} \qquad \mathbf{x}_1 = \begin{bmatrix} 2 \\ 1 \end{bmatrix}, \qquad \mathbf{x}_2 = \begin{bmatrix} 1 \\ 1 \end{bmatrix}$$

$$B: \quad \begin{matrix} \mu_1 = 1, \\ \mu_2 = 3, \end{matrix} \qquad \mathbf{y}_1 = \begin{bmatrix} 1 \\ -1 \end{bmatrix}, \qquad \mathbf{y}_2 = \begin{bmatrix} 1 \\ 1 \end{bmatrix}$$

Therefore the eigenvalues and corresponding eigenvectors of (6.3.3) are

$$\lambda_1\mu_1 = 2, \qquad (2\mathbf{y}_1^T, \mathbf{y}_1^T) = (2, -2, 1, -1)^T$$

$$\lambda_1\mu_2 = 6, \qquad (2\mathbf{y}_2^T, \mathbf{y}_2^T) = (2, 2, 1, 1)^T$$

$$\lambda_2\mu_1 = 3, \qquad (\mathbf{y}_1^T, \mathbf{y}_1^T) = (1, -1, 1, -1)^T$$

$$\lambda_2\mu_2 = 9, \qquad (\mathbf{y}_2^T, \mathbf{y}_2^T) = (1, 1, 1, 1)^T$$

As illustrated by the above example, Kronecker products are sometimes useful in generating large matrices with known eigensystems. Knowing the eigensystem of the 4×4 matrix $A \otimes B$, we can easily compute the eigenvalues of the 16×16 matrix $(A \otimes B) \otimes (A \otimes B)$, and so on.

Kronecker Sums

We next define Kronecker sums. Let A be $n \times n$, and B be $m \times m$. If I_p is the $p \times p$ identity, then $I_m \otimes A$ and $B \otimes I_n$ are both $nm \times nm$, and the $nm \times nm$ matrix

$$(I_m \otimes A) + (B \otimes I_n) \tag{6.3.8}$$

is the *Kronecker* (or tensor) *sum* of A and B. As in the case of Kronecker products, the eigensystem of a Kronecker sum is easily computed.

6.3.2. EIGENSYSTEM OF KRONECKER SUMS. *Let A be an $n \times n$ matrix with eigenvalues $\lambda_1, \ldots, \lambda_n$, and B an $m \times m$ matrix with eigenvalues $\mu_1, \ldots, \mu_m$. Then the nm eigenvalues of the Kronecker sum (6.3.8) are $\lambda_i + \mu_j$, $i = 1, \ldots, n$, $j = 1, \ldots, m$. Moreover, if $\mathbf{x}_1, \ldots, \mathbf{x}_p$ are linearly independent eigenvectors of A, and $\mathbf{y}_1, \ldots, \mathbf{y}_q$ are linearly independent eigenvectors of B, then the Kronecker products*

$$\mathbf{y}_j \otimes \mathbf{x}_i, \qquad i = 1, \ldots, p, \ j = 1, \ldots, q \tag{6.3.9}$$

are linearly independent eigenvectors of the Kronecker sum. In particular, if both A and B have a diagonal Jordan form, then so does the Kronecker sum.

PROOF. Again, let $A = PJ_AP^{-1}$ and $B = QJ_BQ^{-1}$, where J_A and J_B are the Jordan forms of A and B. Then

$$(Q^{-1} \otimes I_n)(B \otimes I_n)(Q \otimes I_n) = J_B \otimes I_n$$

and

$$(I_m \otimes P^{-1})(I_m \otimes A)(I_m \otimes P) = I_m \otimes J_A$$

Moreover, a simple calculation shows that

$$(Q^{-1} \otimes I_n)(I_m \otimes A)(Q \otimes I_n) = I_m \otimes A$$

and, similarly,

$$(I_m \otimes P^{-1})(J_B \otimes I_n)(I_m \otimes P) = J_B \otimes I_n$$

Therefore, with $U = (Q \otimes I_n)(I_m \otimes P)$, we have

$$U^{-1}[(I_m \otimes A) + (B \otimes I_n)]U = (I_m \otimes J_A) + (J_B \otimes I_n)$$

$$= \begin{bmatrix} J_A + \mu_1 I_n & \varepsilon_1 I_n & & \\ & \ddots & \ddots & \\ & & & \varepsilon_{m-1} I_n \\ & & & J_A + \mu_m I_n \end{bmatrix} \tag{6.3.10}$$

where the ε_i are 0 or 1 in accord with J_B. Thus, $(I_m \otimes A) + (B \otimes I_n)$ is similar to the matrix of (6.3.10), which has eigenvalues $\lambda_i + \mu_j$. Moreover, if J_B is diagonal, then (6.3.10) is the Jordan form of $(I_m \otimes A) + (B \otimes I_n)$, and if J_A is also diagonal, then (6.3.10) is diagonal.

Now, let $\mathbf{x}_1, \ldots, \mathbf{x}_p$ and $\mathbf{y}_1, \ldots, \mathbf{y}_q$ be linearly independent eigenvectors of A and B, respectively. Then

$$(I_m \otimes A)\begin{bmatrix} y_{j1}\mathbf{x}_i \\ \vdots \\ y_{jm}\mathbf{x}_i \end{bmatrix} = \begin{bmatrix} y_{j1}A\mathbf{x}_i \\ \vdots \\ y_{jm}A\mathbf{x}_i \end{bmatrix} = \lambda_i \begin{bmatrix} y_{j1}\mathbf{x}_i \\ \vdots \\ y_{jm}\mathbf{x}_i \end{bmatrix}$$

and

$$(B \otimes I_n)\begin{bmatrix} y_{j1}\mathbf{x}_i \\ \vdots \\ y_{jm}\mathbf{x}_i \end{bmatrix} = \begin{bmatrix} \sum b_{1k}y_{jk}\mathbf{x}_i \\ \vdots \\ \sum b_{mk}y_{jk}\mathbf{x}_i \end{bmatrix} = \mu_j \begin{bmatrix} y_{j1}\mathbf{x}_i \\ \vdots \\ y_{jm}\mathbf{x}_i \end{bmatrix}$$

Hence, the vectors (6.3.7) are eigenvectors of the Kronecker sum corresponding to the eigenvalues $\lambda_i + \mu_j$. That they are linearly independent follows as in 6.3.1. □

We next give an example of the use of 6.3.2. The matrix

$$A = \begin{bmatrix} T+2I & -I & & \\ -I & T+2I & \ddots & \\ & \ddots & \ddots & -I \\ & & -I & T+2I \end{bmatrix}, \qquad T = \begin{bmatrix} 2 & -1 & & \\ -1 & 2 & \ddots & \\ & \ddots & \ddots & -1 \\ & & -1 & 2 \end{bmatrix} \tag{6.3.11}$$

arises in the numerical solution of Laplace's equation by finite difference methods. Here, I and T are $n \times n$ and A is $n^2 \times n^2$. It is easy to verify that A is the Kronecker sum

$$A = (I \otimes T) + (T \otimes I) \tag{6.3.12}$$

From Exercise 6.1-16 the eigenvalues of T are $2 - 2\cos[k\pi/(n+1)]$, $k = 1, \ldots, n$, with corresponding eigenvectors

$$\mathbf{x}_k = \left(\sin\frac{k\pi}{n+1}, \sin\frac{2k\pi}{n+1}, \cdots, \sin\frac{nk\pi}{n+1}\right)^T, \qquad k = 1, \ldots, n$$

Hence, by 6.3.2, we can conclude that the eigenvalues of A are

$$4 - 2\left(\cos\frac{k\pi}{n+1} + \cos\frac{j\pi}{n+1}\right), \qquad j, k = 1, \ldots, n \tag{6.3.13}$$

with corresponding eigenvectors

$$\left(\sin\frac{k\pi}{n+1}\mathbf{x}_j^T, \sin\frac{2k\pi}{n+1}\mathbf{x}_j^T, \ldots, \sin\frac{nk\pi}{n+1}\mathbf{x}_j^T\right)^T, \qquad j, k = 1, \ldots, n$$

Circulant Matrices

A *Toeplitz matrix* is an $n \times n$ matrix with constant diagonals

$$T = \begin{bmatrix} a_0 & a_1 & a_2 & \cdots & a_{n-1} \\ a_{-1} & \ddots & \ddots & \ddots & \\ a_{-2} & \ddots & \ddots & \ddots & \\ \vdots & & \ddots & \ddots & \\ a_{-n+1} & & & \ddots & \end{bmatrix}$$

An example is

$$T = \begin{bmatrix} 1 & 2 & 3 & 4 \\ 5 & 1 & 2 & 3 \\ 6 & 5 & 1 & 2 \\ 7 & 6 & 5 & 1 \end{bmatrix}$$

Other examples are the matrices A and T of (6.3.11).

An important and interesting special case of a Toeplitz matrix arises when $a_{-j} = a_{n-j}$, $j = 1, \ldots, n-1$. In this case we write the matrix as

$$C = \begin{bmatrix} c_1 & c_2 & & \cdots & c_n \\ c_n & c_1 & c_2 & \cdots & c_{n-1} \\ \vdots & & & & \vdots \\ c_3 & & & & \\ c_2 & c_3 & & \cdots & c_1 \end{bmatrix} \tag{6.3.14}$$

The matrix C is $n \times n$ and is defined by the n (real or complex) numbers $c_1, \ldots, c_n$. Each row is obtained from the previous one by shifting all elements one place to the right, with the nth element taken back to the first position. Such a matrix is called a *circulant matrix* or, simply, a *circulant*. Some simple examples are

$$C = \begin{bmatrix} 1 & 2 & 3 \\ 3 & 1 & 2 \\ 2 & 3 & 1 \end{bmatrix}, \quad C = \begin{bmatrix} 1 & 0 & 0 \\ 0 & 1 & 0 \\ 0 & 0 & 1 \end{bmatrix}, \quad C = \begin{bmatrix} 1 & 1 & 1 \\ 1 & 1 & 1 \\ 1 & 1 & 1 \end{bmatrix}$$

Circulant matrices have many interesting properties. If C is a circulant, then so is its conjugate transpose C^* (see Exercise 6.3-4). Moreover, it is easy to show that $CC^* = C^*C$, so that C is normal (Exercise 6.3-4). Hence, by 3.1.12, every circulant is unitarily similar to a diagonal matrix, and, surprisingly, it is easy to give explicitly the unitary matrix of this similarity transformation. Let

$$\omega = e^{2\pi i/n} = \cos\left(\frac{2\pi}{n}\right) + i \sin\left(\frac{2\pi}{n}\right) \tag{6.3.15}$$

be an nth root of unity, and define the $n \times n$ matrix

$$U = \frac{1}{\sqrt{n}} \begin{bmatrix} 1 & 1 & 1 & \cdots & 1 \\ 1 & \omega & \omega^2 & \cdots & \omega^{n-1} \\ 1 & \omega^2 & \omega^4 & \cdots & \omega^{2(n-1)} \\ \vdots & \vdots & \vdots & & \vdots \\ 1 & \omega^{n-1} & \omega^{2(n-1)} & \cdots & \omega^{(n-1)(n-1)} \end{bmatrix} \tag{6.3.16}$$

Note that U is symmetric but not Hermitian. The j, k element of UU^* is

$$\gamma_{jk} = \frac{1}{n}(1 + \omega^j \bar{\omega}^k + \omega^{2j} \bar{\omega}^{2k} + \cdots + \omega^{j(n-1)} \bar{\omega}^{k(n-1)})$$

Since $\omega\bar{\omega} = 1$, it is clear that $\gamma_{jj} = 1$, $j = 1, \ldots, n$, while if $j > k$, then, by the identity

$$(1 - \alpha)(1 + \alpha + \cdots + \alpha^m) = 1 - \alpha^{m+1}$$

for any scalar α, we have for $j \neq k$

$$\gamma_{jk} = \frac{1}{n}(1 + \omega^{j-k} + \omega^{2(j-k)} + \cdots + \omega^{(n-1)(j-k)}) = \frac{1}{n} \frac{1 - \omega^{n(j-k)}}{1 - \omega^{j-k}} = 0,$$

since $\omega^n = 1$. Thus U is unitary.

The matrix U turns out to be a diagonalizer for *any* circulant. That is, for any $n \times n$ circulant, the columns of U are n orthonormal eigenvectors. Moreover, the eigenvalues of C are obtained from the polynomial

$$p(\lambda) = c_1 + c_2\lambda + \cdots + c_n\lambda^{n-1} \tag{6.3.17}$$

defined by the entries of C, as shown in the next result.

6.3.3. DIAGONALIZATION OF CIRCULANTS. *Let C be an $n \times n$ circulant* (6.3.14), *and let U be the unitary matrix defined by* (6.3.16) *and* (6.3.15). *Then, if p is the polynomial* (6.3.17),

$$U^*CU = D = \mathrm{diag}(p(1), p(\omega), \ldots, p(\omega^{n-1})) \tag{6.3.18}$$

PROOF. Let C_0 be the special circulant

$$C_0 = \begin{bmatrix} 0 & 1 & & & \\ & 0 & 1 & & \\ & & \ddots & \ddots & \\ & & & \ddots & 1 \\ 1 & & & & 0 \end{bmatrix}$$

and

$$D_0 = \mathrm{diag}(1, \omega, \ldots, \omega^{n-1})$$

Then a calculation analogous to that which showed that U is unitary yields

$$UD_0U^* = C_0 \tag{6.3.19}$$

Now note that

$$C_0^2 = \begin{bmatrix} 0 & 0 & 1 & & \\ & & & \ddots & \\ & & & & 1 \\ 1 & & & & 0 \\ 0 & 1 & & & 0 \end{bmatrix}, \qquad C_0^3 = \begin{bmatrix} 0 & 0 & 0 & 1 & & \\ & & & & 1 & \\ & & & & \ddots & 1 \\ 1 & & & & & 0 \\ 0 & 1 & & & & 0 \\ 0 & 0 & 1 & & & 0 \end{bmatrix}$$

and so on. Thus,

$$C = c_1I + c_2C_0 + c_3C_0^2 + \cdots + c_nC_0^{n-1} = p(C_0)$$

and, therefore, by (6.3.19),

$$C = p(UD_0U^*) = Up(D_0)U^* = UDU^*$$

which completes the proof. □

We give a simple example. Let

$$C = \begin{bmatrix} 1 & 2 & 3 & 4 \\ 4 & 1 & 2 & 3 \\ 3 & 4 & 1 & 2 \\ 2 & 3 & 4 & 1 \end{bmatrix}$$

Then

$$p(\lambda) = 1 + 2\lambda + 3\lambda^2 + 4\lambda^3, \qquad \omega = \cos\frac{\pi}{2} + i\sin\frac{\pi}{2} = i$$

Thus, the eigenvalues of C are

$$p(1) = 10, \quad p(i) = -2 - 2i, \quad p(i^2) = -2, \quad p(i^3) = -2 + 2i$$

and the corresponding eigenvectors are the columns of

$$U = \frac{1}{2}\begin{bmatrix} 1 & 1 & 1 & 1 \\ 1 & i & -1 & -i \\ 1 & -1 & 1 & -1 \\ 1 & -i & -1 & i \end{bmatrix}$$

Exercises 6.3

1. Let A and B be $n \times n$ and $m \times m$ matrices, respectively. Show that $(A \otimes B)^* = A^* \otimes B^*$ and that $\operatorname{rank}(A \otimes B) = \operatorname{rank}(A)\operatorname{rank}(B)$.
2. Let $\mathbf{x}$ be an n-vector and $\mathbf{y}$ an m-vector. Show that $\mathbf{x} \otimes \mathbf{y}^T = \mathbf{x}\mathbf{y}^T$.
3. Let

$$A = \begin{bmatrix} 3 & 1 \\ 1 & 3 \end{bmatrix}, \qquad B = \begin{bmatrix} 1 & 3 \\ 3 & 1 \end{bmatrix}$$

Find the eigenvalues and eigenvectors of $A \otimes B$ and $(I \otimes A) + (B \otimes I)$. Also find $(A \otimes B)^{-1}$.

4. Let C be an $n \times n$ circulant. Show that C^* is a circulant and that $CC^* = C^*C$.
5. Let C be the circulant

$$C = \begin{bmatrix} 2 & 1 & 1 \\ 1 & 2 & 1 \\ 1 & 1 & 2 \end{bmatrix}$$

Find the eigenvalues and eigenvectors of C.

6. Find the eigenvalues and eigenvectors of the matrices of Exercise 3 by means of Theorem 6.3.3.
7. Let B and C be $n \times n$ circulants. Show that BC is a circulant.
8. Let

$$A = \begin{bmatrix} T & -2I & & \\ -2I & \ddots & \ddots & \\ & \ddots & \ddots & -2I \\ & & -2I & T \end{bmatrix}, \quad T = \begin{bmatrix} 4 & -1 & & \\ -1 & \ddots & \ddots & \\ & \ddots & \ddots & -1 \\ & & -1 & 4 \end{bmatrix}$$

where T and I are $n \times n$. Find the eigenvalues and eigenvectors of A.

6.4. Matrix Equations

We will consider in this section matrix equations of the form

$$AX - XB = C \tag{6.4.1}$$

which is sometimes called Sylvester's equation. Here A, B, and C are given $n \times n$ matrices, and we seek $n \times n$ matrices X that satisfy (6.4.1). More generally, we could allow B to be $m \times m$ and C to be $n \times m$ and seek $n \times m$ matrices X as solutions. We shall, however, deal only with the case in which A and B are the same size.

An important special case of (6.4.1) occurs when $B = A$ and $C = 0$, so that (6.4.1) reduces to

$$AX = XA \tag{6.4.2}$$

Clearly, this equation has infinitely many solutions: for example, $X = A^k$, $k = 0, 1, \ldots$. The solutions of (6.4.2) are precisely the matrices that commute with A, and we will treat the problem of finding all such matrices in some detail.

We will consider two approaches to (6.4.1). The first is to write the equations in standard matrix-vector form. Let $\mathbf{x}_i$, $\mathbf{b}_i$, and $\mathbf{c}_i$ be the columns of X, B, and C. Then the ith column of (6.4.1) is

$$A\mathbf{x}_i - X\mathbf{b}_i = \mathbf{c}_i$$

or

$$A\mathbf{x}_i - \sum_{j=1}^{n} b_{ji}\mathbf{x}_j = \mathbf{c}_i$$

These can be combined into the $n^2 \times n^2$ system

$$\begin{bmatrix} A - b_{11}I & -b_{21}I & \cdots & -b_{n1}I \\ -b_{12}I & A - b_{22}I & \cdots & -b_{n2}I \\ \vdots & & & \vdots \\ -b_{1n}I & -b_{2n}I & \cdots & A - b_{nn}I \end{bmatrix} \begin{bmatrix} \mathbf{x}_1 \\ \mathbf{x}_2 \\ \vdots \\ \mathbf{x}_n \end{bmatrix} = \begin{bmatrix} \mathbf{c}_1 \\ \mathbf{c}_2 \\ \vdots \\ \mathbf{c}_n \end{bmatrix} \tag{6.4.3}$$

where I is the $n \times n$ identity. The coefficient matrix (6.4.3) is just the Kronecker sum $(I \otimes A) + (-B^T \otimes I)$ discussed in the previous section. Therefore, (6.4.3) can be written as

$$[(I \otimes A) + (-B^T \otimes I)]\mathbf{x} = \mathbf{c} \tag{6.4.4}$$

where $\mathbf{x}$ and $\mathbf{c}$ are the vectors of (6.4.3). This system of n^2 unknowns will have a unique solution if $(I \otimes A) + (-B^T \otimes I)$ has no zero eigenvalues. By 6.3.2 this Kronecker sum has eigenvalues $\lambda_i - \mu_j$, $i, j = 1, \ldots, n$, where $\lambda_1, \ldots, \lambda_n$ and $\mu_1, \ldots, \mu_n$ are the eigenvalues of A and B, respectively. Hence, $(I \otimes A) + (-B^T \otimes I)$ is nonsingular if and only if $\lambda_i \neq \mu_j$, $i, j = 1, \ldots, n$, and we have the basic result:

6.4.1. UNIQUE SOLUTION OF MATRIX EQUATIONS. *If A, B, and C are given $n \times n$ matrices, Equation* (6.4.4), *and consequently the equation $AX - XB = C$, has a unique solution if and only if A and B have no eigenvalues in common.*

It is of interest to give a direct proof of 6.4.1 that does not rely on the Kronecker sum.

SECOND PROOF OF 6.4.1. It suffices to show that the linear system (6.4.3) with the right-hand side zero has only the zero solution. This is equivalent to showing that

$$AX - XB = 0 \tag{6.4.5}$$

has only the solution $X = 0$. Suppose first that λ is a common eigenvalue of A and B. Then $\bar{\lambda}$ is an eigenvalue of B^*. Let $\mathbf{u}$ and $\mathbf{v}$ be corresponding eigenvectors, $A\mathbf{u} = \lambda\mathbf{u}$, $B^*\mathbf{v} = \bar{\lambda}\mathbf{v}$, and let $X = \mathbf{u}\mathbf{v}^*$. Then

$$AX = A\mathbf{u}\mathbf{v}^* = \lambda\mathbf{u}\mathbf{v}^* = \mathbf{u}\mathbf{v}^*B = XB$$

so that X is a nonzero solution. Conversely, suppose that $X \neq 0$ is a solution of (6.4.5). Then there is an eigenvector or generalized eigenvector $\mathbf{p}$ of B such that $X\mathbf{p} \neq 0$, or else X would take n linearly independent vectors into 0, which would imply that $X = 0$. If $\mathbf{p}$ is an eigenvector corresponding to an eigenvalue λ of B, then, since X is a solution of (6.4.5),

$$0 = (AX - XB)\mathbf{p} = A(X\mathbf{p}) - \lambda X\mathbf{p} \qquad (6.4.6)$$

which shows that λ is an eigenvalue of A. If we cannot find an eigenvector $\mathbf{p}$ of B such that $X\mathbf{p} \neq 0$, let $\mathbf{p}$ be a generalized eigenvector such that $B\mathbf{p} = \lambda\mathbf{p} + \hat{\mathbf{p}}$ with $X\mathbf{p} \neq 0$, $X\hat{\mathbf{p}} = 0$. Then the calculation of (6.4.6) is again valid, so that λ is an eigenvalue of A. □

The Lyapunov Equation

We next consider the application of 6.4.1 to the Lyapunov equation

$$AX + XA^* = C \qquad (6.4.7)$$

treated in Section 5.4. Assume that A is negative stable (all eigenvalues of A have negative real part). Then all eigenvalues of $-A^*$ have positive real part, and A and $-A^*$ have no eigenvalues in common. Hence, by 6.4.1, (6.4.7) has a unique solution X for any C. Moreover, if C is Hermitian, X must also be Hermitian because

$$C = C^* = (AX + XA^*)^* = AX^* + X^*A^*$$

shows that X^* is also a solution. But because the solution is unique, we must have $X^* = X$.

The matrix mapping $X \to AX + XA^*$ is sometimes called the *Lyapunov mapping*. Theorem 6.4.1 shows that if A is negative stable, this is a one-to-one mapping of Hermitian matrices onto Hermitian matrices. Moreover, it can be shown (see number 7 of the References and Extensions) that if C is negative definite, then X must be positive definite. We summarize the previous discussion by the following result, which complements those of Section 5.4.

6.4.2. If the $n \times n$ matrix A is negative stable, then for any $n \times n$ matrix C, Equation (6.4.7) has a unique solution X. If C is Hermitian, then X is also Hermitian; and if C is Hermitian and negative definite, then X is positive definite.

Commuting Matrices

We now turn to the special case of (6.4.1) in which $B = A$ and $C = 0$:

$$AX = XA \tag{6.4.8}$$

We wish to address two related questions:

1. For given A, what are all the solutions of (6.4.8)? That is, what are all the matrices that commute with a given matrix A?
2. Under what conditions do two given matrices A and B commute?

We first prove a result on the eigenvalues of AB and BA. If A is nonsingular, then

$$AB = ABAA^{-1}$$

so that AB and BA are similar and, hence, have the same eigenvalues. The same is true if B is nonsingular instead of A: $AB = B^{-1}BAB$. If both A and B are singular, then AB and BA are not necessarily similar (Exercise 6.4-1), but the following result shows that they still have the same characteristic polynomial and, hence, the same eigenvalues.

6.4.3. If A and B are $n \times n$ matrices, then AB and BA have the same characteristic polynomial and the same eigenvalues.

PROOF. Let μ be the smallest absolute value of the nonzero eigenvalues of A. Then $A + \varepsilon I$ is nonsingular for $0 < \varepsilon < \mu$ and

$$\begin{aligned}\det[(A+\varepsilon I)B - \lambda I] &= \det(A+\varepsilon I)\det[B - \lambda(A+\varepsilon I)^{-1}] \\ &= \det[B - \lambda(A+\varepsilon I)^{-1}]\det(A+\varepsilon I) \\ &= \det[B(A+\varepsilon I) - \lambda I]\end{aligned} \tag{6.4.9}$$

Since the determinant of a matrix is just a sum of products of elements of the matrix, it is a continuous function of the elements of the matrix. Hence, (6.4.9) must hold in the limit as $\varepsilon \to 0$. □

We note that the preservation of eigenvalues of AB and BA is true even if A and B are not square matrices, at least in the following sense.

6.4.4. Let A and B be $n \times m$ and $m \times n$ real or complex matrices, respectively, with $m > n$. Then AB and BA have the same nonzero eigenvalues, counting multiplicities, and BA has $m - n$ more zero eigenvalues than AB.

PROOF. Let $\hat{A}$ and $\hat{B}$ be $m \times m$ matrices obtained by adding zero rows to A and zero columns to B. Then, by 6.4.3, $\hat{A}\hat{B}$ and $\hat{B}\hat{A}$ have the same characteristic polynomial. But since

$$\hat{A}\hat{B} = \begin{bmatrix} AB & 0 \\ 0 & 0 \end{bmatrix}, \qquad \hat{B}\hat{A} = BA$$

we have

$$\det(BA - \lambda I) = \det(\hat{B}\hat{A} - \lambda I) = \det(\hat{A}\hat{B} - \lambda I) = (-\lambda)^{m-n} \det(AB - \lambda I)$$

which proves the result. □

We now return to the problem of commuting matrices. We first show that whenever two matrices commute they must have a common eigenvector.

6.4.5. If A and B are $n \times n$ matrices such that $AB = BA$, then A and B have a common eigenvector.

PROOF. Suppose that λ is an eigenvalue of A with r linearly independent eigenvectors $\mathbf{x}_1, \ldots, \mathbf{x}_r$. Then every eigenvector of A associated with λ is a linear combination of $\mathbf{x}_1, \ldots, \mathbf{x}_r$. Therefore, from

$$AB\mathbf{x}_i = BA\mathbf{x}_i = \lambda B\mathbf{x}_i, \qquad i = 1, \ldots, r$$

it follows that $B\mathbf{x}_i$ is a linear combination of $\mathbf{x}_1, \ldots, \mathbf{x}_r$:

$$B\mathbf{x}_i = \sum_{j=1}^{r} c_{ij}\mathbf{x}_j, \qquad i = 1, \ldots, r$$

Then, for any linear combination $\sum_{i=1}^{r} \alpha_i \mathbf{x}_i$, we have

$$B\left(\sum_{i=1}^{r} \alpha_i \mathbf{x}_i\right) = \sum_{i=1}^{r} \alpha_i \sum_{j=1}^{r} c_{ij}\mathbf{x}_j = \sum_{j=1}^{r} \left(\sum_{i=1}^{r} c_{ij}\alpha_i\right)\mathbf{x}_j \tag{6.4.10}$$

Let μ and $\boldsymbol{\alpha}$ be an eigenvalue and corresponding eigenvector of the $r \times r$ matrix $(c_{ij})^T$. Thus

$$\sum_{i=1}^{r} c_{ij}\alpha_i = \mu\alpha_j, \qquad j = 1, \ldots, r$$

so that (6.4.10) becomes

$$B\left(\sum_{i=1}^{r} \alpha_i \mathbf{x}_i\right) = \mu \sum_{i=1}^{r} \alpha_i \mathbf{x}_i$$

Therefore, $\sum_{i=1}^{r} \alpha_i \mathbf{x}_i$ is an eigenvector of B and also an eigenvector of A. □

The converse of 6.4.5 is not true; A and B may have a common eigenvector but not commute. A simple example is

$$A = \begin{bmatrix} 0 & 1 \\ 0 & 0 \end{bmatrix}, \quad B = \begin{bmatrix} 0 & 1 \\ 0 & 1 \end{bmatrix}$$

in which $(1, 0)^T$ is an eigenvector of both matrices, but $AB \neq BA$. However, if A and B have n linearly independent eigenvectors in common, they must commute.

6.4.6. If A and B are $n \times n$ matrices with n linearly independent common eigenvectors, then $AB = BA$. Moreover, if A has p distinct eigenvalues of multiplicities $m_1, \ldots, m_p$ with $n = \sum_{i=1}^{p} m_i$ and $A = PDP^{-1}$, where D is the diagonal matrix containing the first distinct eigenvalue m_1 times, followed by the second eigenvalue m_2 times, and so on, then every matrix B that commutes with A is of the form $B = PCP^{-1}$, where

$$C = diag(C_1, \ldots, C_p)$$

and C_i is an arbitrary $m_i \times m_i$ matrix. In particular, if A has n distinct eigenvalues, then B commutes with A if and only if $B = P\hat{D}P^{-1}$, where $\hat{D}$ is diagonal.

Proof. If A and B have n linearly independent common eigenvectors, let P be the nonsingular matrix whose columns are these eigenvectors. Then $A = PDP^{-1}$ and $B = PD_1P^{-1}$, where D and D_1 are diagonal, and

$$AB = PDP^{-1}PD_1P^{-1} = PD_1DP^{-1} = BA$$

For the second part, if $A = PDP^{-1}$ and $AB = BA$, then $PDP^{-1}B = BPDP^{-1}$ or

$$DC = CD \tag{6.4.11}$$

where $C = P^{-1}BP$. Equating the elements of (6.4.11) gives

$$\lambda_i c_{ij} = \lambda_j c_{ij}, \qquad i, j = 1, \ldots, n$$

which shows that $c_{ij} = 0$ whenever $\lambda_i \neq \lambda_j$, and otherwise c_{ij} is arbitrary. □

Finally, we consider the general case in which A has an arbitrary Jordan form. Recall (Section 6.3) that a Toeplitz matrix has constant diagonal elements.

6.4.7. COMMUTING MATRICES. *Let $A = PJP^{-1}$, where $J = \text{diag}(J_1, \ldots, J_p)$ is the Jordan form of A, and $\lambda_1, \ldots, \lambda_p$ are the eigenvalues of A associated with the p Jordan blocks. Then every matrix B that commutes with A is of the form $B = PCP^{-1}$, where if C is partitioned according to J, $C_{ij} = 0$ if $\lambda_i \neq \lambda_j$, and each C_{ii} is an arbitrary upper triangular Toeplitz matrix. Finally, if $\lambda_i = \lambda_j$, $i \neq j$, and J_i is $s \times s$ and J_j is $r \times r$, then C_{ij} is $s \times r$ and of the form*

$$C_{ij} = \begin{bmatrix} * & * & * & \cdots & * \\ & \ddots & \ddots & \ddots & \\ & & & & \end{bmatrix}, \quad s \leq r; \qquad C_{ij} = \begin{bmatrix} * & * & \cdots & * \\ & \ddots & \ddots & \\ & & & \\ & & & \end{bmatrix}, \quad s \geq r \tag{6.4.12}$$

where the elements in the first row of C_{ij} are arbitrary, and each corresponding diagonal is constant.

Before proving 6.4.7, we consider some special cases to illustrate the theorem. If the eigenvalues $\lambda_1, \ldots, \lambda_p$ are distinct, then $C = \text{diag}(C_{11}, \ldots, C_{pp})$, where each C_{ii} is an arbitrary upper triangular Toeplitz matrix. If, in addition, $p = n$ so that A has n distinct eigenvalues, then C is a diagonal matrix, and this is the last part of 6.4.6. On the other hand, if A has multiple eigenvalues but a diagonal canonical form, then $p = n$, and each C_{ij} is 1×1. If we group the multiple eigenvalues together on the diagonal of J, we then obtain the rest of 6.4.6. Finally, as an example of the general case, suppose that $J = \text{diag}(J_1, J_2)$, where $\lambda_1 = \lambda_2$, J_1 is $s \times s$, and J_2 is $r \times r$ with $s < r$. Then C has the form

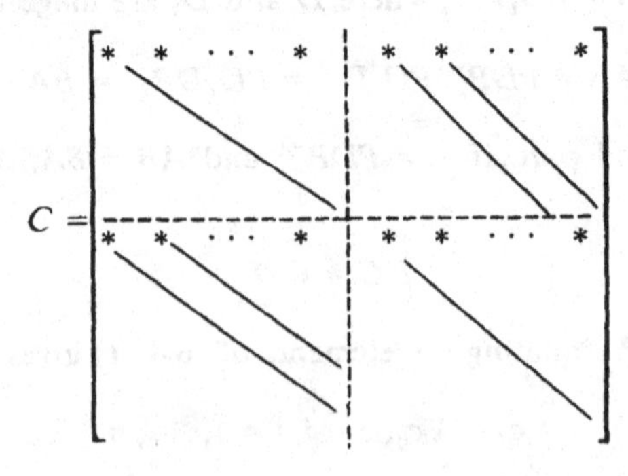

where the diagonals of each block are constant.

PROOF OF 6.4.7. If $AB = BA$, then $JC = CJ$, where $C = P^{-1}BP$. Using the partitioned forms of J and C, we must have

$$J_iC_{ij} = C_{ij}J_j, \qquad i, j = 1, \ldots, p \tag{6.4.13}$$

Assume that J_i is $s \times s$ and J_j is $r \times r$, and set $X = C_{ij}$. Then X is $s \times r$ and (6.4.13) is

$$\lambda_i x_{kl} + x_{k+1,l} = x_{k,l-1} + \lambda_j x_{kl}, \qquad k = 1, \ldots, s-1, i = 1, \ldots, r \tag{6.4.14a}$$

$$\lambda_i x_{sl} = x_{s,l-1} + \lambda_j x_{sl}, \qquad l = 2, \ldots, r \tag{6.4.14b}$$

$$\lambda_i x_{k1} + x_{k+1,1} = \lambda_j x_{k,1}, \qquad k = 1, \ldots, s-1 \tag{6.4.14c}$$

$$\lambda_i x_{s1} = \lambda_j x_{s1} \tag{6.4.14d}$$

We then have two cases:

Case 1. $\lambda_i \neq \lambda_j$. Here, (6.4.14d) shows that $x_{s1} = 0$, and then the Equations (6.4.14c), taken in the order $k = s-1, s-2, \ldots, 1$, show that $x_{k,1} = 0$, $k = 1, \ldots, s-1$. Similarly, Equations (6.4.14b), taken in the order $l = 2, \ldots, r$, show that $x_{sl} = 0$, $l = 2, \ldots, r$. Finally, Equations (6.4.14a), taken in the order $l = 2$, $k = s-1, \ldots, l$; $l = 3$, $k = s-1, \ldots, l$; $\ldots$; $l = r$, $k = s-1, \ldots, l$, show that all $x_{kl} = 0$. Thus, if $\lambda_i \neq \lambda_j$, then $C_{ij} = 0$.

Case 2. $\lambda_i = \lambda_j$. Equation (6.4.14a) shows that the diagonals of X are all constant. By (6.4.14b) the elements $x_{s1}, \ldots, x_{s,r-1}$ in the last row are all zero, and, hence, the corresponding diagonals are zero. If $r \geq s$, there are no other constraints, and C_{ij} has the first form in (6.4.12) with the nonzero diagonals arbitrary but constant. In particular, C_{ii} is an upper triangular Toeplitz matrix. If $s > r$, (6.4.14c) shows that all elements of the first column, except the first, are zero. Hence, the corresponding diagonals are zero, and C_{ij} has the second form in (6.4.12). □

Exercises 6.4

1. Let

$$A = \begin{bmatrix} 0 & 1 \\ 0 & 0 \end{bmatrix}, \qquad B = \begin{bmatrix} 1 & 0 \\ 0 & 0 \end{bmatrix}$$

Show that AB and BA are not similar.

2. If A and B are $n \times n$ Hermitian matrices, show that AB is Hermitian if and only if $AB = BA$.
3. Solve the matrix equation

$$\begin{bmatrix} 1 & 1 \\ 0 & 1 \end{bmatrix} X - X \begin{bmatrix} 2 & 0 \\ 1 & 2 \end{bmatrix} = \begin{bmatrix} 1 & 1 \\ 1 & 1 \end{bmatrix}$$

4. Solve the matrix equation

$$\begin{bmatrix} -1 & 1 \\ 0 & -1 \end{bmatrix} X + X \begin{bmatrix} -1 & 0 \\ 1 & -1 \end{bmatrix} = I$$

and verify that X is symmetric and positive definite.

5. Use 6.4.6 to find all matrices that commute with

$$A = \begin{bmatrix} 1 & 0 & 0 \\ 0 & 2 & 1 \\ 0 & 1 & 2 \end{bmatrix}$$

6. Use 6.4.7 to find all matrices that commute with

$$\text{(a)} \quad A = \begin{bmatrix} 1 & 1 & 0 \\ & 1 & 1 \\ & & 1 \end{bmatrix}$$

$$\text{(b)} \quad A = \begin{bmatrix} 1 & 0 & 0 \\ & 1 & 1 \\ & & 1 \end{bmatrix}$$

Review Questions—Chapter 6

Answer whether the following statements are true or false and justify your assertions. In the following five statements A is assumed to be a positive $n \times n$ matrix.

1. A is stochastic if the elements in each row sum to 1.
2. If A is stochastic, then the trivial solution of the difference equation $x^{(k+1)} = Ax^{(k)}$, $k = 0, 1, \ldots$, is stable.
3. A is irreducible.
4. If A is stochastic, the only eigenvalues of A equal to $\rho(A)$ are 1 and -1.
5. The only eigenvectors associated with the eigenvalue $\rho(A)$ are positive vectors.

In the following four statements A is assumed to be a nonnegative $n \times n$ matrix.

6. A is always reducible unless it is positive.

7. If the diagonal elements of A are all zero, then A is reducible.
8. If A is irreducible, then there are no eigenvalues except $\rho(A)$ equal in absolute value to $\rho(A)$.
9. If A is irreducible, then A^{n-1} is positive.
10. If A has positive diagonal elements and a nonnegative inverse, then A is an M-matrix.
11. If A is an irreducible M-matrix, then $A^{-1} > 0$.
12. A tridiagonal matrix is irreducible if and only if all its off-diagonal elements (in the two diagonals next to the main diagonal) are nonzero.
13. If C is irreducible and $\|C\|_\infty \leq 1$, then $\rho(C) < 1$.
14. If $A \leq B$, then $\rho(A) \leq \rho(B)$.
15. If $A \leq B$, then $A^{-1} \geq B^{-1}$.
16. If A is diagonally dominant and irreducible, then A is nonsingular.
17. If A is irreducibly diagonally dominant, then the Jacobi iteration converges.
18. If A and B are real and symmetric and A is positive definite, then the eigenvalues of $A\mathbf{x} = \lambda B\mathbf{x}$ are real.
19. If A is similar to a Hermitian matrix, then A can be written as a product of Hermitian matrices.
20. If A and B are Hermitian, then AB has real eigenvalues.
21. If A_1 and A_0 are $n \times n$ matrices, then the equation $\det(A_0 + \lambda A_1 + \lambda^2 I) = 0$ has $2n$ roots.
22. The Kronecker product of A and B is nonsingular if A and B are nonsingular.
23. If A and B are Hermitian, then $A \otimes B$ is Hermitian and, hence, has real eigenvalues.
24. If A and B are $n \times n$ matrices, the Kronecker sum $(I \otimes A) + (-B \otimes I)$ is nonsingular if and only if A and B have no eigenvalues in common.
25. If C is a circulant, then its eigenvalues are all real.
26. An $n \times n$ Toeplitz matrix is determined by $2n - 1$ numbers.
27. An $n \times n$ circulant always has n orthonormal eigenvectors.
28. The matrix equation $AX - XB = C$ has a unique solution if and only if both A and B are nonsingular.
29. AB and BA have the same eigenvalues if and only if either A or B is nonsingular.
30. If A has distinct eigenvalues, then every matrix B that commutes with A has the same eigenvectors as A.

References and Extensions: Chapter 6

1. The basic results of Perron and Frobenius on nonnegative matrices were proved in the early part of this century, but the subject has had a renaissance in the last

30 years owing to the many applications of nonnegative matrices in numerical analysis, probability and statistics, economics, and a variety of other areas. For further reading see Berman and Plemmons [1979] and Gantmacher [1959] for overall treatments, Senata [1981] for applications to Markov processes, and Varga [1962] for applications to numerical analysis.

2. An interesting part of the Perron–Frobenius Theorem, which we did not cover, is the nature of the other eigenvalues of absolute value $\rho(A)$. For this and various extensions and ramifications, see Gantmacher [1959], Berman and Plemmons [1979], and Varga [1962]. For further results on nonnegative inverses, diagonal dominance, and Gerschgorian-type theorems, see Varga [1962] and Householder [1964].
3. The set of matrices $\{A - \lambda B\}$, for fixed A and B, is sometimes called a *matrix pencil*, and many results on the generalized eigenvalue problem (6.2.1) will be found in the literature under this term. See, for example, Gantmacher [1959] and Dold and Eckmann [1983].
4. For further results on simultaneous similarity transformations, see Mirsky [1982].
5. If the differential equation (6.2.8) has highest-order term $A_m \mathbf{x}^{(m)}(t)$ and A_m is nonsingular, then the development in the text may be carried out by multiplying the equation by A_m^{-1}, thus replacing the coefficient matrices A_i by $A_m^{-1}A_i$. However, if A_m is singular, the situation is much more difficult. For a detailed discussion of this and the corresponding eigenvalue problem, as well as a number of other results, see Gohberg *et al.* [1982].
6. A wealth of information on circulants is contained in Davis [1979]. For the application of both finite and infinite Toeplitz matrices to trigonometric series and a number of other areas, see Grenander and Szego [1984]. Because of their special structure, linear systems with Toeplitz coefficient matrices can be solved in $O(n^2)$ operations; see Golub and van Loan [1983].
7. We noted in Chapter 5 (see number 9 in the References and Extensions) that the Lyapunov matrix can be represented as a certain integral. More generally, if

$$X = -\int_0^\infty e^{At} C e^{-Bt}\, dt$$

exists, then X is the unique solution of the equation $AX - XB = C$. This is related to the fact that the solution of the matrix differential equation

$$\frac{dX(t)}{dt} = AX(t) - BX(t), \qquad X(0) = C$$

is $X(t) = e^{At} C e^{-Bt}$. In the case of the Lyapunov equation (6.4.7), $B = -A^*$, so that if C is negative definite, then $e^{At} C e^{A^* t}$ is negative definite for all t, and, hence, X is positive definite. For further discussion, see, for example, Bellman [1960].

References

Basilevsky, A. [1983]. *Applied Matrix Algebra in the Statistical Sciences.* New York: Elsevier.

Bellman, R. [1960]. *Introduction to Matrix Analysis.* New York: McGraw-Hill.

Berman, A., and Plemmons, R. [1979]. *Nonnegative Matrices in the Mathematical Sciences.* New York: Academic.

Campbell, S. [1980]. *Singular Systems of Differential Equations.* Marshfield, Mass. Pitman.

Campbell, S., and Meyer, C. [1979]. *Generalized Inverses of Linear Transformations.* Marshfield, Mass.: Pitman.

Coddington, E., and Levinson, N. [1955]. *Theory of Ordinary Differential Equations.* New York: McGraw-Hill.

Davis, P. [1979]. *Circulant Matrices.* New York: John Wiley.

Dennis, J., and Schnabel, R. [1983]. *Numerical Methods for Unconstrained Optimization and Nonlinear Equations.* Englewood Cliffs, New Jersey: Prentice-Hall.

Dold, A., and Eckmann, B., eds. [1983]. *Matrix Pencils.* Lecture Notes in Mathematics, No. 973, New York: Springer-Verlag.

Dunford, N., and Schwartz, J. [1958, 1963]. *Linear Operators.* Vol. 1, *General Theory,* Vol. 2, *Spectral Theory.* New York: Interscience.

Fletcher, R. [1980, 1981]. *Practical Methods of Optimization.* Vol. 1, *Unconstrained Optimization.* Vol. 2, *Constrained Optimization.* New York: Wiley.

Fletcher, R., and Sorensen, D. [1983]. An algorithmic derivation of the Jordan canonical form. *Am. Math. Monthly* 90: 12–16.

Galperin, A., and Waksman, Z. [1980]. An elementary approach to Jordan theory. *Amer. Math. Mon.* 87: 728–732.

Gantmacher, F. [1959]. *The Theory of Matrices,* Vols. 1 and 2. New York: Chelsea.

Gohberg, I., Lancaster, P., and Rodman, L. [1982]. *Matrix Polynomials.* New York: Academic.

Golub, G., and van Loan, C. [1983]. *Matrix Computations.* Baltimore: Johns Hopkins University Press.

Grenander, U., and Szego, G. [1984]. *Toeplitz Forms and Their Applications.* New York: Chelsea (reprint of the 1958 edition).

Gross, H. [1979]. *Quadratic Forms in Infinite Dimensional Vector Spaces.* Boston: Birkhauser.

Hageman, L., and Young, D. [1981]. *Applied Iterative Methods.* New York: Academic.

Halmos, P. [1958]. *Finite Dimensional Vector Spaces.* Princeton, New Jersey: Van Nostrand Reinhold.

Heath, M. [1984]. Numerical methods for large sparse linear least squares problems. *SIAM J. Sci. Statist. Comput.* 5: 497–513.

Householder, A. [1964]. *The Theory of Matrices in Numerical Analysis.* New York: Blaisdell.

LaSalle, J., and Lefschetz, S. [1961]. *Stability by Lyapunov's Direct Method.* New York: Academic.

Lawson, L., and Hanson, R. [1974]. *Solving Least Squares Problems.* Englewood Cliffs, New Jersey: Prentice-Hall.

Lukes, D. [1982]. *Differential Equations: Classical to Controlled.* New York: Academic.

Mirsky, L. [1982]. *An Introduction to Linear Algebra.* New York: Dover (reprint of the 1955 original; Oxford: Clarendon Press).

Mostow, G., and Sampson, J. [1969]. *Linear Algebra.* New York: McGraw-Hill.

Ortega, J. [1972]. *Numerical Analysis: A Second Course.* New York: Academic.

Ortega, J. [1973]. Stability of difference equations and convergence of iterative processes. *SIAM J. Numer. Anal.* 10: 268–282.

Ortega, J., and Poole, W. [1981]. *An Introduction to Numerical Methods for Differential Equations.* Marshfield, Massachusetts: Pitman.

Ortega, J., and Rheinboldt, W. [1970]. *Iterative Solution of Nonlinear Equations in Several Variables.* New York: Academic.

Ostrowski, A., and Schneider, H. [1962]. Some theorems on the inertia of general matrices. *J. Math. Anal. Appl.* 4: 72–84.

Parlett, B. [1980]. *The Symmetric Eigenvalue Problem.* Englewood Cliffs, New Jersey: Prentice-Hall.

Samelson, H. [1974]. *An Introduction to Linear Algebra.* New York: Wiley.

Seber, G. [1977]. *Linear Regression Analysis.* New York: Wiley.

Senata, E. [1981]. *Nonnegative Matrices and Markov Chains.* 2d ed. New York: Springer-Verlag.

Stewart, G. [1973]. *Introduction to Matrix Computations.* New York: Academic.

Strang, G. [1980]. *Linear Algebra and Its Applications.* 2d ed. New York: Academic.

Varga, R. [1962]. *Matrix Iterative Analysis.* Englewood Cliffs, New Jersey: Prentice-Hall.

Young, D. [1971]. *Iterative Solution of Large Linear systems.* New York: Academic.

Index

9 780306 42433